AF251036

# FUNDAMENTAL PHYSICS

## An Introduction

# FUNDAMENTAL PHYSICS

## An Introduction

Sanat Kumar Chatterjee

**Alpha Science International Ltd.**
Oxford, U.K.

**Fundamental Physics**
An Introduction
392 pgs.  |  178 figs.  |  07 tbls.

**Sanat Kumar Chatterjee**
U G C – Emeritus Fellow
Department of Physics
National Institute of Technology
Durgapur

Copyright © 2013

ALPHA SCIENCE INTERNATIONAL LTD.
7200 The Quorum, Oxford Business Park North
Garsington Road, Oxford OX4 2JZ, U.K.

**www.alphasci.com**



ISBN  978-1-84265-821-5

Printed in India

Dedicated to the Everlasting
Memories of
My  Professors

# Preface

Writing a book and that a text book meant for students is probably a more arduous task than writing a book on some advanced research topic or on some specialised subjects. Coming to text book it matters whether the book is meant for undergraduate or for postgraduate students. In the opinion of the author it is a bit more difficult if the book is meant for undergraduate students. The undergraduates have the area of interest much wider and like to read the topics of their interest from one book. Moreover, it must be kept active in the thoughts of the author that the development of the wide spread subject topics must never lose its continuity in development. This is necessary to make the basic understanding of the subject based on stable ground so that the students themselves can move forward in the specialised branch later in their academic career. This zeal to further the knowledge in even any specialised topic depends on how efficiently the students have followed the fundamentals. Now, talking on fundamental Physics, there are innumerable books written by illustrious authors who have excelled in their career either as a researcher or as a teacher or both. The author of this book cannot refrain himself from recollecting from his memory that he was so impressed and sometimes remained spell bound in the classes during his undergraduate student days. The deliberations of most of his teachers during his undergraduate days was so absorbing that it inspired and implanted the idea of writing a book on fundamentals of Physics even in those early days of his career. It has now become a dream realised in practice. While writing this book the undersigned has taken help from a number of books in framing the plan for the development of the book and then develop it in his own way. The book is divided into two parts Part I and Part II. In Part I the Physics whose principles do not require any knowledge of the microstructure to understand are included and so titled as "Macroscopic Physics" In Part II those topics are discussed which essentially require the knowledge of the microstructure of the materials for explanation and understanding and so titled as "Microscopic Physics"

In Part I, Macroscopic Physics the preliminaries like scalar and vectors are included in Chapter 1 along with Newtonian Mechanics, Angular motion, Kepler's laws, Rigid body mechanics, conservation of linear and angular momentum, Friction were discussed giving examples. In Chapter 2 on Special Theory of relativity, the Galilean and Lorentz transformation and relativistic equations of motion have been discussed. In Chapter 3 which is on Elasticity, elastic properties of solids and also fluid mechanics were discussed. From

Chapter 4 to Chapter 7 the topics like Heat and Thermodynamics, Electrostatics, Magnetism, Electrodynamics including Maxwell's equations have been discussed sequentially. From Chapter 8 to Chapter 11 the important topics like Elastic wave, Physical Optics (distributed in two Chapters I and II) to discuss in details interference and diffraction and also polarization have been done and finally in Part I, Chapter 11 is devoted on Electromagnetic wave.

The Part II devoted on topics which require microscopic concepts and state comprising fundamental constituents to explain the phenomena they exhibit. In this Part the topics like Statistical Physics, Quantum Mechanics, Solid state Physics and Semiconductor are discussed in details in Chapters from 12 to 15. The theories of magnetism and magnetic materials are discussed in Chapter 16, where as Super conductivity in Chapter 17 and finally the Nuclear Physics in Chapter 18. Lastly there are two Appendices, Appendix I and Appendix II. Appendix I discusses the interesting aspects of Universe, its creation. As it is experienced that until it is practiced there can never be any solid foundation of understanding and so in Appendix II some solved numerical problems on some topics of Physics are included so that students can gain confidence while they study though there are some theoretical questions which are included as Review Questions at the end of most of the chapters.

In each part the subject has been systematically developed and now all the efforts can be treated as successful if the present book is accepted by the students not merely as a book necessary for examination but more for the understanding of the subject and to cultivate the inner thirst to know more and beyond examination boundaries.

Lastly, the undersigned expresses his sincere thankfulness to his students as without their interest in the class room continued over decades, the idea of taking this massive venture would never appear reasonable and worthwhile. He also expresses his gratefulness to University Grants Commission for sponsoring this project by awarding Emeritus fellowship to him.

**Sanat Kumar Chatterjee**

# Contents

## Chapter 4 : Heat and Thermodynamics — 4.1

# Classical Physics

## 1.1 INTRODUCTION

Science emerged from the quest for finding the relationships that to the human mind appeared to exist between different incidences in the material world. That inter relations of different kinds was also observed to exist between the matter, living or non living and also different phenomena observed to happen in the world before us. To find such inter relations it has been found that the measurement of different parameters is necessary in order to observe and measure different characteristic properties.

When we see a material object, the first aspect of the object that draws our attention is its size i.e. how big it is? There lies the necessity of knowing its dimension and that results in to the knowing a parameter known as "Length or distance". Next that appears important to us its heaviness and that results in to the parameter known as "Mass". Now, when we know or get knowledge of these two parameters of a body, we have the basic idea of the body and when we find that these two parameters of any body may appear different when observed at different intervals, we felt the necessity of measuring and the dependence of different parameters with the "Time". These three parameters are fundamental parameters to know and analyze the matter and different phenomena. All other parameters that we have come across in order to know the more details of material objects are dependent on these three fundamental parameters i.e. Length, Mass and Time.

For example: Area = Length × Length (Breadth), Volume = Length × Length (Breadth) × Length (Height). The secondary parameters which involve any two or all of these fundamental parameters are also the parameters necessary to know the states of material bodies in more details. For example: Density = Mass/Volume,

Speed = Length (distance)/Time, Momentum = Mass × Length (displacement)/Time. Now, irrespective of the fact that the parameters are fundamental or secondary, it is essential to have a quantitative measurement of it in order to have a complete knowledge of the material bodies. When we want to measure the magnitude or the quantity of a parameter we must accept a certain and fixed quantity of it as unit and find how much or how many of such units are present in the material bodies for that parameter. These units are different for different parameters and there are different systems for standardizing such units. Now, before we introduce the different systems of units which are used in the measurement of parameters, it should be said that there must be some "standard" quantity in length, in mass and also in time in any system of measurement which should be universally accepted as units of measurement.

In metric system, the standard of length is taken as the distance between two marks on a platinum-iridium bar kept at International Bureau of Weights and Measures in Paris and kept at 0°C and is known as one Meter. Recently more accurate length of one meter is measured by optical methods and is accepted as 1650763.73 wavelengths of the orange red light emitted by Krypton-86. The unit of mass in metric system is Kilogram and a cylinder of platinum-iridium kept at Paris is said to have mass of one kilogram. It is also equal to the mass of 1000 c.c. of pure water at its maximum density which is at nearly 4°C. The standard time known as Second is said to be 1/86400 of the average length of a solar day. More accurately the duration of one second, the unit of time is taken as the time during which Cesium atom undergoes 9192631770 internal vibrations.

## 1.2 SYSTEMS OF UNITS

There are three established systems of units and they are (i) CGS (ii) MKS and (iii) FPS.

Now for fundamental parameters as Length, Mass and Time, the units in CGS system are respectively as Centimeter for length, Gram for mass and Second for time for MKS they are respectively as Meter, Kilogram and Second and finally for FPS system, foot for length, pound for mass and second for time. The dimensions of the secondary parameters in terms of three fundamental parameters noted respectively as L for length, M for mass and T for time can be derived as follows:

| Parameters | Dimension | CGS | MKS | FPS | SI Units |
|---|---|---|---|---|---|
| Displacement | L | Cm | M | Ft | m |
| Area | $L \times L = L^2$ | Cm² | M² | Ft² | m² |
| Volume | $L \times L \times L = L^3$ | Cm³ (cc) | M³ | Ft³ | m³ |
| Mass | M | Gm | Kg | Lb | Kg |
| Density | Mass/Volume, $ML^{-3}$ | Gm/cc | Kg/m³ | Lb/ft³ | Kg/m³ |
| Velocity | Distance (L)/time, $LT^{-1}$ | Cm/sec | m/sec | Ft/sec | m/s |
| Acceleration | Velocity/time, $LT^2$ | Cm/sec² | m/sec² | Ft/sec² | m/s² |
| Momentum | Mass × Velocity, $MLT^{-1}$ | Gm.cm/sec | Kg m/sec | Lb ft/sec | Kg.ms⁻¹ |
| Force | Mass × Acceleration, $ML\,T^{-2}$ | Gm.cm/sec² (Dyne) | Kg m/sec² (Newton) | Lb ft/sec² (Poundal) | Kg.m.s⁻² Newton |
| Pressure | Force/Area, $ML^{-1}\,T^{-2}$ | Gm./cm.sec² | Kg /m.sec² | Lb /ft.sec² | Kg.m⁻¹s⁻² |
| Work | Force × distance, $ML^2\,T^{-2}$ | Gm. cm²/sec² (erg ) | Kgm²/sec² (joules) | Lb ft²/sec² | m²kgs⁻² joule (J) |
| Power | Work/Time, $M\,L^2\,T^{-3}$ | Gm. cm²/sec³ | Kgm²/sec³ | Lb ft²/sec³ | m²kgs⁻³ watt (W) |
| Energy | $ML^2T^{-2}$ | erg | joule | | m²kgs⁻² |
| Charge | Q | coulomb | A s | | A s |
| Electric Current | I | ampere | s⁻¹ C | | A |

## 1.3 SCALARS AND VECTORS

The physical parameters that have been introduced and many more to be introduced later are classified in to two categories. It can be realized that some of the parameters can fully represent the property of the material or incidence that they represent or stand for, if and only if the physical quantity of the parameters is known. For example, to know fully how heavy an object is, the parameter needed to be known is its Mass and if the quantity of Mass is known then the answer to the question i.e. how heavy the object is, can be known. Such parameter whose magnitude is sufficient to represent it is termed as "Scalar". Like mass we can find out many other parameters which fall under this category like time, density, volume (how large a body is?) and many other. These parameters behave simply like numbers and also follow the basic algebraic relations, like addition, multiplication and division etc. Now, there exist some other kind of parameters and these parameters in addition to having the magnitude, have a sense of direction. Both of these two aspects i.e. magnitude and direction are intimately and inseparably mingled together in the parameters so that without the knowledge of both of these two aspects, the parameter fails to deliver the knowledge of that particular aspect of the incidence or object. For example, suppose a man is for the first time in a city and wants to find the famous church of that locality, finds a man on the street and asks? Where is the Church?

If the man simply replies "It is one Km from here" then it is definitely not the complete answer to the first man's query and so the man must ask him again "In which way?. But if the man to whom the question was asked had replied like " Go nearly half Km towards North and then on the second turn, turn left (West), it will be then another half or so". The man who asked for the church would then be totally satisfied and would express his thankfulness. This explains that not only the 1 Km distance is important, but equally important is the direction of that distance. Such parameters then have a specialty and so they belong to a different category known as "Vectors". The difference between scalars and vectors are, therefore, explicit in their basic properties of the algebra applicable to them. Let $a$ and $b$ arse two scalar quantities and their vector counter parts are say $a$ and $b$.

The Commutative Law of addition is valid for both of two types of quantities:

i.e. $a + b = b + a$ for Scalars and for Vectors $a + b = b + a$.

The Associative Law of addition is valid for both Scalars and Vectors

i.e. $a + (b + c) = (a + b) + c$ and $\quad a + (b + c) = (a + b) + c$.

When $a$ and $b$ are pure numbers, we can easily follow that but a bit more thought is required when we talk about $a$ and $b$ as vectors as $a + b \neq a + b$, $a$ and $b$ will now contain their senses of directions. Suppose a is 5 and b is 3 and so $a + b = b + a = 5 + 3 = 3 + 5 = 8$ but when $a$ and $b$ are vectors having magnitude say 5 and 3 as before, then addition of them will also involve their directions i.e. angle between them and so if the direction expressed by the angle between them is changed, the magnitude of $a + b$ will change even though the magnitude of both $a$ and $b$ remain same. This Vector addition follows the rule of a triangle. $AB$ and $BC$ are respectively as vectors $a$ and $b$ and the vector addition of them is given by the third side $AC$. It is evident that when direction of $b$ is only changed but not in magnitude to $b'$ then the addition of them is changed. The two $AC$ values of the following triangles are not same.

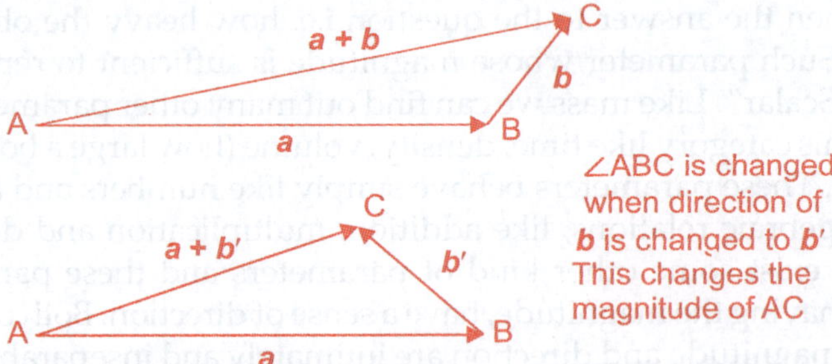


**Fig. 1.1** Vector addition, the law of triangle

The commutative law of Vector addition of more than two vector quantities can be shown graphically and the law $a + b + c = b + c + a$ can also be proved.

Commutative law of addition for more than two vectors

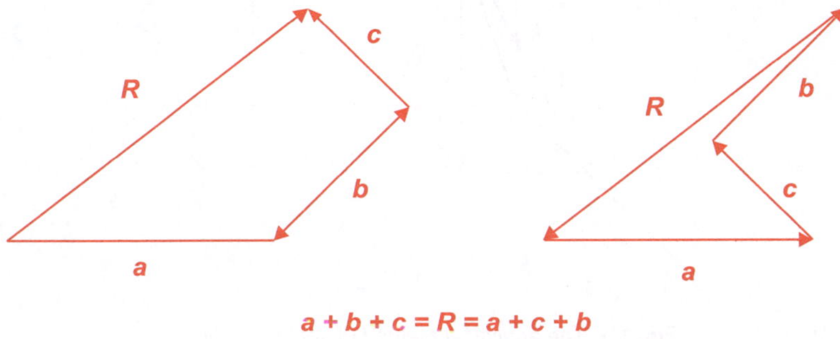

$$a + b + c = R = a + c + b$$

**Fig. 1.2** Commutative law of vector addition

Now, it is not always possible and also not desirable to find the magnitude and also the direction of the resultant vector by this graphical means of triangle (when only two vectors are involved) and of polygons (when any number of vectors is involved).

This can be better done through more analytical way by resolving each vector in to its two mutually perpendicular components. This is a fundamental property of vector that is any vector can be resolved into mutually two perpendicular components.

To explain this, let us now introduce the concept of "Unit Vector". If a vector quantity say $A$, having both the direction and the magnitude is divided by its magnitude only (which is obviously the scalar) which is denoted by $|A|$ then the resultant will maintain the direction of the original vector A but its magnitude will reduce to one. This Vector is now defined as a unit vector.

$$\frac{A}{|A|} = n.$$

This $n$ will retain the directional sense of the vector $A$. In Cartesian coordinate system the directions of $X$, $Y$ and $Z$ axes are given by three unit vectors as $i$, $j$ and $k$ and any vector can be represented in terms of these unit vectors.

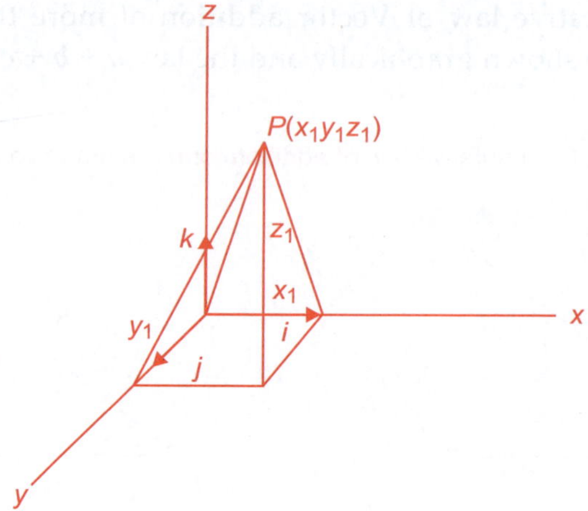



**Fig. 1.3** The analytical method of vector addition

Now, let us consider a point $P$ whose coordinates are $x_1,\ y_1,\ z_1$ and in terms of unit vectors $i$, $j$ and $k$ the position vector of the point $P$ i.e. $OP = x_1\,i + y_1\,j + z_1\,k$.

Again if any other vector say, $OQ$, when drawn from the origin of the coordinate system then if the coordinates of the terminal point $Q$ has coordinates like $x_2$, $y_2$ and $z_2$ then in terms of coordinates of $Q$ the $OQ = x_2\,i + y_2\,j + z_2\,k$. When this two vectors are to be added in the sum could also be obtained analytically as:

$$OP + OQ = x_1\,i + y_1\,j + z_1\,k + x_2\,i + y_2\,j + z_2\,k$$
$$= (x_1 + x_2)\,i + (y_1 + y_2)\,j + (z_1 + z_2)\,k$$

and in this way any number of vectors can be added analytically without going into the drawing as per scale of vectors and following the geometrical polygon rule of addition.

Now, when we come to multiplication it is an interesting point to note that the commutative law of vector multiplication is only valid for one type of multiplication and not for others.

So, there exists two types of multiplication for vectors, a phenomena not observed in the case of Scalars. In first type of multiplication the product is a scalar quantity i.e. it neither carry the sense of directions of the participating vectors, nor it have one of its own. This is called scalar product or "Dot" product. The second type is totally different, here the product is a vector and it has a direction of its own and this direction given by a unit vector is reversed if the order of the product is reversed. Therefore, the commutative law of vector product is not valid for this type of product. This is known as vector product or "Cross" product.

## The Scalar Product ( Dot product )

$a \cdot b = |a||b| \cos \theta$, where $\theta$ is the angle between the vectors a and b and product is a pure scalar quantity or number having value as $|a||b|\cos\theta$ and this is the only thing that exists for the product. Example: Work is defined as the product of force acting on a mass and the displacement of the mass in the direction of the force. The both of on a mass and the displacement of the mass in the direction of the force. The both of these two parameters i.e. force and displacement are Vectors but their product work does not contain any sense of direction and so the product of force and displacement is scalar product or dot product as the multiplication between the vectors are noted by a dot between them.

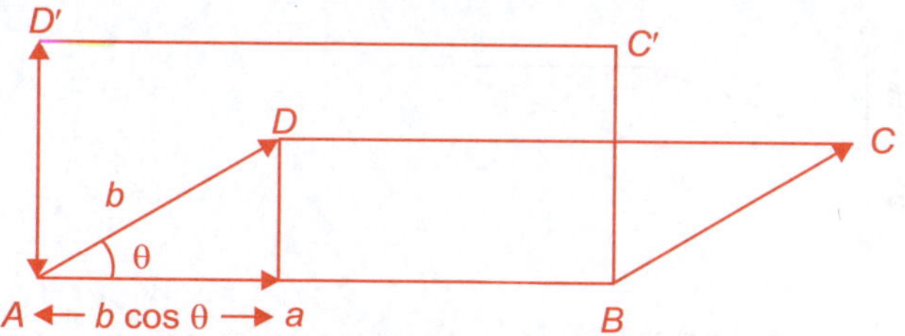


**Fig. 1.4** The scalar product between two vectors

The dot product of $a$ and $b$ give rise to the area $ABC'D'$ when angle between them is $\theta$. This area is obviously different from the area of the parallelogram made by $a$ and $b$.

**Conclusion:**

$$i \cdot i = j \cdot j = k \cdot k = 1 \quad \text{and} \quad i \cdot j = i \cdot k = j \cdot k = 0$$

as $\cos \theta = 1$ and $\cos 90° = 0$.

## The Vector Product (Cross product):

This type of product between two vector quantities $a$ and $b$ are noted as

$a \times b = |a||b|\sin \theta \cdot n$, here $|a||b|\sin\theta$ is the magnitude of the product and its direction is given by the unit vector $n$ and this vector is instead of being parallel to $a$ or $b$, is perpendicular to both $a$ and $b$ i.e. it is perpendicular to the plane containing $a$ and $b$.

The magnitude of this product of $a$ and $b$ is given by $a\,b\,\sin\theta$, which is the area of the parallelogram and of which $a$ and $b$ are the two sides. Therefore, it leads to an interesting conclusion that the area is given by the cross product of two vectors and is a vector quantity the direction of which is given by the unit vector $n$.

The direction of $n$ is not simply perpendicular to the area i.e. the plane of the parallelogram but it follows the right hand screw rule i.e. when it is $a \times b$ the direction of $n$ is vertically "up" (as shown in Fig. 1.5) and when it is $b \times a$ i.e. when the mode of multiplication is reversed i.e. commutated and the $n$ is vertically "down". Therefore, the commutative law of vector multiplication is not valid when it is vector or cross product.

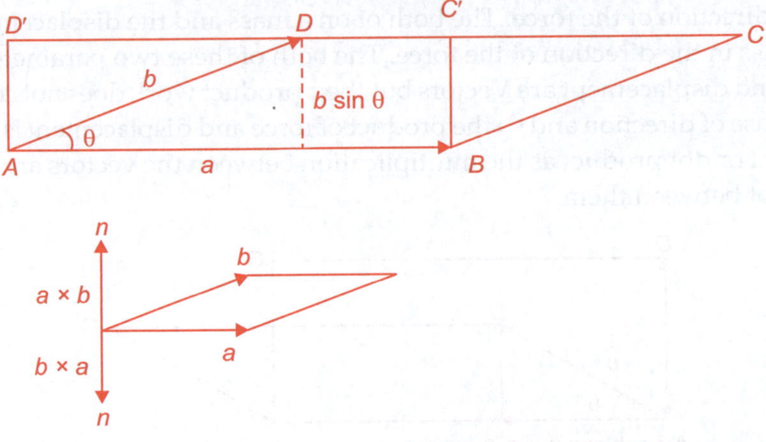

**Fig. 1.5** The vector product between two vectors

Therefore, the area is a vector quantity as for a surface it is the vector (Cross) product of two vectors $a$ and $b$ i.e. lengths on two directions. The directions given by unit vectors $n$ either up or down signify the curvature of the surface i.e. convex or concave. This can however be understood from the following (Fig. 1.6). The volume of any body is however a scalar as it is given by $c \cdot (a \times b)$, the scalar product between the area (vector) and the length in the third direction.

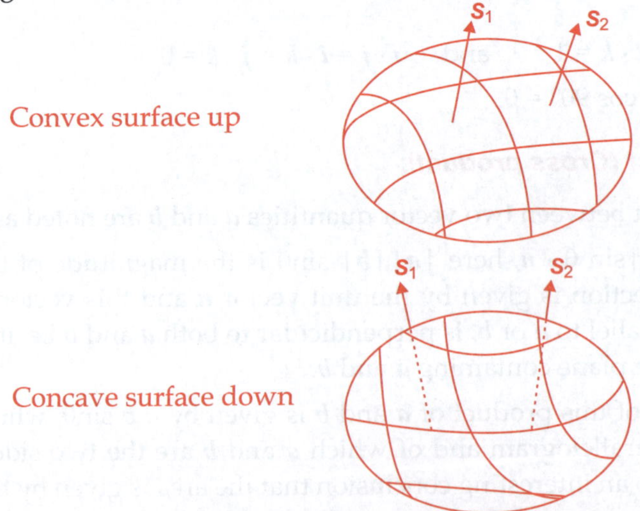


**Fig. 1.6** The surfaces and the directions of the areas of different sections given by unit vectors as $S_1$ and $S_2$. The addition of these vectors $S_1$, $S_2$, $S_3$ ... gives the net direction of the total area

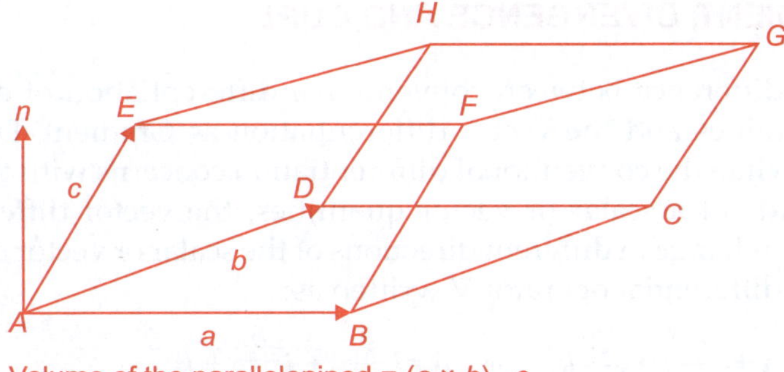



**Fig. 1.7** The volume of any body is the product of three vectors.

$a \times b = a\, b \sin\theta\, n$ where $a\, b \sin\theta$ is the area of the parallelogram *ABCD* and its direction is perpendicular to the area and is given by *n*. Now when $(a \times b)$ is scalar product with *c*, the another dimension of the body then $c \cdot n$ is the perpendicular component of *c* i.e. height of the body and as it is multiplied with the base area, *ABCD* it gives the volume of the body.

**Conclusion:** The dot product between two mutually perpendicular vectors is zero and cross product between two parallel vectors is also be zero and:

$i \times j = k,\ j \times i = -k,\ k \times i = j,$ positive in the order $i \to j \to k$ and in any reversal negative.

### Vector-scalar: classification of some important physical parameters

| Parameters | Classification | | Reason for the Classification |
|---|---|---|---|
| | **Vector** | **Scalar** | |
| Displacement/ Length | Vector | | Has a sense of direction |
| Mass | | Scalar | Has no sense of direction |
| Time | | Scalar | Has only one sense of direction |
| Velocity | Vector | | L/T, Vector/Scalar and so Vector |
| Acceleration | Vector | | $L/T^2$, Vector/Scalar |
| Momentum | Vector | | M.L/T, Scalar × Vector |
| Force | Vector | | $ML/T^2$, Scalar × Vector |
| Moment of force (Torque) | Vector | | $(ML/T^2) \times L$, Cross product between two vectors i.e. force and displacement |
| Pressure | | Scalar | $ML/T^2L^2$, dot product between two vectors, e, force. 1/area: F . n / area i.e. normal component of force / magnitude of area |
| Work | | Scalar | $W = (ML/T^2).\, L$, dot product between two vectors |
| Power | | Scalar | W/T, two scalars are involved |
| Energy | | Scalar | It is work available |
| Temperature | | Scalar | No direction sense |

## 1.4 GRADIENT, DIVERGENCE AND CURL

There is a difference between conventional differentiation of a scalar or vector quantities and the vector differentiation as Gradient, Divergence and Curl. While the conventional differentiation concerns with the change of magnitude of a scalar or vector quantities, the vector differentiation concerns the change in different directions of the scalar or vector quantities. The vector differential operator $\nabla$ written as:

$$\nabla = \frac{\partial}{\partial x} i + \frac{\partial}{\partial y} j + \frac{\partial}{\partial z} k, \quad \text{or} \quad \delta = i \frac{\partial}{\partial x} + j \frac{\partial}{\partial y} + k \frac{\partial}{\partial z}.$$

**The Gradient:**

If $\nabla \varphi(x, y, z)$ is a scalar field which is differentiable at every point of the field $x, y, z$ then

$$\nabla \varphi = \frac{\partial \varphi}{\partial x} i + \frac{\partial \varphi}{\partial y} j + \frac{\partial \varphi}{\partial z} k \text{ and therefore, } \nabla \varphi \text{ defines a vector field.}$$

**The Divergence:**

If $V(x, y, z)$ is a differentiable vector field and is defined as:

$V(x, y, z) = V_1 i + V_2 j + V_3 k$, the divergence of $V$, which is the dot product of $V$ and $\nabla$ is given by:

$$\nabla . V = \left( \frac{\partial}{\partial x} i + \frac{\partial}{\partial y} j + \frac{\partial}{\partial y} k \right) \cdot (V_1 i + V_2 j + V_3 k)$$

$$= \frac{\partial V_1}{\partial x} + \frac{\partial V_2}{\partial y} + \frac{\partial V_3}{\partial z}.$$

**The Curl**

The curl or rotation of the differentiable vector $V$ is the cross product between $\nabla$ and $V$.

$$\nabla \cdot V = \frac{\partial \varphi}{\partial x} i + \frac{\partial \varphi}{\partial y} j + \frac{\partial \varphi}{\partial y} k \cdot (V_1 i + V_2 j + V_3 k)$$

$$\nabla \times V = \left( \frac{\partial}{\partial x} i + \frac{\partial}{\partial y} j + \frac{\partial}{\partial z} k \right) \times (V_1 i + V_2 j + V_3 k)$$

$$= \begin{vmatrix} i & j & k \\ \dfrac{\partial}{\partial x} & \dfrac{\partial}{\partial y} & \dfrac{\partial}{\partial z} \\ V_1 & V_2 & V_3 \end{vmatrix}$$

## 1.5 NEWTONIAN MECHANICS: KINEMATICS

It will be explained later that the dynamics of motion of a particle and of a rigid body having finite shape and of course mass are different, though the same laws of motions are applicable. They are different in the sense that for a particle mass there is only motion of the particle mass on a line which is either straight or in a curve and the motions about its own axis may not be observed and measured because of the fact that it's mass is concentrated in a point having no dimension. The motion of a rigid mass are of two folds one which is similar to that of the motion of the particle and the other the motion of the mass about its own axis resulting into more complicated motion. Therefore, the study of dynamics is classified in to two categories one which is simpler i.e. particle dynamics known as Kinematics and the other more complicated the dynamics of rigid bodies. In the treatment of dynamics it is the first one with which we can safely start and then later we may introduce the finite size effect over the particle nature of mass.

**Motion on a Plane:** A particle can move on a plane in such ways that either its position vector from a fixed origin changes only in magnitude but its direction remains same, under this condition the motion of the particle is straight line motion, if the position vector of the particle only changes its direction but the magnitude remains constant, then the motion is purely circular motion. In these two extremes, the particle in general can move on a plane so that both the magnitude and the direction of its position vector change. This motion on a plane can either be explained by Cartesian coordinates or the polar coordinates of the point.

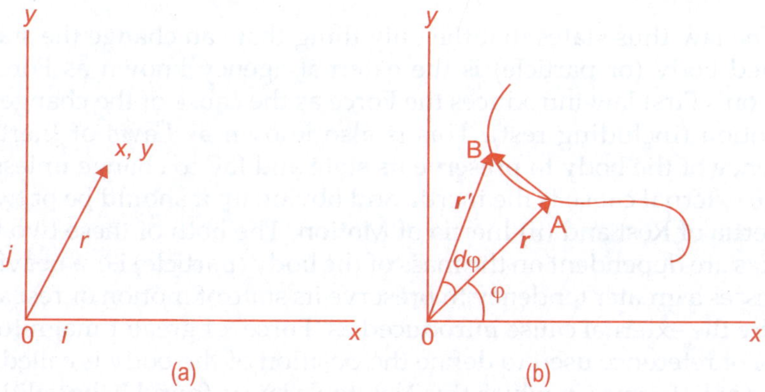


**Fig. 1.8 a and b** A represents the motion of a particle given by the position vector **r** expressed in Cartesian coordinates and B represents the same in polar coordinates

**In Cartesian coordinate:**

$$r = i\,x + j\,y$$

and
$$\frac{dr}{dt} = \dot{r} = i\,\frac{dx}{dt} + j\,\frac{dy}{dt} = i\,v_x + j\,v_y$$

When a particle moves in X, Y plane in random fashion, what changes is its coordinates and the sum total of its such change gives the velocity.

**In Polar coordinates:** When the particle moves, $r$ changes both in magnitude and in direction and its rate of change:

$r = nr$, where $n$ is the unit vector which defines the direction of $r$

$$\frac{dr}{dt} = n\frac{dr}{dt} + r\frac{dn}{dt} \qquad \qquad \dots(1.1)$$

The change in the direction of the unit vector $n$ is in its perpendicular direction given by the component $n_\theta$ and that is given by the change of value of $\theta$ and so

Therefore, $\qquad \frac{d}{dt}n = \lim_{\Delta\theta\to 0}\frac{\Delta\varphi}{\Delta t}\cdot n_\varphi = \varphi n_\varphi \qquad \dots(1.2)$

Therefore, $\qquad v = \frac{dr}{dt} = n\frac{dr}{dt} + rn_\theta = \frac{d\varphi}{dt} \qquad \dots(1.3)$

$v = nv_r + n_\varphi v_\varphi$ where $v_r\left(=\dfrac{dr}{dt}\right)$ and $v_\varphi\left(=r\dfrac{d\varphi}{dt}\right)$ are radial and tangential velocities.

## 1.5.1 Newton's Laws

**First Law: Every body will continue in the state of rest or of uniform velocity unless it is acted on or be influenced externally by an action, known as Force.**

The law thus states that the only thing that can change the state of an isolated body (or particle) is the external agency known as Force. Thus Newton's First law introduces the Force as the cause of the change of state of motion (including rest). This is also known as Laws of Inertia. This tendency of the body to preserve its state and fail to change unless forced by the external cause is the inertia and obviously it should be of two types (i) Inertia of Rest and (ii) Inertia of Motion. The both of these two types of inertias are dependent on the mass of the body (particle) i.e. a heavier body possesses a greater tendency to preserve its state of motion or rest and thus require the external cause introduced as 'Force' of greater magnitude. The frame of reference used to define the position of the body is called Inertial Frame of Reference in which this Newton's law is found to be valid and the frame where it is found to be violated is known as Non Inertial Frame of Reference. If we have a reference plane on which a body is rested and if the plane is moving with uniform velocity then there will be no relative motion between the body and the plane and same thing is obviously observed if the frame is not moving at all. This state of affair will continue until and unless the body is acted on by an external force. But if the frame of reference is given a sudden acceleration in one direction then the body on it will tend to accelerate in reverse direction under the action of a force (provided the generated force exceeds certain limit dependent on the mass of the body and the nature of the surface: limiting Friction) which was not applied directly on it. As this state in an accelerating frame of reference do not obey the Newton's law, the accelerating frame is called Non Inertial Frame of Reference. It has been established that more massive a body is more will be its inertia irrespective of this inertia being either inertia of rest or motion. Therefore this inertia is measured by mass and in same units.

**Second Law: The rate of change of momentum of a body is proportional to the applied force and it takes place in the same direction as that of the force applied.**

Now, as the First Law introduces the concept of force and also defines it, the second law gives its measurement. The momentum is the product of mass of a body (or particle) and its velocity. Now if $v$ is taken as the final velocity after a time t from the state when the velocity of the body was $u$, then the change of momentum is: $mv - mu$, and as this change has taken place in the time interval $t$, the rate of change of momentum is:

$\dfrac{mv - mu}{t}$ and according to the Second Law, if $F$ is the applied force on the body to cause this change of momentum, then:

$$F \propto \frac{mv - mu}{t} \text{ and as } \frac{d}{dt} \text{ where } f \text{ is the acceleration, then } F \propto mf \text{ or, } F = k$$

$mf$, where $k$ is the proportionality constant. Now, as we have already chosen the units of mass say in CGS system as gm and acceleration as cm. $s^{-2}$ and if we take that quantity of force when applied on a mass of 1 gm produces 1 cm. $s^{-2}$ acceleration is a unit of force in CGS system. This is taken as of one unit magnitude and is called 1 dyne, and then 1 dyne is equal to 1 gm. 1 cm. $s^{-2}$. Then the value of the proportionality constant $k$ will be equal to 1 and so, measured in the same units,

$$F = mf \text{ or in differential form:}$$

$$\frac{d}{dt} mv = \frac{d}{dt} p = F.$$

Now, this constant of proportionality $k$ must remain always equal to unity when all the parameters are expressed in the same units. Say for MKS system, acceleration produced of amount 1 ms$^{-2}$ and in a mass of 1 kg is then is to be taken as one unit of measurement of force in MKS system and is known as 1 Newton.

Now, 1 kg = $10^3$ gm, and 1 ms$^{-2}$ = $10^2$ cm. s$^{-2}$ and so,

1 Newton = 1 kg. 1 ms$^{-2}$ = $10^3$ gm. $10^2$ cm.s$^{-2}$ = $10^5$ gm.cm.s$^{-2}$ = $10^5$ dyne.

Like this there is FPS system and the force in that unit is known as poundal.

In addition to these units of force there is one gravitational unit of force. It is the force with which earth attracts a mass towards its centre. So, a gm weight of force in this gravitational unit is the force with which earth attracts a mass of 1 gm towards its centre which is practically the weight of one gram of mass.

Now, a force of appreciable amount acts on a body of mass $m$ for such a short duration of time so that the change of its velocity under the action of force cannot be measured then the force is defined as Impulsive Force or simply Impulse. It is then measured in terms of the change of momentum. When a cricket ball is hit by the bat then the entire force is applied within such a short duration that measurement of acceleration of the ball during the time when it is in contact with the bat (as the entire force is delivered on the ball during the period of its contact with the bat) is not possible and so the impulse measured by the difference of momentum ($mv - mu$) is taken as the representative of force.

### Third Law: To every action there is an equal and opposite reaction.

Now, let us explain what we understand by the word "Action". It can be said that the action is the after effect of the application of any external cause tending to make some changes in the state of rest or motion of the body. When a canon shell say of mass m is fired from the canon with a velocity of $v$, this action will be acted on the canon itself as a reaction and so the canon will also experience a recoil velocity $V$ so that $mv$ is equal to $MV$ where $M$ is the mass of the canon. When a soccer ball is kicked with some velocity, it develops a momentum which the action imparted on the ball by the leg. As a result an equal momentum will be acted as a reaction on the leg. This should be taken in mind always that the action is equal and opposite but both of these two cannot act on the same body. It is thus impossible to move a car by pushing the car while sitting on it.

When a lift with a man of mass $M$ standing on its floor is moving up with an acceleration say, $f$, then net reaction of the floor on the man in the upward direction $R$ will be given as: $R = Mg + Mf$. All the three possible states of motions and the corresponding reactions are given by the following Fig. 1.9.

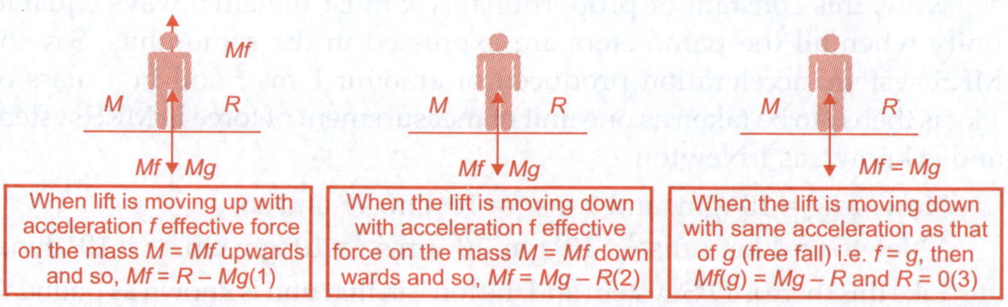


**Fig. 1.9** The different value of reaction R when the lift is moving with acceleration

In the first stage the body will appear heavier, its weight measured under this condition will read more than actual and in the second case the body will weigh lighter and in the third case as the reaction $R$ is zero the body will appear weightless and this will be shown by spring balance or any weighing machine, reading zero.

In the discussion of the Newton's Laws of motion and also in the Newton's Laws, it has to taken in to mind that it is always simpler if we consider a particle mass instead of a rigid body having a particular volume and fixed shape. The treatments to explain the dynamics of these two objects are different and they can thus be classified as (i) Particle Dynamics (ii) Rigid body dynamics. So far we have started with particle dynamics and in due course, we will introduce the rigid body dynamics.

## 1.5.2 Conservation of Linear Momentum and Energy

This is one of the fundamental outcome from the third law. The sum of the momentum of the interacting masses before any action takes place will be equal to the sum total of the momentum after such action is completed. Recalling the same example of firing a shell from the canon, the sum total of the momentum before the shell was fired is

$MV + mv = 0$ as both $V$ and $v$, the velocities of the canon and the shell were zero. Now, after the shell was fired,:

$M \times 0 + m \times 0 = 0 = MV + mv$, and so if the direction of $v$ is from left to right, the direction of $V$ is opposite and is from right to left and is of magnitude:

$V = -mv/M$. Now, if the canon is moving in either left to right or from right to left then after the action of firing the shell from left to right its velocities will as per:

**Left to Right:**

$$MV_1 + m \times 0 = MV + mv$$

and so, $\qquad V = (MV_1 - mv)/M$ and

**Right to Left:**

$$-MV_1 + m \times 0 = MV + mv$$

and so, $\qquad V = -(MV_1 + mv)/M.$

Now, conservation of energy is one of the most important fundamental principle. It says that no energy can be either created out of nothing or destroyed to nothing. It can only be changed or transformed from one existing form to other. When a moving wheel is stopped by applying breaks, its mechanical energy (Kinetic energy) will cease to exist and the energy will then is transformed in to heat energy which is manifested by the increase of temperature of the break shoes. When a body is thrown up with a certain velocity (it cannot be done without it), then as the body moves up its velocity decreases as it gains another mechanical energy, known as potential energy. At the highest point of its movement of its velocity becomes equal to zero and all the energy is then transformed in to potential energy. An opposite thing goes on happening when the body descends its potential energy decreases and it is then converted to the equal amount of kinetic energy. It is an interesting note that at every position of its motion either up or down, the sum total of its potential and kinetic energy will remain constant. Whatever example we can think of, there is no violation of this fundamental principle and the principle of conservation of energy.

## 1.6 ANGULAR MOTION AND ANGULAR MOMENTUM

In equn. 1.1 if the vector $r$ given by $n_r r$ changes only in direction and not in magnitude then $\dfrac{dr}{dt} = 0$ and the equation then results in to

$$\frac{d}{dt} r = r \frac{d}{dt} n = r \frac{d\varphi}{dt} \cdot n_\varphi,$$

where $r \dfrac{d\varphi}{dt}$ is the tangential velocity ..... whose direction is given by the unit vector $n_\varphi$ perpendicular to $n_r$.

Now, $\dfrac{d\varphi}{dt}$ is known as angular velocity and as it is a velocity its direction should determine whether the rotation of the body is clockwise or anti-clockwise and to specify the direction of rotation right hand screw rule is taken to give the direction of this vector.

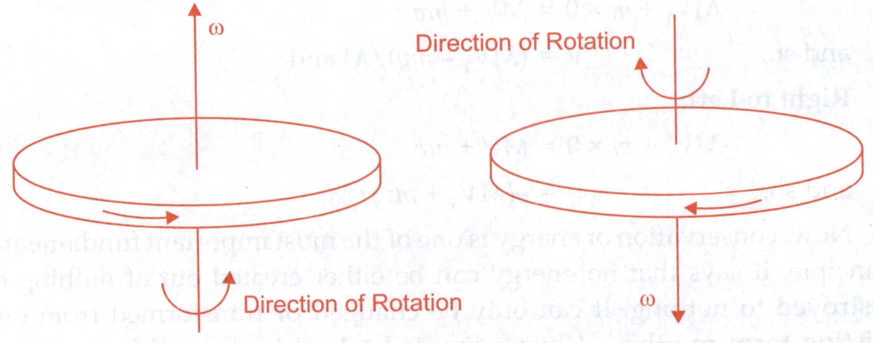


**Fig. 1.10** Rotation and the directions of the angular velocity vector ω

Now, when a body is acted on by a constant force it will go on accelerating and conversely if a body is observed to be constantly accelerating it is definitely under the action of a constant force. Consider a body is rotating in a circle with constant angular velocity. Its tangential velocity changes in its direction at every point of existence of the body on the circle. As the change of velocity or associated momentum can only be made possible if the body is acted on by a force, then at every point of its motion on a circular path of constant radius and constant angular velocity, a force must be constantly acting on the body and that force needed to change direction of tangential velocity is known as Centripetal force.

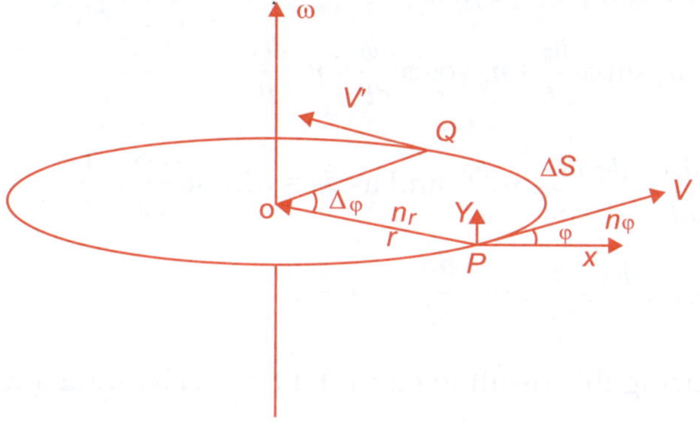

Fig. 1.11 Centripetal and centrifugal acceleration and the forces

Now, if $\Delta\theta$ is the angular displacement in $\Delta t$ time the angular velocity:

$$\omega = \lim_{\Delta t \to o} \frac{\Delta\varphi}{\Delta t} \text{ and its relation with tangential velocity } v = r \times \omega$$

$$= r \cdot \omega \sin 90° n_\varphi.$$

Now, through the magnitude of the product $r \cdot \omega$ which is $v$ for a constant $\omega$ and $r$ do not change but its direction constantly changes with $n_\varphi$. This cannot be achieved without the action of a force and that force acting constantly on the body moving on the perimeter of the circular path must not have any of its component along tangent as in that case, the movement of the body with constant angular velocity $\omega$ would not be possible. Therefore, this force must always act along the radius of the circular path so that it cannot have any component along the tangent and increase the tangential velocity and the angular velocity also. This radial force is the Centripetal force and it acts on the body towards the centre along the radius. As every action has its equal and opposite reaction, so this centripetal force is also balanced by its reaction centrifugal force acting on the centre and directed towards the body along the radius.

Now, this centripetal force causes an acceleration called centripetal acceleration say $a$.

The **a** is then given by: $\quad a = \lim\limits_{\Delta t \to 0} \dfrac{\Delta v}{\Delta t} = \dfrac{dv}{dt}$

$$a = \frac{dv}{dt} = \frac{d}{dt}(n_\varphi^v) = n_\varphi \frac{dv}{dt} + v\frac{dn_\varphi}{dt} \qquad \ldots(1.4)$$

$$n_\varphi = n_x \cos\varphi + n_y \sin\varphi \text{ and } n_r = n_x \cos(\varphi + \pi/2) + n_y \sin(\varphi + \pi/2)$$

$$= -n_x \sin\varphi + n_y \cos\varphi$$

Differentiating $n_\varphi$ we get:

$$\frac{dn_\varphi}{dt} = -n_x \sin\varphi \frac{d\varphi}{dt} + n_y \cos\varphi \frac{d\varphi}{dt} = n_r \frac{d\varphi}{dt}$$

Now, $\dfrac{d\varphi}{dt} = \dfrac{d\varphi}{ds}\dfrac{ds}{dt} = v\dfrac{d\varphi}{ds}$ and as $ds = rd\varphi$, so $\dfrac{d\varphi}{ds} = \dfrac{1}{r}$

Therefore, $\dfrac{d\varphi}{dt} = \dfrac{v}{r}$ and $\dfrac{dn_\varphi}{dt} = n_r \dfrac{v}{r}$.

Introducing this result in equn. 1.4 for acceleration a we get,

$$a = n_\varphi \frac{dv}{dt} + vn_r \frac{v}{r} \text{ and when the body is moving round in circular orbit}$$

with constant angular velocity, $\dfrac{dv}{dt} = 0$ and so the acceleration directed towards the centre along the radius is known as centripetal acceleration and is given by:

$$a = n_r \frac{v^2}{r} \qquad \ldots(1.5)$$

The force $F$ acting on the body is $F = ma = m\,n_r \dfrac{v^2}{r}$ this is centripetal force and its reaction acts on the centre. Now to prevent the movement of the body under the action of this centripetal force towards the centre, a force is imagined to act on the body which is equal and opposite to this centripetal force and as this can only be measured on the moving system (body) and not outside of it, this force is a pseudo force and is known as Centrifugal force. To develop this centripetal force the roads having sharp turn is banked towards the turn so that a component of the weight of the body may act as the necessary centripetal force.

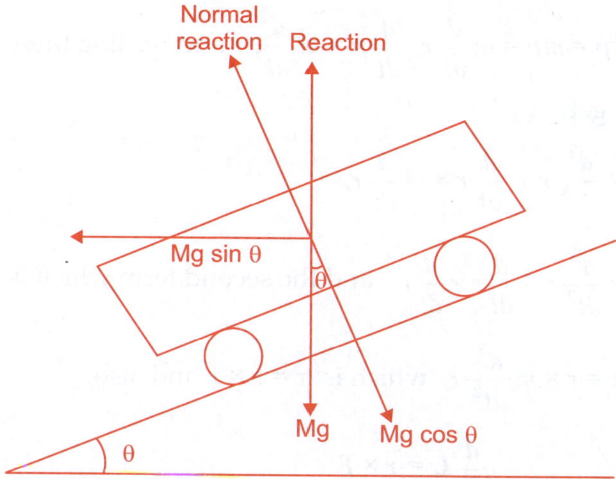



**Fig 1.12** Banking of the road at the turn. Mg sin θ component serves as the necessary centripetal force for a smooth turn. When the road at the turn is not banked then frictional force serves as the necessary centripetal force

Now, for a smooth turn on the road the centripetal force equals to $Mg \sin \theta = Mv^2/R$.

Where, $R$ is the radius of the turning path and the velocity $v$ is the velocity permissibly during turn for effective and smooth turn.

## 1.7 ANGULAR MOMENTUM CONSERVATION

The angular momentum $L$ is defined as $L = r \times p$, where $p$ is the linear momentum i.e. $m \cdot v$ and $r$ is the distance from the axis around which the particle is rotating.

Therefore, as $L = r \times p$, $L = r \times mv = mr \times v$

Now, as the rate of change of linear momentum $p$ is introduced in the Newton's second law as force, the rate of change of this angular momentum $L$ is also the force responsible for rotational motion and acceleration and this force is known as Torque. If $\tau$, a vector designate this torque, then:

$$\tau = \frac{dL}{dt} = \frac{d}{dt}(r \times p) = r \times \frac{d}{dt}p + \left(\frac{d}{dt}r\right) \times p$$

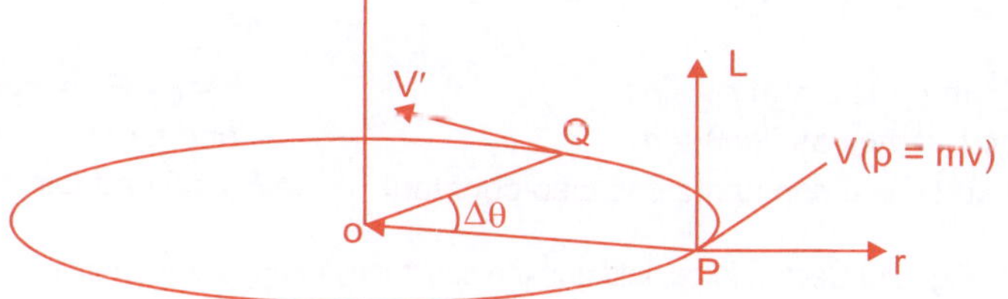

Now, as $p = mv = m\dfrac{d}{dt}r$, $\dfrac{d}{dt}p = m\dfrac{d^2}{dt^2}r$ and putting these in the above equation we get:

$$\tau = r \times m\,\frac{d^2}{dt^2}r + \frac{d}{dt}r \times m\frac{d}{dt}r.$$

$$= m\left( r \times \frac{d^2}{dt^2}r + \frac{d}{dt}r \times \frac{d}{dt}r \right) \text{ and the second term which is } v \times v \text{ is zero.}$$

and so, $\tau = r \times m\dfrac{d^2}{dt^2}r,$ which is $\tau = r \times F$ and also,

$$\frac{d}{dt}L = r \times F \qquad\qquad \dots(1.6)$$

Now, $L = 0$, when $L$ is constant and does not vary with time. This can be achieved for two conditions i.e.

1. When $F = 0$, no force acting on the particle and the particle is then moving with constant velocity in a straight line. This particle which is not under the action of a force is a free particle and we can conclude that for a "Free Particle" the angular momentum $L$ is constant.

2. When $r = 0$. As $r$ is a vector defining the normal distance of the direction of applied force $F$ from a fixed point and so $r = 0$ or $r \times F = 0$ means either the force $F$ passes through the point concerned or its direction is parallel to the direction $r$.

This second condition is then achieved when the force is called a central force and the motion of the particle will then have its angular momentum constant or conversely, when the angular momentum is constant the motion of the particle is then executed under the action of a central force.

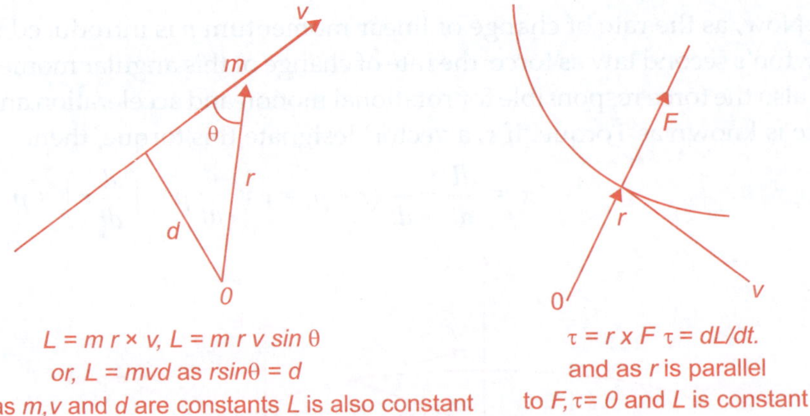


**Fig 1.13** Central force field and the constancy of angular momentum

## 1.8 MOTION UNDER CENTRAL FORCE

If a force $F$ acts on a particle at a distance $r$ from the origin of a reference frame at which another particle is located and if $f(r)$ is a function of $r$ defining the interaction, then

$$F = f(r).r/r \text{ or, } F = f(r).n_r$$

then the potential energy $U$ which is the work that is to be done on a particle is:

$$U(r) = -\int_{r_1}^{r} f(r)dr$$

Now, if there are two such point masses $m_1$ and $m_2$ at distances from the origin as $r_1$ and $r_2$ so that the distance between them is given by $r = r_1 - r_2 = rn_r$.

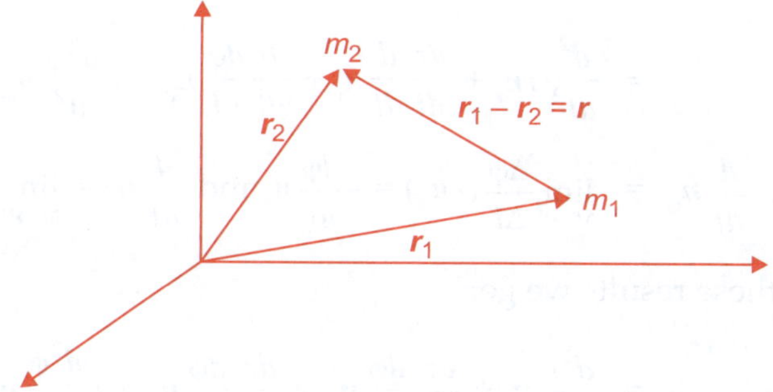


The force on mass $m_1$ is: $m_1 . \dfrac{d^2}{dt^2} r_1 = f(r)n_r$, where $r_1$ is a vector.

And the force on mass $m_2$ is: $m_2 . \dfrac{d^2}{dt^2} r_2 = -f(r).n_r$, ($r_2$ is also a vector)

the function which determines the nature of force between masses is $f(r) \leq 0$ if the force is attractive and $f(r) \geq 0$ if it is repulsive. These two equations for masses $m_1$ and $m_2$ can be combined in terms of "reduced mass" $\mu$, where $\mu =$ so that

$$\mu \frac{d^2}{dt^2} r = f(r).n_r.$$

As the motion is under central force, the torque of this force $f(r)$ and so the cross product of this force with $r$ should vanish, which is the condition for angular momentum to be conserved.

$$f(r)(r \times n_r) = r \times \mu \frac{d^2}{dt^2} r = \frac{d}{dt}\left(r \times \mu \frac{d}{dt} r\right) = L = 0$$

**[Note:**  $\dfrac{d}{dt}\left(r\times\mu\dfrac{d}{dt}r\right) = \mu\left(\dfrac{d}{dt}r\times\dfrac{d}{dt}r + r\times\dfrac{d^2}{di^2}r\right) = r\times\mu\dfrac{d^2}{dt^2}r$

$$= \dfrac{d}{dt}(r\times F) = \dfrac{d}{dt}L.$$

Since the direction of $L$ lies in space, the position vector $r$ can move only in a plane perpendicular to $L$ through the centre of force and so the motion is planar. We then require only two dimensional polar coordinates to define the position of the system the radial and tangential equations of motion can be written by recalling the equns. (1.3) derived before a:

$$v = \dfrac{dr}{dt} = n_r\dfrac{dr}{dt} + r\dfrac{d\varphi}{dr}n\varphi$$

$$a = \dfrac{d}{dt}v = \dfrac{d}{dt}\left\{n_r\dfrac{dr}{dt} + r\dfrac{d_\varphi}{dt}n_\varphi\right\}$$

$$= \dfrac{d^2}{dt^2}rn_r + \dfrac{dr}{dt}\dfrac{d}{dt}n_r + \dfrac{dr}{dt}\dfrac{d\varphi}{dt}n_\varphi + r\dfrac{d^2\varphi}{dt^2}n_\varphi + r\dfrac{d\phi}{dt}\dfrac{d}{dt}n_\varphi.$$

Now,  $\dfrac{d}{dt}n_\varphi = \lim\limits_{\Delta t\to 0}\dfrac{\Delta\varphi}{\Delta t}(-n_\varphi) = -\dfrac{d\varphi}{dt}n_r$, and $\dfrac{d}{dt}n_r = \lim\limits_{\Delta t\to 0}\dfrac{\Delta r}{\Delta t}n_\varphi = \dfrac{dr}{dt}n_\varphi.$

Using these results we get:

$$a = \dfrac{d^2r}{dt^2}n_r + \dfrac{dr}{dt}\dfrac{d\varphi}{dt}n_\varphi + \dfrac{dr}{dt}\dfrac{d\varphi}{dt}n_\varphi + r\dfrac{d^2\varphi}{dt^2}n_\varphi - r\left(\dfrac{d\varphi}{dt}\right)^2 n_r$$

$$= \left[\dfrac{d^2r}{dt^2} - r\left(\dfrac{d\varphi}{dt}\right)^2\right]n_r + \left[\dfrac{d^2\varphi}{dt^2}r + 2\dfrac{dr}{dt}\cdot\dfrac{d\varphi}{dt}\right]n_\varphi$$

and

$$\mu\left[\dfrac{d^2r}{dt^2} - r\left(\dfrac{d\varphi}{dt}\right)^2\right] = f(r) \qquad\qquad \dots(1.7\ a)$$

$$\mu\left[r\dfrac{d^2\varphi}{dt^2} + 2\dfrac{dr}{dt}\cdot\dfrac{d\varphi}{dt}\right] = 0 \qquad\qquad \dots (1.7\ b)$$

which is

$$\dfrac{\mu}{r}\dfrac{d}{dt}\left(r^2\dfrac{d\varphi}{dt}\right) = 0, \qquad\qquad \dots(1.7\ c)$$

and $\dfrac{\mu}{r}\dfrac{dL}{dt} = 0$ so, L is conserved and constant. Now, eliminating $\varphi$ from equation 1.7 b and writing it in terms of $L$ we get:

$$\mu \frac{d^2}{dt^2} r - \frac{L^2}{\mu r^3} = f(r) \qquad \qquad \ldots(1.8)$$

$$\text{or, } \mu \frac{d^2}{dt^2} r = -\frac{d}{dr}\left\{ U(r) + \frac{L^2}{2\mu r^2}\right\}, \text{ where } U(r) = -\int f(r)dr \qquad \ldots(1.9)$$

Now, multiplying both sides by $\dfrac{dr}{dt}$

$$\mu \frac{dr}{dt}\frac{d^2}{dt^2} r = -\frac{d}{dr}\left\{ U(r) + \frac{L^2}{2\mu r^2}\right\} = \left\{ U(r) + \frac{L^2}{2\mu r^2}\right\}$$

which is equivalent to:

$$\frac{d}{dt}\left\{ \frac{1}{2}\mu\left(\frac{dr}{dt}\right)^2 + \frac{L^2}{2\mu r^2} + U(r)\right\} = 0 \text{ and this concludes that the total}$$

energy $E_{total}$ which is $E_{total} = \dfrac{1}{2}\mu\left(\dfrac{dr}{dt}\right)^2 + \dfrac{L^2}{2\mu r^2} + U(r)$ under central force is also constant.

The effective potential energy $U_{eff}(r) = U(r) + \dfrac{L^2}{2\mu r^2}$ $\qquad\qquad \ldots(1.10)$

where $U(r)$ is a constant and the second term $\dfrac{L^2}{2\mu r^2}$ is the centrifugal

potential energy as it represents an additional force in the motion under central force. Therefore,

$$E_{total} = \frac{1}{2}\mu\left(\frac{dr}{dt}\right)^2 + U_{eff}(r) \qquad\qquad \ldots(1.11)$$

Now, recalling equn. (1.8)

$$\mu \frac{d^2 r}{dt^2} = f(r) + \frac{L^2}{\mu r^3} = f(r) + \mu r\varphi^2$$

$f(r) = -\dfrac{C}{r^2}$ i.e. $U(r) = -\dfrac{C}{r}$ for positive $C$ the interaction is attractive.

Now, the following Fig. 1.14 gives the variation of potential energy for central force for a specific case.

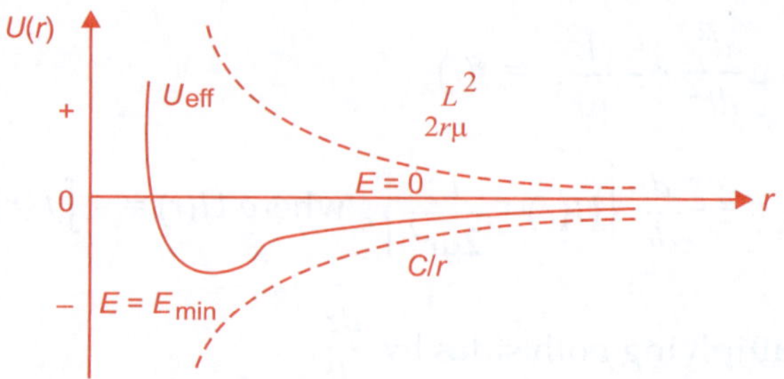



**Fig. 1.14** The potential energy variation for an attractive central force

Now, the motions of planets and the motions of electrons round nucleus of an atom are the most important motions of masses under the action of central force.

For motions of electron:

$$f(r) = -\frac{1}{4\pi\varepsilon_0}\frac{Ze^2}{r^2}\, ,\ U(r) = -\frac{1}{4\pi\varepsilon_0}\frac{Ze^2}{r} = -\frac{C}{r}$$

For planetary motions:

$$f(r) = -G\frac{m_1 m_2}{r^2}\, ,\ U(r) = -G\frac{m_1 m_2}{r} = -\frac{C}{r}$$

Now, $U_{\text{eff}} = -\dfrac{C}{r} + \dfrac{L^2}{2\mu r^2}$ and in the above figure $U_{\text{eff}}$ is given by the solid line and the curves given in broken lines represent $-\dfrac{C}{r}$ and also $\dfrac{L^2}{2\mu r^2}$.

Now, recalling Equn. 1.11

$$E_{\text{total}} = \frac{1}{2}\mu\left(\frac{dr}{dt}\right)^2 + U_{\text{eff}}(r)$$

Now, for a mass $m$ moving under central force

$$E_{\text{total}} = \frac{1}{2}mv^2 - G\frac{mm'}{r}\ \text{and as}$$

$$\frac{mv^2}{r} = G\frac{mm'}{r^2}$$

and therefore, $\quad \dfrac{1}{2}mv^2 = G\dfrac{mm'}{2r}$

so, $\qquad E_{\text{total}} = -G\dfrac{mm'}{2r}$

Now, the kinetic energy $T$ is given by $T = E_{total} - U_{eff}$ and the motions of the mass under central force is determined by total energy and the path of the mass is determined by the conditions i.e. $E > 0$, $E < 0$ or $E = 0$.

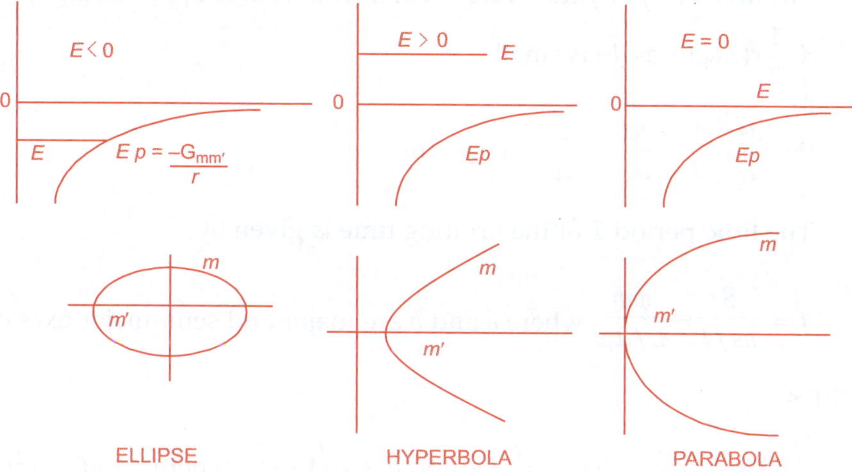


**Fig. 1.15** The total energy $E_{total}$ ($E$) and the path of motion of mass under inverse square central force

## 1.9 KEPLER'S LAW

Kepler's three laws of planetary motions though restricted to the inverse square force but in general it is an important formulation of central force field.

**Kepler's First Law:** The planets move in elliptical orbits with sun at one of its foci.

**Kepler's Second Law:** The vector from the sun to the planet describe equal areas in equal times.

**Kepler's Third Law:** The square of the period of a planet is proportional to the cube of the semi major axis of its orbit.

**Discussions on Kepler's Law:**

The path of planets are elliptical as the total energy in negative and with sun at one of the foci.

**Recalling equn. (1.7 c) we get:**

$$\frac{\mu}{r}\frac{d}{dt}\left(r^2\frac{d\varphi}{dt}\right) = 0$$

$$\text{or, } \frac{1}{r}\frac{d}{dt}\left(\mu r^2\frac{d\varphi}{dt}\right) = \frac{1}{r}\frac{d}{dt}L = 0 \text{ as } L = \mu r^2\frac{d\varphi}{dt}$$

and so $\dfrac{L}{2\mu} = \dfrac{1}{2} r^2 \dfrac{d\varphi}{dt}$ = constant.

The area swept by the radius vector in time $dt$ is approximately

$ds \dfrac{1}{2} r(rd\varphi) =$ as $d\varphi$ is small.

So, $\dfrac{ds}{dt} = \dfrac{1}{2} r^2 \dfrac{d\varphi}{dt} = \dfrac{L}{2\mu}$ = constant.

The time period $T$ of the orbiting time is given by

$$T = \frac{S}{ds/t} = \frac{\pi ab}{L/2\mu}$$ where $a$ and $b$ are major and semi-major axes of the

ellipse.

And $a = \dfrac{p}{1-e^2}$, $b = \dfrac{p}{\sqrt{1-e^2}}$ and $b = a\sqrt{1-e^2}$, where $p = a(1-e^2)$.

Now, as $\pi ab$ is its area.

Therefore, $T^2 = \dfrac{4\mu^2}{L^2} \pi^2 a^4 (1-e^2)$ again as $L = \mu C p = \mu C a (1-e^2)$

So we get: $T^2 = \dfrac{4\pi^2 \mu}{C} a^3$, which is the third law of Kepler.

## 1.10 RIGID BODY DYNAMICS

A "Rigid body" means a mass having definite finite volume and shape and its dynamics is more complicated than that of a "Particle mass" having definite mass but no volume. When a force acts on a particle it may execute a translational motion or a rotary motion around an axis lying out side the body. These motions are discussed in this chapter. When a body is a rigid body which does not change its shape on the application of a force is acted on by a force, whether it will execute a pure translational motion or also a rotary motion about an axis passing through its own body depends on the point of application of the said force. This requires an introduction of the concept of "Centre of Mass". In the following Fig. 1.16, the rigid body is under the action of a series of parallel forces acting on the point masses with which the entire body may be assumed to be made up.

Now, as the forces applied $F_1, F_2, F_3, \dots$ etc. are all parallel to each other the resultant force $F$ will be given by $F = F_1 + F_2 + F_3 + F_4 + \dots$ as they act on the point masses $m_1$, $m_2$ and $m_3$ etc. they will produce accelerations which will be all equal as they body is a rigid body and there is no movement of any part of it relative to the other. Now, suppose relative to a Cartesian coordinate system the point masses have coordinates respectively as $x_1 y_1 z_1$, $x_2 y_2 z_2$, $x_3 y_3 z_3$ etc. Therefore, we can write:

For the entire mass of the body:

$$m_1 + m_2 + m_3 + \dots = \sum_i m_i = M(\text{say}) \text{ and}$$

For all the parallel forces:

$$F_1 + F_2 + F_3 + \dots = \sum_i F_i = F(\text{say})$$

Again as these forces acting on different point masses generate same acceleration say **a** then:

$$a(m_1 + m_2 + m_3 + \dots) = a \sum_i m_i = F.$$

Now, the concept of centre of mass is that through this specific point the resultant force passes and this remains the sum total of the component actions unaltered.

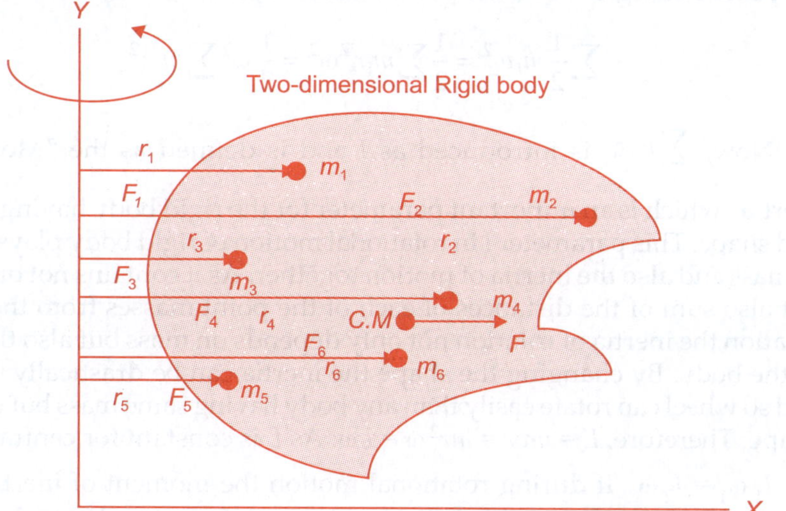


**Fig. 1.16** A two-dimensional rigid body made up of point masses

Taking moment about Y-axis for all system of forces acting point masses we get:

$$a\, m_1 x_1 + a\, m_2 x_2 + a\, m_3 x_3 + a\, m_4 x_4 + \dots = a \sum_i m_i x_i = FX = aX \sum_i m_i$$

and then the $x$-coordinate of the centre of mass $X = \dfrac{\sum\limits_i m_i x_i}{\sum\limits_i m_i}$ and similarly

taking moment of the component forces about X-axis we get the Y-coordinate of the centre of mass:

$$Y = \frac{\sum\limits_i m_i y_i}{\sum\limits_i m_i}.$$

Therefore, the centre of mass which is an important concept of rigid body dynamics is a definite position which either lies inside the body or sometimes outside can only change if the shape of the body is changed. Now, as given in the above figure if the body rotates round the Y-axis with an angular velocity of $\omega$ then the kinetic energy of the point masses $m_1, m_2, m_3, m_4$ etc are given as:

$$\sum_i \frac{1}{2} m_i v_i^{\,2} = \sum_i \frac{1}{2} m_i r_i^2 \omega^2 \text{ for the point masses } m_1, m_2, m_3 \text{ etc. The total}$$

kinetic energy of the entire body can be obtained by simply summing over the kinetic energies of all the point masses and so,

$$\sum_i \frac{1}{2} m_i v_i^2 = \frac{1}{2} \sum_i m_i r_i^2 \omega^2 = \frac{1}{2} \omega^2 \sum_i m_i r_i^2$$

Now, $\sum\limits_i m_i r_i^2$ is introduced as $I$ and is defined as the "Moment of Inertia" which is an important parameter for the rigid body having volume and shape. This parameter I in rotational motion of rigid body plays the role of mass and also the inertia of motion together. As it contains not only mass but also sum of the distances of each of the point masses from the axis of rotation the inertia of rotation not only depends on mass but also the shape of the body. By changing the shape the inertia can be drastically changed and so wheel can rotate easily than any body having same mass but different shape. Therefore, $L = mvr = mr^2 \omega = I \omega$. As $L$ is constant for central force,

$I_1 \omega_1 = I_2 \omega_2$. If during rotational motion the moment of inertia of the rotating body decreases either due to decrease of mass or due to the change of shape, then the angular velocity of the body will increase. In short the moment of inertia plays the same role in rotational motion that mass plays in linear motion.

Now, when a force is applied on such rigid body, if the applied force passes through the centre of mass then the motion of the body will be purely translational but when a force acting through a point on the body other than the centre of mass of the body will execute both the translational and the rotation motion about the centre of mass. The translational motion is due the force transformed through the centre of mass and parallel to the applied force and rotational due to the moment and the applied force through the actual point of application of this moment will be equal to the force, which is multiplied by the perpendicular distance of the applied force from the centre of mass.

**Example:** When we throw a marble it will move in a trajectory, but when we throw a brick it will execute both trajectory motion and also it will go on rotating about its centre of mass. The movement of boomerang used as weapon by aborigines of Australia is a good example of the role of centre of mass in the dynamics of rigid bodies.

## 1.11 FRICTION

Friction being a resistive force generates, when a body tries to move over the surface. It acts through the point of contact of the body along the surface in the opposite direction to the direction of motion of the body. It always opposes the motion of the body and is applied by the surface and so depends on the nature of the surface of contact and the normal reaction of the surface on the body. It does not exist until a force is applied on the body tending to move the body and increases with the increase of the applied force until it reaches an upper limit and if the applied force is increased beyond that then the body starts its motion under the action of net force equal to the difference of applied force and the limiting frictional force.

Frictional force $f \propto N$, where $N$ is the normal reaction of the surface.

Then $f = \mu_s N$, this is the limiting frictional force. In the following (Fig. 1.17), the resulting frictional force is plotted against the applied force. It should be noted that the frictional force increases equally with the applied force up to the limiting friction and until that limit is crossed by the applied force the body cannot move. After the limit is crossed the frictional force decreases slightly as the kinetic friction coefficient $\mu_k$ is smaller than static friction coefficient $\mu_s$ and then remains constant.

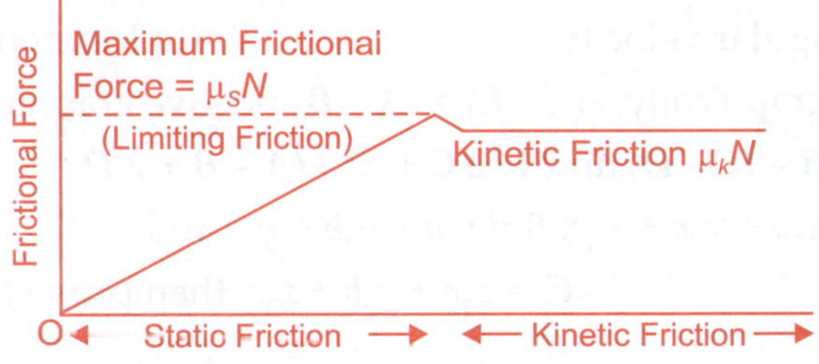


**Fig. 1.17** The variation of frictional force and the applied force

When a force more than the static friction $\mu_s N$, is applied continuously on a body of mass $m$, it will be under the action of a resulting force $F - \mu_s N$ and this will cause of an acceleration is equal to:

$$\frac{F - \mu_s N}{m} = a$$ and after this force is withdrawn, the kinetic energy attained by the body will be spent in doing work against kinetic friction and ultimately the body will stop after travelling through a distance say $S$, which is equal to:

$$\mu_k N \cdot S = \frac{1}{2} mv^2.$$

There are certain advantages and disadvantages of friction and among the advantages the most important is that it helps to make rolling possible. The rolling friction is less than sliding friction and so a body capable of rolling will continue to move longer distance. The following Fig. 1.18 gives the sliding friction down an inclined plane and the rolling friction.

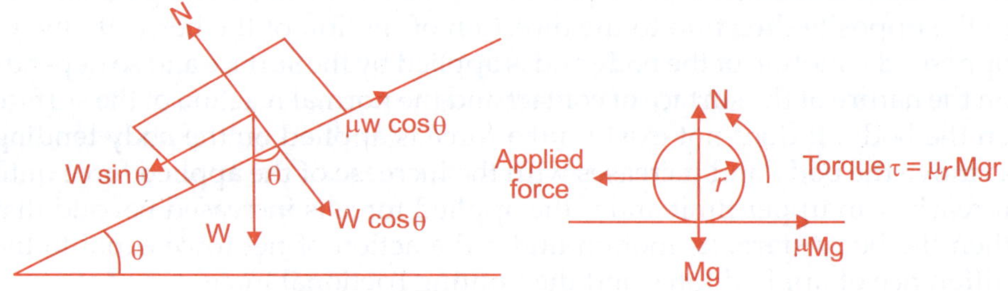


**Fig. 1.18**  The sliding friction down an incline and the rolling friction of a ball over a plane surface

## REVIEW QUESTIONS

1. Which of the following are scalars and which are vectors?

   (a) Kinetic energy  (b) Temperature  (c) Charge

   (d) Work             (e) Time          (f) Frequency

   (g) Angular velocity                   (h) Moment of inertia

2. Show graphically, $-(A - B) = -A + B$ and give graphical construction of:

   $3A - 2B - (C - D)$ and $1/2\,C + 2/3\,(A - B + 2D)$.

3. If $A = x_1 a + x_2 b + x_3 c$, $B = y_1 a + y_2 b + y_3 c$ and

   $$C = z_1 a + z_2 b + z_3 c,$$ then prove that

   $$A \cdot B \times C = (a \cdot b \times c) \begin{vmatrix} x_1 & x_2 & x_3 \\ y_1 & y_2 & y_3 \\ z_1 & z_2 & z_3 \end{vmatrix}$$

4. Prove that $(A \times B) \cdot (C \times D) + (B \times C) \cdot (A \times D) + (C \times A) \cdot (B \times D) = 0$.

5. If A is a constant vector, then prove that $\nabla(r \cdot A) = A$.

6. $\nabla f(r) = \dfrac{f'(r)}{r}\, r.$

7. Prove $\nabla^2 (\varphi \psi) = \varphi \nabla^2\psi + 2 \nabla\varphi \cdot \nabla \psi + \psi\nabla^2 \varphi.$

8. Describe how a driver of a car steer a car travelling at constant velocity so that (a) the acceleration is zero (b) the magnitude of acceleration remains constant.

9. Describe the trajectory of a large body of finite size and of irregular shape, when thrown in horizontal direction from a certain height.

10. How and when a solid body of rectangular cross section having sufficient weight that can rest motionless on an inclined plane.

## REFERENCES


1. H. L. Anderson–A Physicist's Desk Reference, American Institute of Physics, New York, 1989.

2. H. G. Jerrard, D. B. McNuill – Dictionary of Scientific Units, Chapman and Hall, London, 1992.

3. H. Goldstein – Classical Mechanics, (2nd Ed.), Addison-Wesley Publishing Co., Reading, Massachusetts, 1980.

4. M. Alonso and E. J. Finn – Fundamental University Physics, (Vol. 1), Addison-Wesley Pub. Co., Reading, Mass, 1968.

5. R. P. Feynman and R. B. Leighton-Feynman Lectures on Physics, Addison-Wesley Pub. Co. Reading, Mass.

6. M. R. Speigel-Vector Analysis, Schaum's outline series, Mc-Graw Hill Book Co. NY, 1974.

7. R. A. Serway and J. W. Jewett (Jr.)-Principles of Physics, Thomson Books/ Cole, 2002.

# Special Theory of Relativity

## 2.1 INTRODUCTION: GALILEAN TRANSFORMATION

There is nothing like absolute motions or rest. The dynamic state of every object is measured with respect to a frame of reference. To a person on a moving train on a straight track all the objects like poles, trees and also the person standing on the platform of a passing station appear to move in the opposite direction and to the person standing on the platform the person seating in the train will appear to move with the train. Let us consider the frame of references 1 and 2 of the following figure with 2 moving with constant velocity with respect to frame 1.

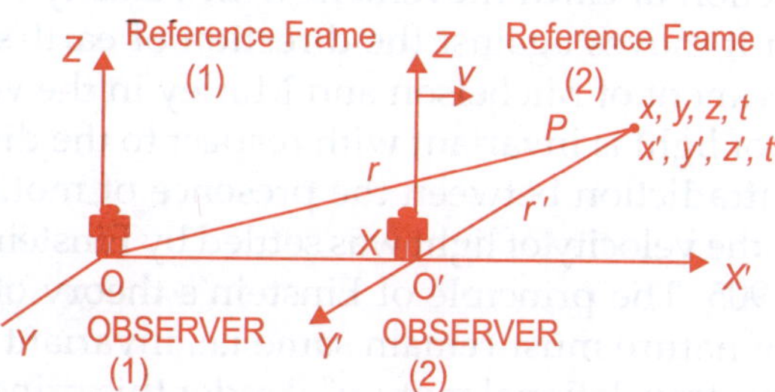


**Fig. 2.1** Relative motion and velocities. At $t = 0$, $OO' = 0$ and at $t = t$, $OO' = Vt$. The coordinates of the point of reference P as per frame 1 are $x, y, z, t$ and for frame 2 $x'$ $y'$ $z'$ $t'$

Let when $OO' = 0$, $t = 0$ and it is then measured equal in both frames, so that $t = t'$. Now, for the point $P$:

We get $X' = X - vt$ and as the frame 2 is moving along X-axis, $Y = Y'$ and $Z = Z'$.

$$\frac{dX'}{dt} = \frac{dX}{dt} - v$$ and $V' = V - v$ and if we reverse the direction of $v$, then $V' = V + v$.

The set of equations i.e.

$X' = X - v\,t$, $Y' = Y$ and $Z' = Z$ are known as **Galilean Transformations.**

The relative velocities for linear motions of frames of reference are shown in the following (Fig. 2.2)

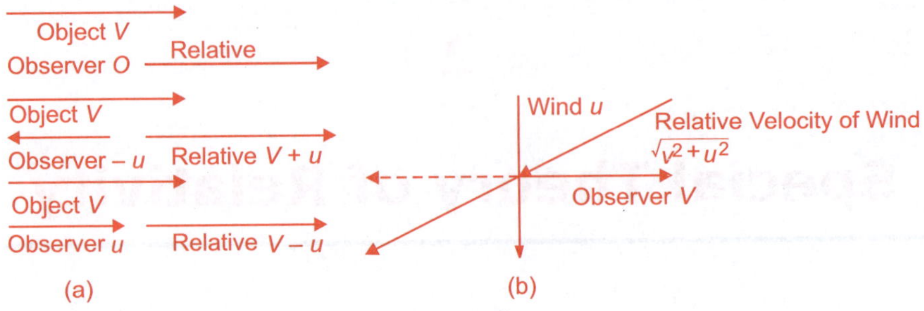


**Fig. 2.2** The vector representations of relative motions. (a) represent the relative velocities when both frames move along the same line and (b) when the directions are perpendicular to each other

## 2.2 LORENTZ TRANSFORMATION

Till the end of nineteenth century it was believed that the empty space above earth was filled up by a medium known as ether and earth moves within this medium without disturbing it and keeping it stationary. Therefore, the velocity of light will depend on the direction of light propagation relative to earth motion. It should then be equal to $c' = c - v$ (where $c' > c$) if it is directed in the same direction of earth movement with velocity $v$ and would be $c' = c + v$ ($c' > c$), if it moves against the direction of earth's movement. But the famous experiment of Michelson and Morley in the year 1881 showed that the velocity of light is invariant with respect to the direction of earth's motion. This contradiction between the presence of motionless ether and the invariance of the velocity of light was settled by Einstein's special theory of relativity in 1905. The principle of Einstein's theory of relativity states that "All Laws of nature must remain same i.e. invariant for all observers in uniform relative translational motion". Under this principle the Galilean transformation is not valid as inferred under this transformation that the velocity of light cannot remain invariant.

Recalling the (Fig. 2.1), let us assume that in the two frames $t = t' = 0$ when both of the observers' positions coincide. At this moment, let a splash of light is emitted from $O$ and according to Observer 1 it reaches the point $P$ after time $t$ so that $r = c\,t$ where $c$ is the velocity of light. As $x, y$ and $z$ are the position coordinates of point $P$ then

$$x^2 + y^2 + z^2 = r^2 = c^2 t^2 \qquad \qquad \ldots(2.1)$$

the Observer 2 at $O'$ will find that light emitted from $O$ reaches the point $P$ in time $t'$ so that:

$$x'^2 + y'^2 + z'^2 = r'^2 = c^2 t'^2 \qquad \qquad \ldots(2.2)$$

We are now to find the transformation relations between equns. (2.1) and (2.2).

Now as $y = y'$, $z = z'$ and $OO' = vt$, then we can write

$x' = k(x - v\,t)$ where $k$ is a constant and $t' = a(t - bx)$ and $a$ and $b$ are also constants. Now, from equns. (2.1) and (2.2) we get:

$$k^2 (x^2 - 2vxt + v^2t^2) + y^2 + z^2 = c^2a^2 (t^2 - 2bxt + b^2 x^2)$$

or, $(k^2 - b^2 a^2 c^2) x^2 - 2(k^2v - ba^2/c^2) x\,t + y^2 + z^2$

$$= (a^2 - k^2 v^2/c^2)\, c^2 t^2 \qquad \qquad \text{...(2.3)}$$

Comparing equn. (2.3) with equn. (2.1) we get:

$k^2 - b^2 a^2 c^2 = 1$, $k^2v - ba^2 c^2 = 0$ and $a^2 - k^2 v^2/c^2 = 1$ and solving these set of equations we get:

$$k = a = \frac{1}{\sqrt{1 - v^2/c^2}} \text{ and } b = v/c^2.$$

The transformations possessing the compatibility with the invariance of velocity of light lead to:

$$x' = k(x - vt) = \frac{x - vt}{\sqrt{1 - v^2/c^2}}$$

$$y' = y$$

$$z' = z \qquad \qquad \text{... (2.4)}$$

$$t' = k(t - bx) = \frac{t - vx/c^2}{\sqrt{1 - v^2/c^2}}.$$

This set of transformation equations are known as Lorentz Transformation. These primed values of coordinates of point $P$ as per frame 2 are in terms of the coordinates of the same point measured from frame 1 when frame 2 moves along X with velocity $v$. Now, the inverse Lorentz Transformation equations are those of the unprimed parameters measured in terms of frame 2 when frame has relative velocity $-v$ in the $-X'$ direction. They can be calculated by replacing $v$ by $-v$ and $x$ by $x'$ and they are:

$$x = k(x + vt') = \frac{x - vt}{\sqrt{1 - v^2/c^2}}$$

$$y = y'$$
$$z = z'$$

$$t = k(t' + bx') = \frac{t - vx/c^2}{\sqrt{1 - v^2/c^2}} \qquad \text{... (2.5)}$$

As commonly the velocity of any body is much less than the velocity of light i.e. as $v \ll c$, so $v^2/c^2 \approx 0$ and also $vx/c^2 \approx 0$ and then the set of Lorentz transformation equations revert back to Galilean transformation equations. In that general condition the time $t$ in one frame is equal to time $t'$ measured in the other frame moving relative to each other. The incidence observed in both of the frames will then appear to be simultaneous and the situation is then under classical relative motion. But if the velocity $v$ is very high and $v/c$ cannot be neglected, then there will be same change in the dimension of length of any body and also the time measured in two frames moving relative to each other.

## 2.3 CONSEQUENCES OF LORENTZ TRANSFORMATION

As the value of $k = \left(1 - \dfrac{v^2}{c^2}\right)^{-\frac{1}{2}}$ determines the difference between the Galilean and Lorentz transformations and as $k$ changes from $1(v \ll c)$ to $(v \approx c)$, these differences become more and more prominent. We then have the following consequences like contraction of length of a body in the direction of relative motion of the two frames and a dilation of time.

A. **Length Contraction**: Consider the Fig. 2.1 and let the observer in Reference frame 2 moving along X-axis with constant velocity $v$ parallel to X-axis has measured the length of a rod parallel to X-axis and stationary in his frame of reference. The readings at the two ends of rod read as $x_2'$ and $x_1'$ so that the length parallel to X-axis reads as:

$$L' = x_2' - x_1'$$

The readings $x_2'$ and $x_1'$ measured by the observer in Reference frame 1 are respectively as $x_2$ and $x_1$ $(L = x_2 - x_1)$ and the relations between them from equn. (2.4) are as:

$$x_2' = \frac{x_2 - vt}{\sqrt{1 - v^2/c^2}} \quad \text{and} \quad x_1' = \frac{x_1 - vt}{\sqrt{1 - v^2/c^2}}$$

$$\text{Therefore, } L' = x_2' - x_1' = \frac{x_2 - x_1}{\sqrt{1 - v^2/c^2}} = \frac{L}{\sqrt{1 - v^2/c^2}} \qquad \ldots(2.6)$$

Now, as $\sqrt{1 - v^2/c^2} < 1$ and $\dfrac{1}{\sqrt{1 - v^2/c^2}} > 1$, so, $L < L'$.

Therefore, the length of the rod measured by the observer in Reference frame 1 will appear shorter than the length measured by the observer in Reference frame 2 moving relative to 1 with constant velocity $v$. This is contraction of length in Lorentz transformation.

**B. Time dilation**: Let the time duration of an incidence which happens in Reference frame 2 is measured by the observer 2 and the difference between beginning and end time of the incidence at a place $x'$ in frame 2 measured as $t_1'$ and $t_2'$ so that, $t_2' - t_1' = \nabla t'$.

Now, the time intervals of the same incidence as measured from the Reference frame 1 will appear as $t_1$ and $t_2$ and $t_2 - t_1 = \nabla t$. From the above equn. (2.5) we get:

$$t_1 = \frac{t_1' + vx'/c^2}{\sqrt{1 - v^2/c^2}} \text{ and } t_2 = \frac{t_2' + vx'/c^2}{\sqrt{1 - v^2/c^2}}$$

Therefore, $t_2 - t_1 = \nabla t = \dfrac{t_2' - t_1'}{\sqrt{1 - v^2/c^2}} = \dfrac{\Delta t'}{\sqrt{1 - v^2/c^2}}$ ...(2.7)

and as $\dfrac{1}{\sqrt{1 - v^2/c^2}} > 1$.

So, $\nabla t > \nabla t'$ that shows that completion of the incidence is delayed in Frame 1 compared to the Frame 2.

## 2.4 RELATIVISTIC TRANSFORMATION OF VELOCITIES

Referring to the coordinates of the point $P$ from Reference frames 1 and 2 respectively as: $x, y, z, t$ and $x'\ y'\ z'\ t'$ and differentiating them

$$\frac{dx}{dt} = v_x, \frac{dy}{dt} = v_y, \frac{dz}{dt} = v_z$$

and

$$\frac{dx'}{dt'} = v'_x, \frac{dy'}{dt'} = v'_y, \frac{dz'}{dt'} = v'_z \qquad \ldots (2.8)$$

are the velocities in their respective frames of reference.

Now, taking differentials of the equn. (2.4) we get

$$dx' = \frac{dx - vdt}{\sqrt{1 - v^2/c^2}}, \ dy' = dy, dz' = dz \ \text{ and } \ dt' = \frac{dt - vdx/c^2}{\sqrt{1 - v^2/c^2}}.$$

From the above equations we get:

$$v'_x = \frac{dx - vdt}{\sqrt{1 - v^2/c^2}} \times \frac{\sqrt{1 - v^2/c^2}}{dt - vdx/c^2} = \frac{dx - vdt}{dt - vdx/c^2}$$

$$= \frac{dx/dt - v}{1 - v(dx/dt)/c^2} = \frac{v_x - v}{1 - vv_x/c^2}$$

And similarly, $v_x = \dfrac{v'_x + v}{1 + vv_x/c^2}$ and

$$v'_y = \frac{v_y\sqrt{1 - v^2/c^2}}{1 - vv_x/c^2} \quad \text{or} \quad v_y = \frac{v'_y\sqrt{1 - v^2/c^2}}{1 + vv_x/c^2}$$

$$v'_z = \frac{v_z\sqrt{1 - v^2/c^2}}{1 - vv_x/c^2} \quad \text{or} \quad v_z = \frac{v'_z\sqrt{1 - v^2/c^2}}{1 + vv_x/c^2}. \qquad \ldots (2.9)$$

The above equations are the relativistic additions of velocities and these equations will reduce to those given by Galilean transformation when the velocity $v \lll c$ so that $v/c \approx O$.

We can also see that it is not possible to obtain a velocity greater than the velocity of light i.e. $c$ even by changing the reference frame because if in the limiting case $v'_x = c$, then from equn. (2.9) velocity of the moving system

$v$ will be $v = \dfrac{c + v}{1 + vc/c^2} = c$. This gives the most important conclusion that

the velocity of light is same for all the case of inertial frames.

These mathematical results are the verification of the two postulates of Special Theory of Relativity and these are:

### Postulates Of Special Relativity

**First:** The laws of Physics take the same form in all inertial systems.

**Second:** The velocity of light in empty space is a universal constant c, and remains same for all observers and is independent of the state of motion of the emitting source.

### An imaginary example of the consequence of relativity

All the above results are from transformation equations and if we imagine a situation to visualize the outcome of special theory of relativity then the following imaginary situation and observation may be cited. The time dilation compared to length contraction is more difficult to conceive as this can never be experienced and is beyond our expectation. In the following Fig. 2.3 let two frames of references $O$ and $O'$ coincide at $t = 0$ i.e. $OO' = 0$ before the second frame starts moving relative to the first with velocity $v$ and along axis $x'$. A mirror parallel to $x'$ is fixed on $y'$ at a distance $L$ from the common origin (at $t = 0$) $O$ and $O'$. The time taken by a light beam to travel from $O$ and $O'$ to the mirror at a distance of $L$ get reflected and return to $O$, $O'$ at $t = 0$ will be $2L/c$ where $c$ is the velocity of light.

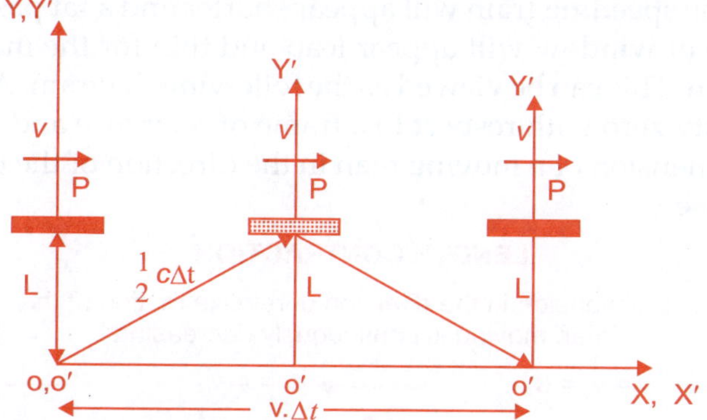


**Fig. 2.3** Hypothetical experimentation of time dilation in special theory of relativity

Now, suppose the frame $2(O')$ is moving with velocity $v$ and after an interval of time measured in frame $1(O)$ $\nabla t$, the new position of $O'$ will be $v \cdot \nabla t$. The light beam emitted from $O$, $O'$ (at $t = 0$) reaches the new position of $O'$ after the time interval $\nabla t$ as measured from frame $1(O)$ and after traveling a distance of $c. \nabla t$ ($OPO'$). But for the frame of reference 2 the light beam appear to travel a distance $O'$ $PO'$ and the time taken will be as: $\nabla t'$ and that will be equal to $2L/c$ again as velocity of light remains same in both of the frames. Now, from simple geometry:

$$\left(\frac{1}{2}c \cdot \nabla t\right)^2 = \left(\frac{1}{2}c \cdot \nabla t\right)^2 + L^2$$

$$\Delta t^2 = 4L^2/(c^2 - v^2) = \frac{4L^2}{c^2(1 - v^2/c^2)} \quad \text{but as} \quad \frac{2L}{c} = \Delta t'$$

Therefore, $\Delta t = \dfrac{\Delta t'}{\sqrt{1 - v^2/c^2}}$ , Now, as $\nabla t'$ and $\nabla t$ are the time intervals

between two incidences i.e. emitting light and receiving it back after reflection as observed respectively with reference to frame 2 which is moving with velocity $v$ with respect to frame 1 and the reference frame 1.

As $\dfrac{1}{\sqrt{1 - v^2/c^2}}$ is more than 1, $\nabla t > \nabla t'$.

So, the time interval between these two incidences will be prolonged with respect to frame 1 compared to frame 2.

Another hypothetical incident may be cited which can only be experienced if velocity $v$ is very large or we consider a world where the velocity of light $c$ is small so that $v/c$ cannot be neglected. Suppose a man is standing on a railway platform and a train is passing through with a high velocity, so that $v/c$ can not be neglected. Under that condition the original length of the speeding train will appear shorter and a fat passenger seating by the side of window will appear lean and thin for the man standing on the platform. This can be viewed as the following diagram. A man standing with velocity zero with respect to a frame of reference and with respect to him the dimension of a moving man in the direction of the relative motion will decrease.

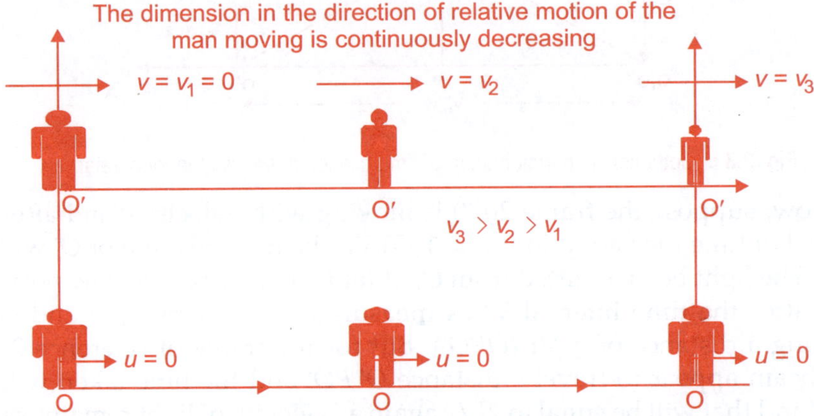


**Fig. 2.4** The imaginary situation of length contraction in the moving frame

The duration between two incidences say striking a match and lighting a cigar by a passenger viewed through the window of the moving train will be different as observed by the man standing on the platform and by any co-passenger seating beside the passenger. To the man standing on the platform this duration of the two incidences will appear much longer than it is actually and as observed by the co-passenger. This thinning of the figure of the passenger in the moving train and also the lengthening of the duration of the two incidences mentioned above depends on how $v$, the velocity of the train is close to the velocity of light c and if $v$ coincides with c then according to the out come of the special theory of relativity the dimension of any body in the direction of motion of the train will vanish and the time duration will be prolonged to infinity.

## 2.5 RELATIVISTIC MOMENTUM AND ENERGY

From Newtonian Mechanics the momentum $p = mv$, mass is considered a constant but in relativistic motion let us begin with an assumption that mass m is a function of velocity i.e. $m = m(v)$. Let us consider an elastic collision problem between two particles of same masses (1) and (2) respectively in two reference frames 1 and 2 denoted by $X_1$, $Y_1$ and $X_2$, $Y_2$. The frame 2 is moving with respect to 1 with velocity $v$ and correspondingly frame 1 with $-v$ with respect to 2. Consider the following Fig. 2.4 where in Frame 1 which is the rest frame for particle 1 it is colliding elastically moving before collision with velocity $v_0$ and colliding back with velocity $v'$.

In this frame particle 2 is moving with velocity $w$ and moving with $w'$ after the collision.

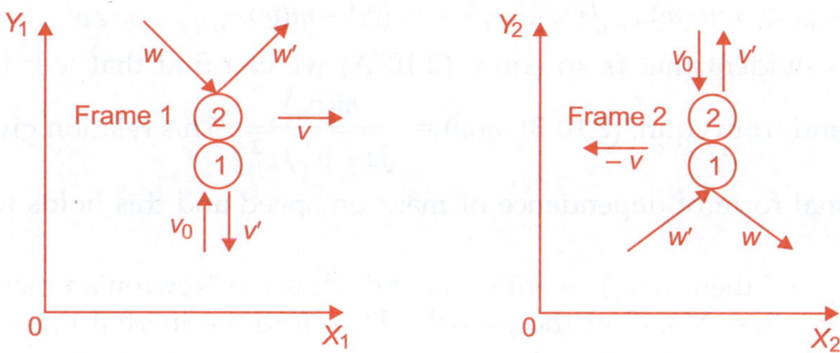


**Fig. 2.5** The diagram at left is that of frame 1 with respect to which frame 2 is moving with velocity $v$ so that for particle 1 this is a rest frame along X and diagram at right which is for frame 2 and is a rest frame for particle 2

In frame 2 which is rest frame for particle 2 it is found to move with velocity $v_0$ and $v'$ before and after the collision, where as particle 1 is moving with $w$ before and $w'$ after the collision. That is both of the two particles have same velocity $v_0$ in their respective rest frames in the Y-axis direction before collision and $v'$ after collision and the Y-component of the oppositely moving particles in their respective rest frames are $v_y$ and $v'_y$ before and after collision i.e. with respect to particle 1 in its rest frame the Y-component of velocity of the particle 2 in its own rest frame before and after collision are $v_y$ and $v'_y$ and vice versa. Therefore, with respect to frame 1 and using the transformation (equn. 2.9) we get:

$$v_y = \frac{v_0\sqrt{1 - v^2/c^2}}{1 - vv_x/c^2} = v_0\sqrt{1 - v^2/c^2}$$

$$v'_y = \frac{v'\sqrt{1 - v^2/c^2}}{1 - vv_x/c^2} = v'\sqrt{1 - v^2/c^2}$$

as $v_x = 0$ for particle in before collision does not have any X-component of its velocity.

Now, the speed of the second particle $w$ and $w'$ before and after collision can be obtained from the considerations that they are the resultant of the X and Y-components.

$$w = \left[v^2 + v_0^{\,2}(1 - v^2/c^2)\right]^{1/2} \quad \text{and} \quad w' = \left[v^2 + v'^{\,2}(1 - v^2/c^2)\right]^{1/2}$$

Applying the principle of the conservation of momentum for the X-component, in frame 1 we get:

$$m(v_0).O + m(w).v = m(v').O + m(w').v \qquad \qquad \text{...(2.10 A)}$$

and for Y-component of momentum in frame 1:

$$- m(v_0) \cdot v_0 + m\,(w).v_0\sqrt{1 - v^2/c^2} = m\,(v_0) - m(w).v_0\sqrt{1 - v^2/c^2} \quad \text{...(2.10 B)}$$

It is evident that from equn. (2.10 A) we can find that $w = w'$ and $v' = v_0$ and from equn. (2.10 B) $m(w) = \dfrac{m(v_0)}{\sqrt{1 - v^2/c^2}}$. This relation gives the functional form of dependence of mass on speed and this holds for any value of $v_0$.

If $v_0 \to 0$ then $m\,(v_0) \to m(0) = m_0$, which is the Newtonian mass or is called as "Rest Mass" of the particle. Therefore the inertial mass of the particle depends on its speed as:

$$m(v) = m = \frac{m_0}{\sqrt{1 - v^2/c^2}} \qquad \qquad \text{...(2.11)}$$

writing an arbitrary speed $v$ in terms of velocity, the momentum $p$ of a mass $m$ moving with velocity $v$ can then be written as:

$$p = mv = \frac{m_0}{\sqrt{1 - v^2/c^2}}\, v. \qquad \qquad \text{...(2.12)}$$

As energy $E$ is defined as the work done against a force $F$ and can be written as:

$$dE = F \cdot dr = \frac{d}{dt}p. \, dr = v.dp \qquad \qquad \text{...(2.13)}$$

Using equn. (2.12) $dp = d\left[\dfrac{m_0 v}{\sqrt{1 - v^2/c^2}}\right] = \dfrac{m_0 dv}{(1 - v^2/c^2)^{3/2}}$ and putting

this in equn. (2.13) $dE = \dfrac{m_0 v dv}{(1 - v^2/c^2)^{3/2}} = -\dfrac{m_0 c^2}{2}\dfrac{d(1 - v^2/c^2)}{(1 - v^2/c^2)^{3/2}}$

$$= m_0 c^2 d\left[\frac{1}{\sqrt{1 - v^2/c^2}}\right]$$

Now, abbreviating $\dfrac{1}{\sqrt{1 - v^2/c^2}}$ as $\gamma$ we can write $dE = m_0 c^2 d\gamma$ ...(2.14)

Now, if we integrate the above equation from $\gamma = 1(v = 0)$ to $\gamma = 1(v = v)$, we will get the kinetic energy of the body $T$.

$$T = m_0 c^2 \int_1^\gamma d\gamma = m_0 c^2 (\gamma - 1) = m_0 c^2 \left( \frac{1}{\sqrt{1 - v^2/c^2}} - 1 \right).$$ Now for non relativistic.

Condition, i.e. $v \ll c$, the above equation can be written as:

$$T = m_0 c^2 \left[ (1 - v^2/c^2)^{-1/2} - 1 \right] \text{ and if } v \ll c$$

$$T = m_0 c^2 \left[ 1 + \frac{1}{2} v^2/c^2 - 1 \right]. \qquad \ldots(2.15)$$

Therefore, Kinetic energy, $T$ reduces to the classical expression $T = \frac{1}{2} m_0 v^2$.

For relativistic case integrating equn. (2.14) and setting the integration constant as zero, we get:

$$E = m_0 c^2 \gamma = \frac{m_0 c^2}{\sqrt{1 - v^2/c^2}}$$

or,    Total energy, $E = T + m_0 c^2$. $\qquad \ldots(2.16)$

Now, integration over can be done in some other way as:

$$dE = c^2 dm_0 \gamma = c^2 d \frac{m_0}{\sqrt{1 - v^2/c^2}} \text{ but } m = \frac{m_0}{\sqrt{1 - v^2/c^2}} \text{ from equn. (2.11)}$$

$dE = c^2 dm$ which after integration and equating the integration constant to zero

$$E = mc^2 \qquad \ldots(2.17)$$

This is famous Einstein's mass energy relation.

On the other way it can be said also that when an energy $\Delta E$ is given to a mass it results in an increase of its inertial mass by $\Delta m$ and as this is true and so it may be considered as the origin of mass. As this is again true irrespective of the form of energy, it goes beyond the classical conservation of mechanical energy.

## 2.6 RELATISTIC EQUATIONS OF MOTION

Recalling the Lorentz transformation equn. (2.4) as:

$$x' = k(x - vt) = \frac{x - vt}{\sqrt{1 - v^2/c^2}}$$

$$y' = y$$

$$z' = z$$

$$t' = k(t - bx) = \frac{x - vt}{\sqrt{1 - v^2/c^2}} \text{ and introducing the abbreviations as:}$$

$\beta = \dfrac{v}{c}$ and $\gamma = \dfrac{1}{\sqrt{1 - v^2/c^2}} = \dfrac{1}{\sqrt{1-\beta^2}}$ , the above transformation

equations may be simplified as:

$$x' = \gamma(x - \beta ct) \quad \text{and inverse relations as} \quad x = \gamma(x' + \beta ct)$$
$$y' = y \qquad\qquad\qquad\qquad\qquad\qquad\quad y = y'$$
$$z' = z \qquad\qquad\qquad\qquad\qquad\qquad\quad z = z'$$
$$t' = \gamma(t - \beta x/c) \qquad\qquad\qquad\qquad t = \gamma(t + \beta x'/c) \qquad ...(2.18)$$

Now, as $i = \sqrt{-1}$ we can rewrite the above equations in terms of $i$ as:

$$x_1' = \gamma(x_1 + i\beta ct) \qquad \text{and} \qquad x_1 = \gamma(x_1' - i\beta ct')$$
$$x_2' = x_2' \qquad\qquad\qquad\qquad\qquad\quad x_2 = x_2'$$
$$x_3' = x_3 \qquad\qquad\qquad\qquad\qquad\quad x_3 = x_3'$$

and multiplying the time relations by $ic$ on both sides.

$ict' = \gamma(ict - i\beta x_1)$ and writing $ict$ as $x_4$

$$x_4' = \gamma(-i\beta x_1 + 4x) \qquad \text{and} \qquad x_4 = \gamma(i\beta x_1' + x_4') \qquad ...(2.19)$$

This formulation of Lorentz transformation is then in four dimensional space with $x_1$, $x_2$, $x_3$ and $x_4$ one variables. The common characteristics of all these sets of quantities is that they have "four components" and for this reason they are called "four vectors" and can be supposed to describe a four dimensional representative space.

Like these equations a four dimensional vector may be specified in terms of components $A_\mu = (A_1, A_2, A_3, A_4)$ which are equivalent to: $A_\mu = (A_x, A_y, A_z, A_t)$ and these components must obey the transformation rule analogous to equn. (2.18) and can be in compact notation as:

$$A_v' = \sum_{\mu=1}^{4} a_{v\mu} A_\mu \text{ the coefficients } av\mu \text{ define the matrix elements in the}$$

direct Lorentz transformation as:

$$\begin{bmatrix} \gamma & 0 & 0 & i\beta\gamma \\ 0 & 1 & 0 & 0 \\ 0 & 0 & 1 & 0 \\ -i\beta\gamma & 0 & 0 & \lambda \end{bmatrix} \text{ and changing the sign of } i\beta\gamma \text{ we get the corresponding}$$

coefficients for inverse Lorentz transformation.

Now, from Newton's equation of motion which is invariant under the Galilean transformation, the force with its components and defined as:

$\dfrac{dp_i}{dt} = F_i$ where $i$ = 1, 2, 3 may be generalized in the form which is invariant under Lorentz transformation as:

$\dfrac{dp_\mu}{dt} = F_\mu$ where $\mu$ = 1, 2, 3, 4 the spatial components 1, 2, 3 must reduce to that from Newton's for very low velocity i.e. when $v/c \approx 0$.

The four vector $p_\mu$ can then be written as the product of rest mass $m_0$ and the corresponding four vector velocity $v_\mu$ as:

$p_\mu = m_0 v_\mu = \gamma (m_0 v, im_0 c) = (mv, imc)$. The spatial components of $p_\mu$ is the momentum of the particle as $p = \dfrac{m_0}{\sqrt{1 - v^2/c^2}} v$ and the time like components $p_4$ is:

$p_4 = \dfrac{i}{c}(mc^2) = \dfrac{i}{c}E$ and the four momentum is given as: $p_\mu = \left( p, \dfrac{i}{c}E \right)$.

This is a combined relation of both momentum and energy and is called "Momentum-Energy four-vector". The transformation rule stated above leads to:

$$p_x{}' = \gamma\left[ p_x - \frac{\beta}{c}E \right]$$

$$p_y{}' = p_y$$

$$p_z{}' = p_z$$

$$E' = \gamma(E - \beta c p_x) \qquad \qquad \ldots(2.20)$$

Equation (2.19) shows that the energy and momentum are inter related, like time and longitudinal position in Lorentz transformation. The norm of the four momentum is a Lorentz invariant:

$$\sum_\mu p_\mu{}^2 = p^2 - \frac{E^2}{c^2} = -m_0{}^2 c^2 . \qquad \qquad \ldots(2.21)$$

This is a result which is equivalent to momentum-energy relation as below:

$$E = m_0 c^2 \sqrt{1 + \frac{p^2}{m_0{}^2 c^2}} \quad \text{or,} \qquad E^2 = p^2 c^2 + (m_0 c^2)^2$$

or, $\dfrac{E^2}{c^2} = p^2 + m_0{}^2 c^2$. This can be geometrically expressed as a right angled triangle and sine of the angle is

$\dfrac{p}{E/c} = \dfrac{mv}{mc^2/c} = \dfrac{m\beta c}{mc} = \beta$. (as $\beta = v/c$) shown in Fig. 2.6.

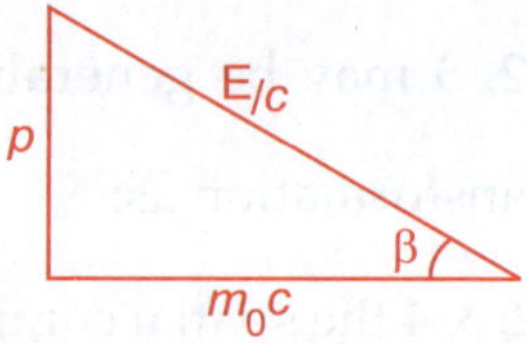

**Fig. 2.6** The relativistic and non relativistic change of energy

Where, from we see that if $\beta \to 0$, $E/c \to m_0 c$ i.e. $E \to m_0 c^2$ situation transforms in to non relativistic and if $\beta \to 1$, $p \to E/c$ i.e. $E \to mc^2$ which is the extreme relativistic case. Now, considering the equn. (2.15) we get:

$$\frac{T}{m_0 c^2} = \left[1 + \frac{1}{2} v^2 / c^2 + \ldots - 1\right]$$ and if the left hand is plotted against the

velocity $v$ then we get a comparison between the energy per mass with velocity for relativistic and non relativistic change of mass. This is shown in the following Fig. 2.7.

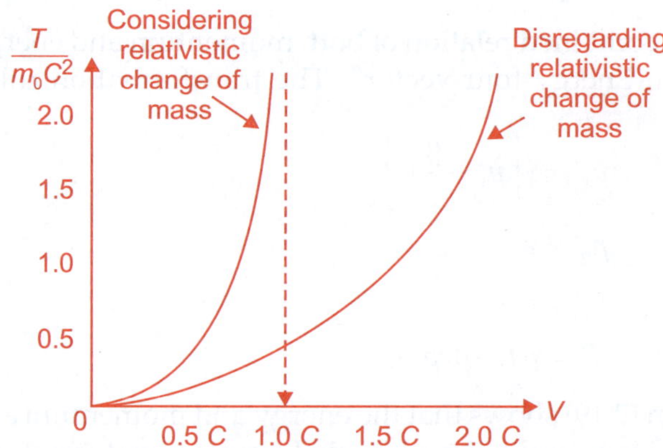


**Fig. 2.7** Comparison between energies in relativistic and non relativistic change of mass. For velocities less than 0.2 $C$ both yield the same result

## REVIEW QUESTIONS

1. Show that all laws of Physics are same in all inertial reference.

2. Show that the speed of light in a vacuum has the same value in all inertial frame, regardless of the velocity of the observer or the velocity of the light emitting source.

3. In the relativistic motion of a proton. Calculate the rest mass energy and if the total energy of the proton is three times of its rest energy then find the speed of the proton. ($m_0 = 1.67 \times 10^{-27}$ kg and $c = 3.00 \times 10^8\ m/sec$).

4. An electron has a kinetic energy five times greater than its rest energy, find the total energy and its speed.

5. Imagine a situation in which one of the twin brothers of same age travels by say rocket with speed 0.5 $c$, after 10 years of travel in space craft, the brother returns to earth and found his brother left on earth looks older than him. Comment on this.

## REFERENCES


1. W. Rindler – Introduction to Special Relativity, Clarendon Press, Oxford, 1982.

2. R. E. Turner – Relativity Physics, Routledge and Kegan Paul, London, 1984.

3. R. Resnick and D. Halliday – Basic concepts in Relativity and Early Quantum Theory, 2nd Ed., Macmillan Publishing Co., New York, 1992.

4. M. Alonso and E. J. Finn – Fundamental University Physics (Vol. 1), Addison Wesley Pub. Co., Reading, Mass, 1968.

5. S. B. Palmer and M. S. Rogalski – Advanced University Physics, Gordon and Breach Pub. USA, 1996.

6. R. P. Feynman, R. B. Leighton and M. Sands – The Feynman Lectures on Physics, Addison Wesley Pub. Co., Reading, Mass, Narosa Pub. House New Delhi, 2003.

# Elasticity

## 3.1 INTRODUCTION

The molecules or atoms in pure state of solids are not free to move from each other as it's in the case of gaseous state. Though the molecule/atoms are not static or fixed to their positions and they oscillate about their respective mean positions. They can also be displaced from their positions resulting in a deformation either in any dimension of the bulk, change in shape of the bulk or change in volumes. As this state characteristic of the solid state of matter is universally true, it may also be mentioned here that the deformations resulted by the application of external force are resisted within certain limit by the solid. As an outcome of these resistive force which is not a reaction of the applied force, the body is deformed and may recover either fully or partially after the said force is withdrawn. As action (applied force) and the resistive forces (not as reaction) act and develop in the same body, the resistive force cannot be a reaction force of the applied force. This resistive force will henceforth be called as "Stress" and the deformation that the stress has created is known as "Strain". The stress and the corresponding strain determine the physical nature known as "Elasticity". This important property of elasticity classifies the solids into three categories namely perfectly elastic, partially elastic and perfectly plastic. No matter is either perfectly elastic or plastic; it remains within these two extreme boundaries. Hence the study of elastic properties of a solid matter is of immense importance. The strain is the precondition of the development of stress. When a solid is subjected to an external force it gets deformed i.e. strained and within the body a stress i.e. resistive force develops and this stress remains proportional to the strain within certain limit. When the applied deforming force within this limit is withdrawn the body relieves the stress by returning back to its state before deformation and thereby the strain is eliminated. This is valid within a limit of stress, different for different materials of the solid and is known as elastic limit. If the applied stress crosses this limit the material then suffers permanent deformation. The following Fig. 3.1 shows the nature of change of stress developed against strain created by the external deforming force.

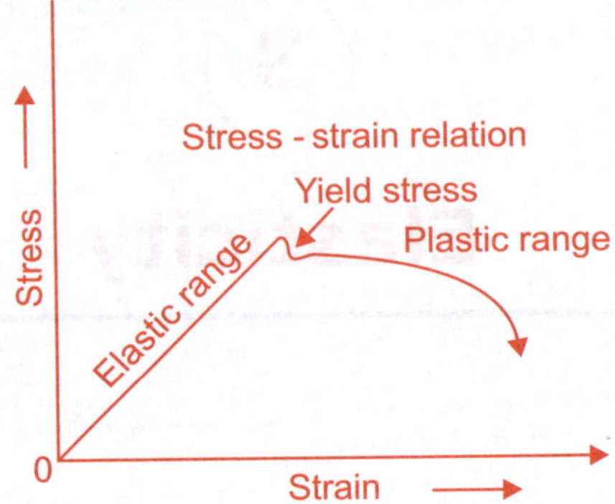


**Fig. 3.1** The Stress *vs.* Strain curve for a crystalline solid. There are sometimes for some crystalline structures, there are two yield stresses

Within the elastic limit i.e. yield stress the ratio of stress and strain remains constant and is known as elastic constant or elastic modulus. Depending on the nature of strain and stress i.e. depending on the intending effect i.e. change of length, change of shape and change of volume the elastic module are known respectively as Young's Modulus, Shear Modulus and Bulk Modulus. The following linear relationship between stress and strain up to the limit of yield stress is known as Hooke's Law.

They are:

$$\frac{\text{Longitudinal Stress}}{\text{Longitudinal Strain}} = Y \text{ (Young's Modulus)}$$

$$\frac{\text{Volume Stress}}{\text{Volume Strain}} = k \text{ (Bulk Modulus)}$$

$$\frac{\text{Shear Stress}}{\text{Shear Strain}} = n \text{ (Modulus of Rigidity)}$$

$$\frac{\text{Lateral Strain}}{\text{Longitudinal Strain}} = \sigma \text{ (Poisson's Ratio)}$$

$$\frac{\text{Longitudinal load per unit area of cross section}}{\text{Increase in length per unit length}} = \chi \text{ (Axial Modulus)}.$$

It may be mentioned that the Stress is defined as Resistance force per unit area of cross section and strain is the ratio of amount of deformation per original dimension or volume.

It appears that both of Young's modulus and Axial modulus are same. The $\chi$ however is defined in terms of the principal stress needed to produce a simple elongation without lateral change. While $y$ is the complete stress which includes two perpendicular stresses of such magnitude as will prevent lateral contraction, the $\chi$ represents the extensional stress to increase length per unit length.

## 3.2 DETERMINATION OF MODULI OF ELASTICITY

The different moduli of elasticity can be determined by various methods and some of them are described below:

### A. Young's Modulus by bending of beam

A bar having rectangular cross section of area a and of length $l$ is supported at the ends and is centrally loaded with weight $W$. Let us first consider only one half of the beam of length $l/2$. In the element of length shown below in the Fig. 3.2 (a). The line taken through the axis of the beam may be assumed to remain undistorted, the upper part is elongated where as the lower is contracted. In Fig. 3.2 (b), let $S$ denotes a section of that neutral axis. Then from diagram $dS/z = \varphi = S/r$, where $r$ is the radius of curvature of the bent section. So that $dS/S = z/r$.

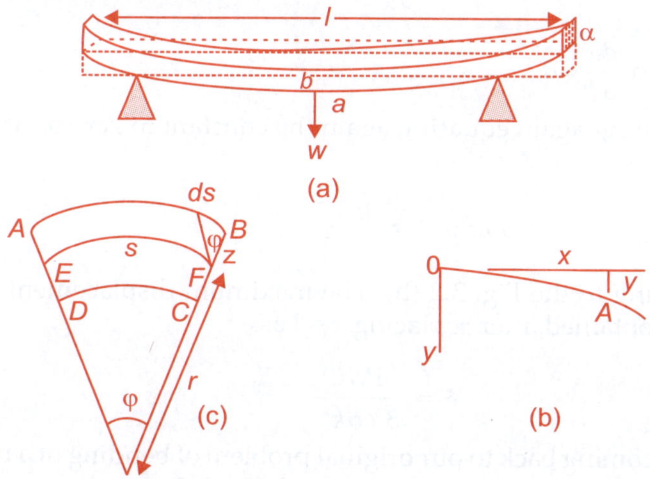

**Fig. 3.2** Bending of a beam supported at the ends and loaded at the centre. In (a) the original beam loaded on the centre, (b) shows the half of the total beam acting like a cantilever and (c) the section of the beam

If $p$ is the magnitude of the internal force and a is the cross section then

$$\frac{p}{\alpha} = Y\frac{dS}{S} = Y\frac{z}{r} \quad \text{or,} \quad p = \frac{Y}{r}z\alpha$$

The moment of $p$ around $F$ is $pz = \frac{Y}{r}\alpha z^2$.

The internal bending moment of resistance is the sum of all such terms and so, $\sum pz = \dfrac{Y}{r}\sum \alpha z^2$ but the quantity $\sum \alpha z^2$ is analogous to the moment of inertia of the cross section about that axis and is equal to $ak^2$, where $k$ is the radius of gyration.

Then 'Internal bending moment' $= \dfrac{Y\alpha k^2}{r}$

The curvature at the point $A$ Fig. 3.2 (b) having coordinate $x, y$ is given by:

$$\frac{1}{r} = \frac{d\psi}{dS} = \frac{d}{dS}(\tan\psi) = \frac{d^2 y}{dx^2}.$$

Hence equating the internal bending moment with that of external bending moment at the point $A$, we get:

$$W(l-x) = \frac{Y}{r}\alpha k^2 = Y\alpha k^2 \frac{d^2 y}{dx^2}$$

Integrating we get $Y\alpha k^2 \dfrac{dy}{dx} = W\left(lx - \dfrac{x^2}{2}\right) + c_1.$ The constant $c_1 = 0$ at

$x = 0$ as also $\dfrac{dy}{dx} = 0.$

Integrating again equating again the constant to zero as at $x = 0$, $y = 0$, we get:

$$y\alpha k^2 y = W\left(l\frac{x^2}{2} - \frac{x^3}{6}\right)$$

Referring to the Fig. 3.2 (b), The maximum displacement $y = d$ at the point B is obtained after replacing $x = l$ as:

$$\delta = \frac{Wl^3}{3Y\alpha k^2} \qquad\qquad \ldots(3.1)$$

Now, coming back to our original problem of bending of a beam loaded on the centre, we must replace the length $l$ by $l/2$, as so long we considered a cantilever of length l and a total beam here can be considered of two such cantilevers making a beam of length l and each is subjected to a load at the end by $W/2$.

Then the relative elevation of the original centre after the loaded centre $\delta$ is given by:

$$\delta = \frac{Wl^3}{48Y\alpha k^2} \qquad\qquad \ldots(3.2)$$

The equn. (3.2) can be used to determine the Young's by measuring $\delta$ and knowing $W, l$ and the cross section, $\alpha$.

## B.  Modulus of Rigidity by Torsion of a cylinder or Torsional pendulum

The experiment appropriate for the determination of modulus of rigidity has been shown in the following Fig. 3.3 (a). The wire of cylindrical shape of radius $r$ and of length $l$ is mounted firmly at A and at the other end a thread is wound around a cylinder and two equal weights are hung to apply torsion. Two mirrors are fixed on the ends of the wire to measure relative twist of the wire by lamp and scale method. The Fig. 3.3 (b) shows the magnified section of the cylindrical wire. The section $ABCD$ is twisted to the position $ABED$ due to torsion applied. Figure 3.3 (c) shows the base of the section both in undistorted and twisted configurations. $FGIH$ a such a section element taken of element of width $dr$ which is sheared through the angle $f$ and takes the position $JKML$ in Fig. 3.3 (d) the distorted situation of the element of area $dr \cdot dx$ is sheared through the distance $r\varphi$. Now, the force $f$ acting tangentially on the face $FGIH$, $f$ which produces a shear of an angular displacement $\theta$ is given by $\dfrac{f}{dr.dx}$ and equating this shearing stress with the modulus of rigidity $n$, we get:

$$n = \frac{f}{\theta\, dr.dx} \qquad \qquad \dots(3.3)$$

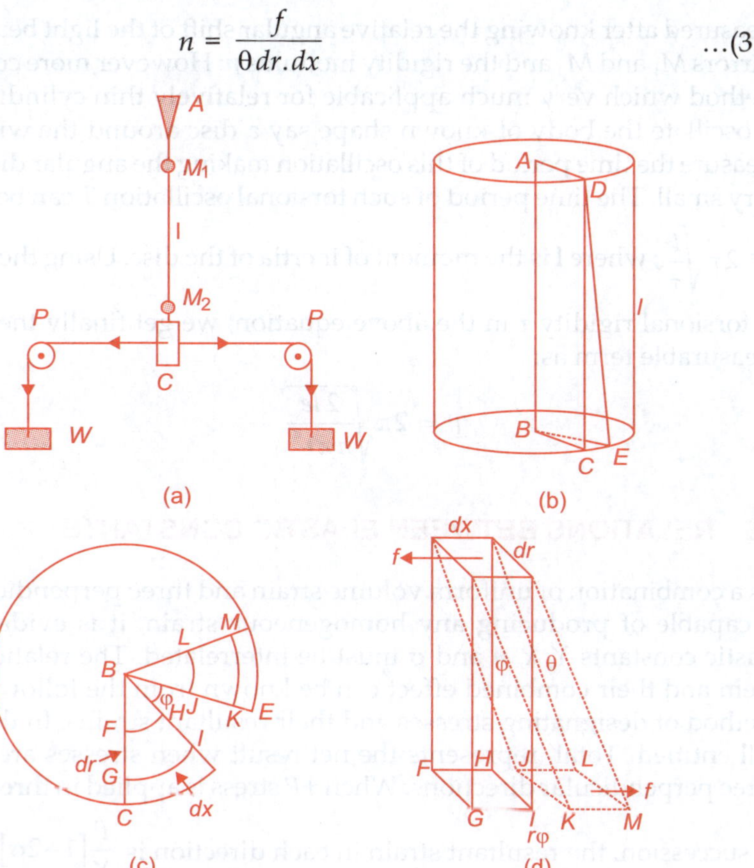



**Fig. 3.3**  Torsion of a cylindrical wire and the experiment to determine the modulus of rigidity

Now, since, $l\theta = r\varphi$, we get:

$$f = ndr \cdot dx \cdot \frac{r\varphi}{l}$$

This force $f$ has a moment $fr$ about the axis about the axis of the cylinder and thus total moment is $\Gamma = \dfrac{n\varphi}{l} \iint r^2\, drdx$ and taking integration on $x$ around the circumference of the cylindrical cross section of radius $r$,

$$\Gamma = \frac{2\pi n\varphi}{l} \int r^3\, dr.$$

$\Gamma = \dfrac{n\pi r^4 \varphi}{2l}$. Now, writing $\dfrac{\Gamma}{\varphi} = \tau = \dfrac{2WR}{\varphi}$ where $t$ is the torsional rigidity

and $R$ is the radius of the cylinder attached to the end of the torsion wire.

Equating this relation, we get finally $\tau = \dfrac{n\pi v^4}{2l}$. The torsional rigidity can be measured after knowing the relative angular shift of the light beam from the mirrors $M_1$ and $M_2$ and the rigidity modulus $n$. However more conventional method which very much applicable for relatively thin cylindrical wire is to oscillate the body of known shape say a disc around the wire axis and measure the time period of this oscillation making the angular displacement very small. The time period of such torsional oscillation $T$ can be equated to

$T = 2\pi \sqrt{\dfrac{I}{\tau}}$, where I is the moment of inertia of the disc. Using the expression

of torsional rigidity $\tau$ in the above equation, we get finally the relation in measurable term as:

$$T = 2\pi \sqrt{\frac{2\,Ie}{n\pi r^4}} \qquad\qquad \ldots(3.4)$$

## 3.3   RELATIONS BETWEEN ELASTIC CONSTANTS

As a combination of uniform volume strain and three perpendicular shears is capable of producing any homogeneous strain, it is evident that the elastic constants $Y$, $k$, $n$ and $\sigma$ must be interrelated. The relation between them and their combined effect can be known from the following tabular method of designating stresses and their resultant strains. In this table the cell entitled 'Total' represents the net result when stresses are applied in three perpendicular directions. When $+P$ stress is applied in three directions

in succession, the resultant strain in each direction is $\dfrac{P}{Y}[1-2\sigma]$.

The resultant volume strain is $\dfrac{3P}{Y}[1-2\sigma]$ but this is equal from Bulk modulus of elasticity to $\dfrac{P}{k}$. Therefore, we can write:

$$\frac{P}{k} = \frac{3P}{Y}(1-2\sigma) \qquad \text{or,} \qquad Y = 3k(1-2\sigma) \rightarrow$$

| Stress Directions | | | Strains produced along | | | |
|---|---|---|---|---|---|---|
| X | Y | Z | X | Y | Z | |
| +P | 0 | 0 | $+\dfrac{P}{Y}$ | $-\dfrac{\sigma P}{Y}$ | $-\dfrac{\sigma P}{Y}$ | |
| 0 | +P | 0 | $-\dfrac{\sigma P}{Y}$ | $+\dfrac{P}{Y}$ | $-\dfrac{\sigma P}{Y}$ | |
| 0 | 0 | +P | $-\dfrac{\sigma P}{Y}$ | $-\dfrac{\sigma P}{Y}$ | $+\dfrac{P}{Y}$ | |
| +P | +P | +P | $\dfrac{P}{Y}[1-2\sigma]$ | $\dfrac{P}{Y}[1-2\sigma]$ | $\dfrac{P}{Y}[1-2\sigma]$ | Total |
| +P | 0 | 0 | $+\dfrac{P}{Y}$ | $-\dfrac{\sigma P}{Y}$ | $-\dfrac{\sigma P}{Y}$ | |
| 0 | −P | 0 | $+\dfrac{\sigma P}{Y}$ | $-\dfrac{P}{Y}$ | $+\dfrac{\sigma P}{Y}$ | |
| +P | −P | 0 | $\dfrac{P}{Y}[1+\sigma]$ | $-\dfrac{P}{Y}[1+\sigma]$ | 0 | Total |
| +P | 0 | 0 | $+\dfrac{P}{Y}$ | $-\dfrac{\sigma P}{Y}$ | $-\dfrac{\sigma P}{Y}$ | |
| 0 | +P1 | 0 | $-\dfrac{\sigma P_1}{Y}$ | $+\dfrac{P_1}{Y}$ | $-\dfrac{\sigma P_1}{Y}$ | |
| 0 | 0 | +P1 | $-\dfrac{\sigma P_1}{Y}$ | $-\dfrac{\sigma P_1}{Y}$ | $+\dfrac{P_1}{Y}$ | |
| +P | +P1 | +P1 | $\dfrac{1}{Y}\left[P-2\sigma P_1\right]$ | $\dfrac{1}{Y}\left[P_1-\sigma(P+P_1)\right]$ | $\dfrac{1}{Y}\left[P_1-\sigma(P+P_1)\right]$ | Total |

Now, from the second, when two perpendicular stresses one extensional and other compressional given respectively as +P and −P are combined, they produce extensional along X and equal compressional along Y and are equivalent to a shear strain of magnitude $\dfrac{2P}{Y}[1+\sigma]$ but this shear is equal to $\dfrac{P}{n}$. Therefore,

$$\frac{P}{n} = \frac{2P}{Y}[1+\sigma] \text{ or, } Y = 2n(1+\sigma)$$

Now, from third, the result of applying a set of stresses which if produces no strain in $Y$ and also in $Z$ direction, then $P_1 = s(P + P_1)$ and in this case the extension along $X$ direction is $\dfrac{P}{Y}\left(1 - \dfrac{2\sigma^2}{1-\sigma}\right) = \dfrac{P}{Y}\dfrac{(1+\sigma)(1-2\sigma)}{(1-\sigma)}$. Now, this is equal to $\dfrac{P}{\chi}$ and so we get:

$$Y(1 - \sigma) = \chi(1+\sigma)(1-2\sigma)$$

From these relations we may write:

$$Y = \frac{9nk}{3k+n}, \quad \sigma = \frac{3k-2n}{6k+2n}, \quad \chi = \frac{3k+4n}{3}. \qquad \qquad …(3.5)$$

## 3.4 STRESS, STRAIN AND MODULUS OF ELASTICITY

In a solid which is deformable the description of motion of the constituents which are not fixed at a position is influenced by various factors which include in addition to the applied force, the presence of the neighbouring constituents in the continuous medium and their mutual effect are also important. Though there exists local variations but as long as these variations are microscopic, the property of the bulk may be considered from macroscopic point view as that which is averaged over the entire volume element. Therefore, when a force is applied on the bulk, it is considered to be same as applicable on an elementary volume $dv = dx \cdot dy \cdot dz$, irrespective of the size of the element. If we consider any property of the bulk as $F$ it is a function of $x$, $y$, $z$ and also $t$ as the property may vary with position and time i.e. $F(x, y, z, t)$. This time dependence of the property is an interesting phenomenon of the solids.

$$\text{Now,} \quad \frac{dF}{dt} = \frac{\partial F}{\partial x}\frac{\partial x}{\partial t} + \frac{\partial F}{\partial y}\frac{\partial y}{\partial t} + \frac{\partial F}{\partial z}\frac{\partial z}{\partial t} + \frac{\partial F}{\partial t}$$

$$= v_x \frac{\partial F}{\partial x} + v_y \frac{\partial F}{\partial y} + v_z \frac{\partial F}{\partial z} + \frac{\partial F}{\partial t} = (v \cdot \Delta)\, F + \frac{\partial F}{\partial t}$$

Considering $F$ to represent an elastic parameter in general let us consider first the strain.

### A.  Strain in the volume element (continuum)

Consider the extension (elongation) of the volume element $dv(dx \cdot dy \cdot dz)$ in the $x$-axis direction at a point $x$, $y$, $z$ within the element at $t = 0$ and consider that the point concerned is displaced at $t = t$. The displacement vector $\xi$ is defined as a "property" which is a function of position and time i.e. $\xi = (\xi_x, \xi_y, \xi_z)$, at time $t = t$.

The strain (extensional) in the $X$ direction $\varepsilon_{xx}$ is the displacement in the $X$ direction per unit change in $x$-coordinate and is given by :

$\varepsilon_{xx} \dfrac{\partial \xi_x}{\partial x}$ and similarly the strain in $Y$ and $Z$ directions are given as:

$\varepsilon_{yy} \dfrac{\partial \xi_y}{\partial y}$ and $\varepsilon_{zz} \dfrac{\partial \xi_z}{\partial z}$ . These strains are also called longitudinal strain.

The shear strain is due to the change in shape of the element and at a point $x$ is given by the displacement in $Y$ direction with unit change in $x$ direction i.e. that is change in $\xi_y$ per change in $x$.

Therefore, the shear strain along $X$ direction, $\xi_{yx} = \dfrac{\partial \xi_y}{\partial x}$ and similarly along $y$ as: $\varepsilon_{xy} = \dfrac{\partial \xi_x}{\partial y}$ . The angle of shear $\gamma_{xy}$ in the $xy$ plane is given

by: $\gamma_{xy} = \dfrac{\partial \xi_y}{\partial x} + \dfrac{\partial \xi_x}{\partial y}$ but for shear in rigid element $\dfrac{\partial \xi_y}{\partial x} = \dfrac{\partial \xi_x}{\partial y}$ and so,

$\varepsilon_{xy} = \dfrac{1}{2}\gamma_{xy} = \dfrac{1}{2}\left( \dfrac{\partial \xi_x}{\partial y} + \dfrac{\partial \xi_y}{\partial x} \right)$, which is the shear strain in $xy$ plane and

similarly we can also get shear strain in $xz$ and $yz$ planes. We then have six components representing the state of strain in terms of three displacement components $\xi_x,\ \xi_y,\ \xi_z$ and they can be written as:

$$\varepsilon_{xx} = \dfrac{\partial \xi_x}{\partial x},\ \varepsilon_{yy} = \dfrac{\partial \xi_y}{\partial y},\ \varepsilon_{zz} = \dfrac{\partial \xi_z}{\partial z}$$

$$\varepsilon_{xy} = \varepsilon_{yx} = \dfrac{1}{2}\left[ \dfrac{\partial \xi_x}{\partial y} + \dfrac{\partial \xi_y}{\partial x} \right]$$

$$\varepsilon_{yz} = \varepsilon_{zy} = \dfrac{1}{2}\left[ \dfrac{\partial \xi_z}{\partial y} + \dfrac{\partial \xi_y}{\partial z} \right]$$

$$\varepsilon_{zx} = \varepsilon_{xz} = \dfrac{1}{2}\left[ \dfrac{\partial \xi_x}{\partial z} + \dfrac{\partial \xi_z}{\partial x} \right].$$

These strain components when expressed in the form of a matrix is called "Strain Matrx"

$$\hat{\varepsilon} = \begin{bmatrix} \varepsilon_{xx} & \varepsilon_{xy} & \varepsilon_{xz} \\ \varepsilon_{yx} & \varepsilon_{yy} & \varepsilon_{yz} \\ \varepsilon_{zx} & \varepsilon_{zy} & \varepsilon_{zz} \end{bmatrix}. \qquad \ldots(3.6)$$

The diagonal element of the above matrix represents strains in the specified directions and it will be elongation (dilation) if they are positive and compression if they are negative. The other elements represent shear strains.

The expansion of the unit volume, called dilation Δ or volume expansion is

$$X = \frac{dV}{V} = \varepsilon_x + \varepsilon_y + \varepsilon_z = \varepsilon_{xx} + \varepsilon_{yy} + \varepsilon_{zz} = \frac{\partial \xi_x}{\partial x} + \frac{\partial \xi_y}{\partial y} + \frac{\partial \xi_x}{\partial z}$$

Taking the time derivative of the above equation

$$\frac{1}{V}\frac{dV}{dt} = e_{xx} + e_{yy} + e_{zz},$$ where $e's$ are the components of rate of strain.

Now, using conservation of mass $\rho V$, we get

$$\frac{d}{dt}(\rho V) = \rho \frac{dV}{dt} + V\frac{d\rho}{dt} = 0 \quad \text{or,} \quad \frac{1}{V}\frac{dV}{dt} = -\frac{1}{\rho}\frac{d\rho}{dt}$$

Now, expressing the rate of strain with velocities as

$$e_{xx} + e_{yy} + e_{zz} = \frac{\partial v_x}{\partial x} + \frac{\partial v_y}{\partial y} + \frac{\partial v_z}{\partial z} = \Delta \cdot v.$$

Therefore, $$\Delta \cdot v = \frac{1}{V}\frac{dV}{dt} = -\frac{1}{\rho}\frac{d\rho}{dt}$$

So, $$\frac{1}{\rho}\frac{d\rho}{dt} + \Delta \cdot v = 0. \qquad \qquad \text{...(3.7)}$$

This is the 'Equation of continuity' for a fluid of density, $r$.

### B. Stress in the volume element (continuum)

It has been stated earlier that whenever a solid is strained (elongation, shear and volume) a resistive force develops within the solid and that resistive force per unit area is defined as stress designated by δ. Therefore,

$$\delta_s = \lim_{\Delta S \to 0} \frac{\Delta F}{\Delta S} = \frac{d}{dS}F$$

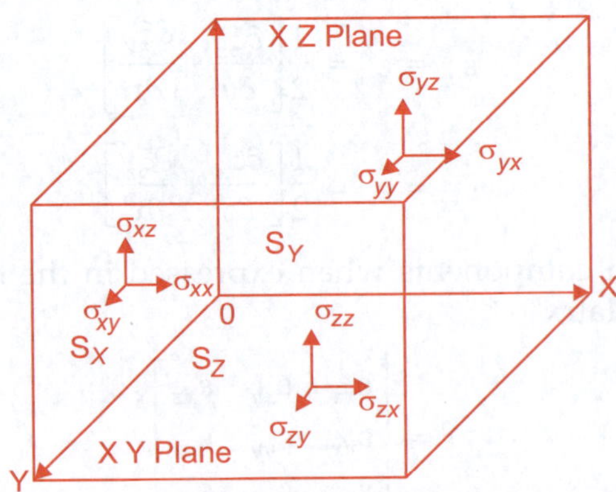


Stress field in a volume element within the solid

**Fig. 3.4** The stress field in a volume element $dV = dx \cdot dy \cdot dz$,. The stress components on three independent planes $S_X$, $S_Y$ and $S_Z$ i.e. YZ, XZ and XY planes are shown. $\sigma_{xx}$, $\sigma_{yy}$ and $\sigma_{zz}$ are elongation (dilation) stresses as they are perpendicular and $\sigma_{xy}$, $\sigma_{xz}$ etc. which are parallel to the planes are shear stresses

Therefore, on the surfaces $S_X$, $S_Y$ and $S_Z$, the stress components are:

$$\sigma_{SX} = \sigma_{xx}, \sigma_{xy}, \sigma_{xz}$$
$$\sigma_{SY} = \sigma_{yx}, \sigma_{yy}, \sigma_{yz}$$
$$\sigma_{SZ} = \sigma_{zx}, \sigma_{zy}, \sigma_{zz}$$

These nine components of stresses are positive if the direction of the axes are positive, otherwise they are negative. These stress components are expressed in the following form are known as "Stress Matrix"

$$\hat{\sigma} = \begin{bmatrix} \sigma_{xx} & \sigma_{xy} & \sigma_{xz} \\ \sigma_{yx} & \sigma_{yy} & \sigma_{yz} \\ \sigma_{zx} & \sigma_{zy} & \sigma_{zz} \end{bmatrix} \qquad \ldots(3.8)$$

Similar to equn. (3.6) the diagonal elements of the above matrix represent elongation stress if they are positive and compression if they are negative and all the other elements are represent shear stress.

## 3.5 ELASTIC SOLIDS

Normal polycrystalline metals have practically same elastic properties in all directions because this uniformity results due to distribution of very small crystals at random and in chaos. The discontinuity between these small crystals is the cause of the brittleness of the gross material, which increases continuously with continuous variation of strain. It is thus to be expected that the properties, especially those of elasticity, of large metal crystals are greatly influenced by the internal regularity of orientation of molecular fields. It is interesting to study this elastic behaviour for specially fabricated large metal single crystals, grown by using some special methods namely that due to Czochralski. As is to be expected, practically all the normal physical properties including elasticity show a directional sense in mono crystalline state and follow the law:

$$P = P_1 \cos^2 \alpha = P_2 \sin^2 \alpha$$

where $P$, $P_1$, and $P_2$ are the measure of elasticity at an angle a to the symmetry axis, parallel to that axis and perpendicular to it respectively.

In general, single crystals have smaller elasticity and markedly greater plasticity than its polycrystalline state and these reactions to stresses vary markedly with directions. Because of existence of symmetry in single crystalline state, there are directional variation in tensile strength, and notable tendencies for fracture to occur along fixed cleavage planes. These cleavage planes naturally tend to lie on where molecular spacing is of maximum magnitude. If various planes are drawn in the metal, the breaking stress normal to the planes will exhibit defined minima of different magnitudes and so the actual direction of fracture depends upon the ratio of stress component, perpendicular to one of the cleavage planes, to its particular breaking stress, and break occurs at that plane for which the ratio is greatest.

In connection with plastic deformation, the metal crystals show some extraordinary properties. Ordinary Cadmium polycrystalline wire has a high value of rigidity modulus, $n$ but single crystalline Cadmium wire shears into a permanent set under its own weight. Just as there are minimum breaking stress planes (cleavage planes), so there are minimum shearing stress planes, and plastic deformation occurs as a result of mutual gliding of these planes over one another. This is the cause of glide deformation. This glide deformation occurs as a series of jerks and the same time a striated appearance can be observed on the surface of the material as the 'glide packets' are parallel slabs of metal of a few microns thick. This glide of one plane over the other is commonly known as 'slip' and this slip is found to occur at the slip plane in the direction called slip direction and together they are called slip system.

The following Fig. 3.3 shows an example of striated appearance on the surface of Cadmium wire under stress.

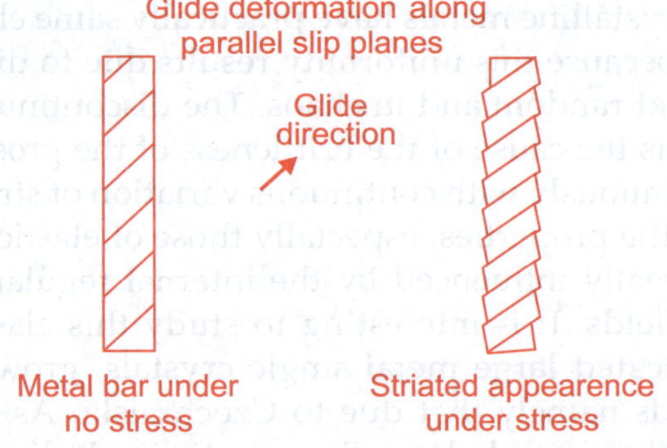


**Fig. 3.5** Deformation by slip on parallel slip planes in the glide or slip direction

If this slip is arrested during the process of deformation, then it cannot continue up to breaking point with that slip system and then it will 'climb' to a parallel plane and the phenomena is normally called work hardening as more stress is required to allow the deformation to continue by this process of climb. After a limiting degree of this kind of deformation an increasing strain with time at constant stress occurs and this called 'creep'. We can however apply a general equation to this effect of creep for many substances metallic, non-metallic, crystalline and amorphous in which the detailed atomic processes must vary. This onset of plasticity can be studied by stress-strain curve and also by X-ray diffraction, whereas the slip bands can be viewed by transmission electron microscopy.

### 3.5.1 Elastic Constants and Ultrasonic

By the transmission of ultrasonic waves, it is possible to measure the elastic constants. When ultrasonic waves are passed through small crystals, it moves by creating adiabatic compression and rarefaction through the material. This method is applicable for materials like glass and plastics. Either shear or longitudinal waves may be used and from the deduced velocities, $V_s$ and $V_d$ respectively and their ratio, which is $k$, then Poisson's

ratio is given by: $\sigma = \dfrac{k^2 - 2}{2(k^2 - 1)}$.

### 3.5.2 Elastic Constants and Temperature

As a general rule elastic moduli decrease with temperature and for a small range of temperature the relation is approximately linear. It has been established that the relation between $Y$ and $T$ is in exponential form

$$Y = Y_1\, e^{-bT}.$$

The constant $b$ has one value up to about one half the absolute temperature of the melting point and another value at higher temperature. The rigidity modulus also depends on temperature. Kohlrausch and Loomis observed this dependence of rigidity modulus of solids on temperature between 15°C and 100°C and proposed a relation as under:

$$n_T = n_0(1 - \alpha T - \beta T^2).$$

Though it is observed by many physicist that elastic moduli decreases with temperature but the exact dependence and the values of the constants depends on the purity of the metals.

### 3.6 FLUID MECHANICS

The elastic behaviour of liquids and fluids in general is widely different from that of solids, crystalline or non-crystalline. While a solid posses both rigidity and bulk moduli, but a fluid has no rigidity and so cannot permanently resist a tangential stress. In solid the stress at any point on a given element of area may have any direction with reference to that area but in a fluid at rest it must act along the normal to the plane, and it follows that in liquids at rest, the pressure at a point is independent of direction and thus a function of the position of the point alone. In a perfect fluid, whether at rest or in relative motion, no tangential stress ideally can exist, but in practice, relative motion is accompanied by tangential forces tending to prevent that motion and they persist as long as the motion lasts. Thus the fluid may be regarded as yielding to these stresses, different liquids yielding at very different rates. The condition at which yield takes is determined by a property known as viscosity and the latter may be regarded as a transient type of rigidity. The word transient is used to clarify that this property exists as long as the liquid flows. This can be easily experienced from the fact that when we pour liquid say water from a glass to another glass, we can feel the transportation of water even without seeing it. This may related to the transient resistance that flowing water exerts as it changes its shape at the brim of the edge of the glass pouring out water. Though no liquid is incompressible, but changes of volume at moderate increase of pressure is very small and can be practically neglected unless the pressure applied is very large. Therefore, a liquid can be used in hydraulic compressors or other devices and as due to Pascal's law of distribution of applied stress, the undiminished applied force is distributed equally in all directions. Therefore, the quantitative study of the bulk modulus of liquids and its relation to the other properties of the liquid is a matter of great practical importance.

However, a distinction between a viscous fluid and a solid is that the viscous flow corresponds to large displacements of the particles, whereas in elasticity they are regarded as small. The equations of fluid mechanics can be derived from a postulated linear relationship between stress and rate of strain. The shear per unit velocity gradient is called the dynamic coefficient of viscosity, $h$ which is defined as:

$$\sigma_{xy} = 2\eta\, e_{xy} = \eta\left[\frac{\partial v_x}{\partial y} + \frac{\partial v_y}{\partial x}\right]$$

$$\sigma_{yz} = 2\eta\, e_{yz} = \eta\left[\frac{\partial v_y}{\partial z} + \frac{\partial v_z}{\partial y}\right]$$

$$\sigma_{xz} = 2\eta\, e_{xz} = \eta\left[\frac{\partial v_x}{\partial z} + \frac{\partial v_z}{\partial x}\right] \qquad \ldots(3.9)$$

The equn. (3.9) above, show the dependence of the stress components $s$ with the time derivatives of strain components $\mu$, involving velocities of the element as $e$.

The significance of this equn. (3.9) can be illustrated by the following Fig. 3.6.

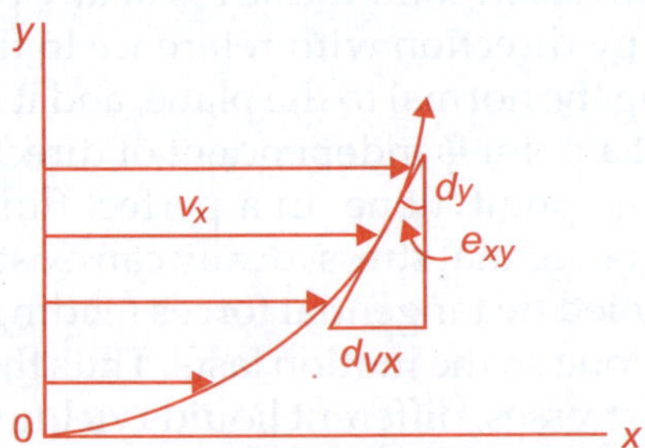


**Fig. 3.6** Velocity distribution in *x-y* plane in a viscous fluid flow parallel to *x*

From equn. (3.9), it follows that in a fluid which has zero velocity there are no shear forces i.e. $\sigma_{xy} = \sigma_{yz} = \sigma_{xz} = 0$. We can now define a scalar quantity, pressure, $p$ which from Pascal's law: $p_x = p_y = p_z = p$ and as stress force is independent of orientation in a fluid at rest, we have

$$\sigma_{xx} = \sigma_{yy} = s_{zz} = -p \text{ and } \sigma_{xy} = s_{yz} = s_{xz} = 0.$$

Therefore, in a viscous fluid the direct stress components will be given in terms of rate of strain components as follows:

$$\sigma_{xx} = -p + 2\eta\, e_{xx} = -p + 2\eta\frac{\partial v_x}{\partial x}$$

$$\sigma_{yy} = -p + 2\eta\, e_{yy} = -p + 2\eta\frac{\partial v_y}{\partial y}$$

$$\sigma_{zz} = -p + 2\eta\, e_{zz} = -p + 2\eta\frac{\partial v_z}{\partial z} \qquad \ldots(3.10)$$

Now, introducing $f$ as body force per unit mass, the dynamic equations with viscous forces are obtained by substituting the foregoing relationships between stress and rate of strain the above equation into the equations of motion for a continuum. The resulting equation can be derived for $y$ direction from the following equation:

$$\rho\frac{dv_y}{dt} = \rho f_y + \frac{\partial \sigma_{xy}}{\partial x} + \frac{\partial \sigma_{yy}}{\partial y} + \frac{\partial \sigma_{zy}}{\partial z}, \text{ where } \rho \text{ is the density and } \frac{dv_y}{dt} \text{ is}$$

the $y$-component of acceleration. The left hand side (LHS) represents the total body force per unit mass. The other directions follow the similar equations. Therefore, using the equn. (3.6) we get:

$$\rho\frac{dv_y}{dt} = \rho f_y - \frac{\partial p}{\partial y} + \eta\frac{\partial}{\partial y}\left(\frac{\partial v_x}{\partial x} + \frac{\partial v_y}{\partial y} + \frac{\partial v_z}{\partial z}\right)_{+\eta} \left(\frac{\partial^2 v_x}{\partial x^2} + \frac{\partial^2 v_y}{\partial y^2} + \frac{\partial^2 v_z}{\partial z^2}\right)$$

$$= \rho f_y - \frac{\partial p}{\partial y} + \eta\frac{\partial}{\partial y}\, \Delta \cdot v + h\, \Delta^2 v_y \qquad \ldots(3.11)$$

As for an incompressible fluid of constant density, $\dfrac{d\rho}{dt} = 0$, the equation of continuity (3.7), we get:

$$\Delta \cdot V = 0 \qquad \ldots(3.12)$$

This equation allows us to simplify equation of motion (3.7)

$$\rho\frac{dv_x}{dt} - \rho\, f_x - \frac{\partial p}{\partial x} + \eta\nabla^2 v_x$$

$$\rho\frac{dv_y}{dt} = \rho f_y - \frac{\partial p}{\partial y} + \eta\nabla^2 v_y$$

$$\rho \frac{dv_z}{dt} = \rho f_z - \frac{\partial p}{\partial z} + \eta \nabla^2 v_z \qquad \ldots(3.13)$$

These three components can be made in compact form as:

$$\rho \frac{d}{dt} v = \rho f - \Delta p + \eta \nabla^2 v \qquad \ldots(3.14)$$

These equns. (3.10) and (3.14) are known as 'Navier - Stokes equations' for incompressible viscous fluids.

Let us now apply these equations for the viscous flow of a liquid through a horizontal tube of radius $R$ and let the length of the tube be taken as $x$-axis. There are no body forces i.e. $f_x = f_y = f_z = 0$ and also the velocity components, $v_y = v_z = 0$. The equation of continuity (3.7) reduces to $\frac{\partial v_x}{\partial x} = 0$, which signifies that the liquid flows along $x$-axis with constant velocity and so, $v_x$ is not a function of $x$ but is only function of $y$ and $z$-axes. Therefore, imagining the liquid to be divided into different layers in $y$ and $z$-axes, the liquid flows with different velocities i.e. a velocity gradient exists in $y$ and $z$-directions only. Moreover, as we consider a steady state flow velocity is independent of time i.e. $\frac{\partial v_x}{\partial t} = 0$. We then can obtain:

$$\frac{dv_x}{dt} = v_x \frac{\partial v_x}{\partial x} + v_y \frac{\partial v_x}{\partial y} + v_z \frac{\partial v_x}{\partial z} + \frac{\partial v_x}{\partial t} = 0$$

The $x$-component of equn. (3.13) then reduces to:

$$\frac{dp}{dx} = \eta \left( \frac{\partial^2 v_x}{\partial y^2} + \frac{\partial^2 v_x}{\partial z^2} \right) \qquad \ldots(3.15)$$

As there exists a cylindrical symmetry, transforming this equn. (3.15) in cylindrical coordinates, we get

$$\frac{d^2 v_x}{dr^2} + \frac{1}{r} \frac{dv_x}{dr} - \frac{1}{\eta} \frac{dp}{dx} = 0 \qquad \ldots(3.16)$$

Here $r$ is the radial distance from the axis of the cylindrical tube.

Integrating the above equation, we get

$$v_x = \frac{1}{4\eta} \frac{dp}{dx} r^2 + A \ln r + B .$$ Here $A$ and $B$ are integrating constants but

$A = 0$ since $v_x$ has a finite magnitude for $\rho = 0$.

The viscous flow of liquid through this tube of radius $R$ and the velocity gradient of the liquid flow are demonstrated in the following Fig. 3.7.

There is no velocity of the layer in contact with the inner wall of the tube and so for this no slip condition, $v_x = 0$ for $\rho = R$ in all direction around the $x$-axis. Under this condition of laminar flow of the viscous liquid through the tube we get

$$v_x = \frac{1}{4\eta}\frac{dp}{dx}(r^2 - R^2) \qquad \qquad \dots (3.17)$$

This equn. (3.17) is known as Poiseuille's equation for steady state of laminar flow.

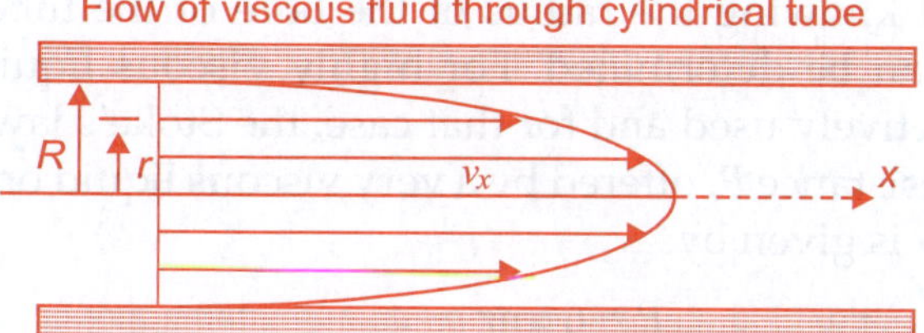


**Fig. 3.7** Laminar viscous flow in a horizontal tube

In the above Poiseuille's equn. (3.17) as pressure decreases with the increase of $x$ the factor $\dfrac{dp}{dx}$, the pressure gradient along $x$-axis is negative and so the above equation considering the sign of this pressure gradient can be written as:

$$v_x = \frac{1}{4\eta}\frac{dp}{dx}(R^2 - r^2).$$

The above equation may be transformed into more practical measurable form by transforming in terms of measurable volume of liquid that flow out of the tube. For doing this, if we multiply both sides of the above equation with cross sectional area of the area element i.e. $2\,prdr$, then we will get the volume of liquid that flow and calling it $dv$, we get:

$$2\pi r\, dr\, v_x = \frac{1}{4\eta}\frac{dp}{dx}(r^2 - R^2)2\pi r\, dr.$$

As $x$ is taken along the length of the tube writing $\dfrac{dp}{dx} = \dfrac{P}{l}$, where $P$ is the pressure difference at the two ends of the tube of length $l$ and equating the left side with $dv$, we can write:

$$dv = \frac{P\pi}{2\eta l}(r^2 - R^2)r\, dr.$$

Integrating the above equation for the entire are of cross section of the tube and equating the left side to the volume, $V$ of liquid flown per unit time, we get

$$V = \int_0^R \frac{P\pi}{2\eta l}(r^2 - R^2)\,r\,dr = \frac{P\pi R^4}{8\eta l}. \qquad \ldots (3.18)$$

The above equation can be easily used for the determination of viscosity of liquids having low or moderate viscosity coefficient. The simple experimental requirement is to set the flow of the liquid at one end of a horizontal narrow tube at constant pressure and collect the liquid flowing out from the other end. Measuring the volume of liquid pouring out per unit time and knowing the liquid height from the inlet end, the constant pressure difference and knowing the radius of the bore of the tube, the viscosity coefficient $h$ can be determined. For highly viscous liquid this method cannot be effectively used and for that case, the Stoke's law. According to this law, the resistance $P$, offered by a very viscous liquid on a falling small spherical body is given by:

$$P = 6\,\pi\eta r\,v$$

where $r$ is the radius of the small sphere and $v$ is the velocity of fall through the liquid. This resistance force on the spherical body will be added with the buoyancy force and these two together equal the downward force due to gravity, the spherical body assumes a constant velocity called terminal velocity $v_t$. The condition can be stated as follows:

$$6\,\pi\eta r\,v_t + \frac{4}{3}\pi r^3\,\rho_l g = \frac{4}{3}\pi r^3\,\rho_s g$$

where $\rho_1$ and $\rho_S$ are the densities of the liquid and material of the spherical body respectively. A sinking ship in ocean experiences this viscous drag and can almost safely land and settle on the ocean bed. The wreckage of ship can be recovered much later.

If we assume that the coefficient of viscosity is negligible i.e. $\eta = 0$, we then get a 'perfect fluid'. The perfect fluid is characteristic of the fact that it does not support any shear. For this case the Navier-Stokes equn. (3.10) reduces to:

$$\rho\frac{d}{dt}v = \rho f - \Delta p \qquad \ldots(3.19)$$

This equation is called 'Euler's equations of hydrodynamics'.

The motion of a perfect fluid follows a 'tube of flow' consisting of a family of 'streamlines', which are the imaginary curves in the fluid and are tangential at every point to the velocity vector, as illustrated in the following Fig. 3.8. So that

$$v \times dr = 0 \qquad \ldots(3.20)$$

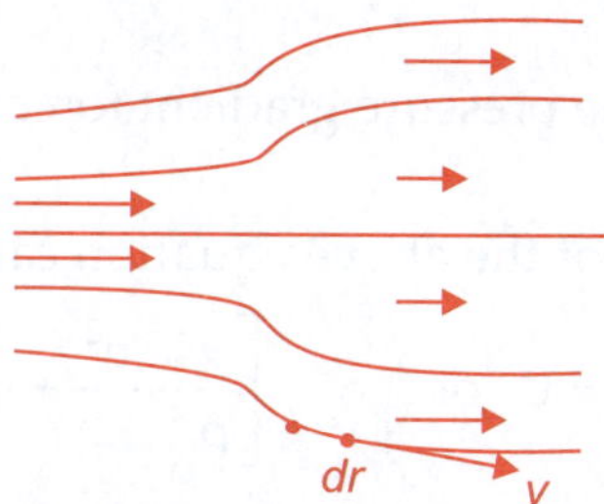

**Fig. 3.8** A family of streamlines in a perfect fluid

Now, recalling the equation known as Euler's equn. (3.19), the equation can also be written in the form considering $v$ as function of time and displacement:

$$\frac{\partial}{\partial t}v + v\,(\Delta \cdot v) = f - \frac{1}{\rho}\Delta p$$

But for a perfect fluid acted on by conservative force like gravity, the body force $f$ can be written as $f = -\Delta\varphi$, where $\varphi$ denotes the gravitational potential and also in steady flow we must consider $\frac{\partial}{\partial t}v = 0$. We then get from the above equation

$$V(\Delta \cdot v) = -\Delta\left(\frac{p}{\rho} + \varphi\right)$$

Using vector identities we write:

$$\frac{1}{2}\Delta v^2 - v \times (\Delta \times v) = -\Delta\left(\frac{p}{\rho} + \varphi\right)$$

or, $\Delta\left(\dfrac{v^2}{2} + \varphi\right) - v \times (\Delta \times v) = -\Delta\left(\dfrac{p}{\rho}\right).$

Taking scalar product of the above equation with a line element $dr$, we get

$$dr \cdot \Delta\left(\frac{v^2}{2} + \varphi\right) = dr \cdot \frac{\partial \xi_y}{\partial x}$$

As $\quad dr \cdot [v \times (\Delta \times v)] = (\Delta \times v) \cdot (dr \times v) = 0$ and then we can write,

$$d\left[\frac{v^2}{2} + \varphi\right] = dr \cdot \left\{-\Delta\left(\frac{p}{\rho}\right)\right\}. \qquad \qquad \dots(3.21)$$

The equn. (3.21) above states the energy conservation law for a perfect fluid as this equation equates the change in energy per unit mass of the fluid with the work done by the pressure gradient force per unit mass i.e. $\Delta\left(\dfrac{p}{\rho}\right)$ along a stream line.

The equivalent form of the above equation can be written as:

$$d\left[\frac{p}{\rho}+\frac{v^2}{2}+\varphi\right]=0 \quad\text{or,}\qquad \left[\frac{p}{\rho}+\frac{v^2}{2}+\varphi\right]=\text{const.} \qquad \dots(3.22)$$

This equn. (3.22) is known as 'Bernoulli Equation' along a stream line of a perfect fluid.

As an example of this famous Bernoulli Equation, if we neglect external forces, the velocity increases as pressure is lowered and vice versa. If a liquid flows through a pipe having constriction, the velocity at the constricted part has to increase to maintain same rate of flow all through the pipe, the pressure at the constricted part is diminished accordingly. This is shown in the following (Fig. 3.9). The wings of a speeding aircraft on the runway experience an upward force and thus can lift its nose up and take off. Another example of the application of this equation is the cause for the roof of house to fly off during a strong gale. The wind moving at the top of the roof suffers decrease of pressure where as below the roof inside the house the pressure becomes much high and this difference forces the roof to get detached and fly off with the wind.

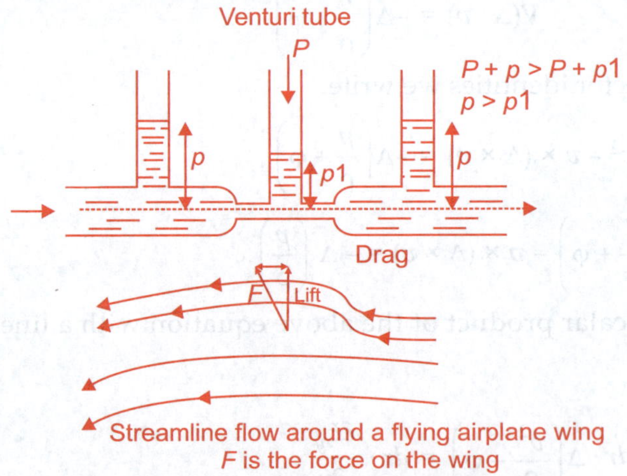


**Fig. 3.9** Applications of Bernoulli's Theorem. The top is the venturi tube and bottom the wing of airplane helping the craft to move up

### 3.6.1 Dependence of Viscosity on Temperature

The viscosity of liquids depends markedly on temperature. There are various proposition but none of them satisfy with the experimental results. However, Andrade's theory based on some assumptions can be applied to first approximation. The relation proposed is that: $\eta = Ae^{\frac{c}{T}}$ can be applied. However, it can be stated that with the increase of temperature the viscosity of liquids decreases. For water if temperature is varied from 0°C to 100°C, the viscosity of water changes from 0.01785 to 0.00282 in CGS unit.

Again it can be stated here that the velocity of moving liquid if remains within certain limit called 'Critical Velocity', the flow of liquid is found to be independent of density but if it exceeds then the liquid flow depends largely on density than on viscosity. This explains the comparatively rapid flow of very viscous lava during volcanic eruptions.

## REVIEW QUESTIONS

1. Explain why "Stress" is defined as a resistance force and not as a reaction to the applied force.

2. In practice all types of stresses are applicable but under what condition only one type of stress and the associated elastic modulus is considered?

3. Establish a relation between different type of stresses and the associated moduli of elasticity.

4. A 95 kg load tied with a rope of 15 m long, when allowed to fall from the point of suspension is found to be dangling from the end. Find the strain and stress if the rope is of 9.5 mm in diameter and has have an extension by 2.8 cm.

5. A horizontal beam is supported at its two ends and loaded in the centre. Show that the upper part of the beam is under compression while the lower part under tension.

## REFERENCES


1. F. H. Newman and V. H. L Searle - The General Properties of Matter, (5th Ed.), Edward Arnold (Publishers) Ltd., London, 1962.

2. A. J. M Spencer - Continuum Mechanics, Longman, London, 1980.

3. R. J. Atkin and N. Fox – An introduction to the theory of Elasticity, Longman, London, 1080.

4. S. B. Palmer and M. S. Rogalski – Advanced University Physics, Gordon and Breach Publishers, U. K., 1996.

5. R. P. Feynman, R. B. Leighton and M. Sands – The Feynman Lectures on Physics, Addison Wesley Pub. Co., Reading, Mass, Narosa Pub. House New Delhi, 2003.

# Heat and Thermodynamics

## 4.1 INTRODUCTION

The word "Thermo" means heat and dynamics is the motion. The thermodynamics is the macroscopic Physics which deals with the conversion of heat i.e. thermal energy into motion representing mechanical energy. It is within the domain of bulk science or macroscopic Physics as this branch of science does not concern with the constituents and their behaviour under different conditions and rather it concerns a matter as solid, liquid or gas, the state in which it belongs. It is based on logical inferences observed considering the matter as a whole and its interaction with the surrounding. Though thermodynamics is a macroscopic science, heat cannot be explained macroscopically as this form of energy depends on the nature of microscopic constituents and their motions in fluids where they can move rather freely and the vibratory motions in solids. We will discuss the origin of this microscopic science heat later and presently let us begin with the macroscopic manifestation of this heat energy. Thermodynamics while failing to give the definition of heat, it gives the independent definition of temperature from macroscopic point of view. We have already come across two other parameters pressure and volume, which are very important particularly when the matter, introduced in thermodynamics as "System" is a gas. Pressure of course has microscopic origin as it depends like heat on the motions (velocity) and the change of momentum of the constituent molecules/atoms on their collisions with container wall. There exist two relations one that relates pressure of a gas with its volume at constant temperature, known as Boyle's Law and the other relates volume of a gas with temperature when pressure remains constant, known as Charles's Law.

**Boyle's Law**: $P \propto 1/v$ i.e. $P = k/V$, where $k$ is a constant.

Therefore, $PV = k$(constant) for a certain mass of gas. Taking differentials $VdP + PdV = 0$ i.e. $\dfrac{dP}{dV} = -\dfrac{P}{V}$. This shows that the slope of $P$ vs. $V$ curve is negative. This condition remains valid for an ideal gas when temperature is maintained constant.

**Charles's Law**: $V \propto V_0 t$, where $t$ is the temperature in Celsius scale and pressure remains constant and $V_0$ is the initial volume for a certain mass of a gas. Therefore, $V = kV_0 t$, where $k$ is a constant.

Taking differential $dV = kV_0\, dt$ i.e. $\dfrac{dV}{V_0} = kdt$. This constant of proportionality $k$ is introduced as volume expansion coefficient and is noted as $\gamma$. Which is defined in finite form as:

$$\frac{V_2 - V_1}{V_1(t_2 - t_1)} = \gamma$$

For finite change of volume due to finite change of temperature, from 0 to $t$, we get: $V_t - V_0 = V_0{}^3 t$, where $V_0$ is the volume at $t = 0°C$

$V_t = V_0[1 + \gamma t]$, Now, if temperature is changed from $0°C$ to $-\dfrac{1°C}{\gamma}$.

The volume of the gas $V_t = 0$. No further decrease of temperature is possible, as that would make the volume negative. Therefore temperature $-1/\gamma°C$ is regarded as absolute zero of the scale of temperature and the new scale of temperature introduced is known as Kelvin Absolute Scale. It can be found that $-1/\gamma°C$ is $-273°C$ from the slope of the linear plot of $V_t$ vs. $t$. We will however, introduce this Kelvin absolute scale from thermodynamical point of view having relevance to all the physical states of the matter later in this Chapter.

## 4.2 HEAT AND TRANSPORTATION OF HEAT

Heat is transported in system by three processes i.e. Conduction, Convection and Radiation. Conduction process is that when there is no mass movement of the medium, convection when there is and radiation is the process when there is no requirement of the medium. Though three processes act simultaneously but one of the three dominates over other depending on the physical state of the system. If it is solid, then conduction process dominates as mass movement is not possible and in liquid or gas the convection is the process of transportation of heat.

**Conduction and conductivity:** Let us consider a rectangular bar of cross section S and at the steady state i.e. when there is no change of temperature of any longitudinal section, $\theta_1$ and $\theta_2$ are the steady temperatures in Celsius of two sections $C$ and $D$ within the bar. Let $Q$ be the heat transported during the time interval $t$.

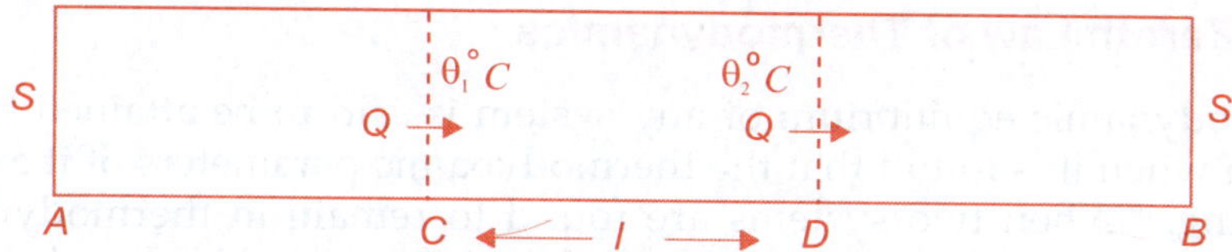

**Fig. 4.1** A rectangular bar of cross section (uniform) S, C and D are two cross sections within the bar at a distance of l

Then $Q \propto S$

$$\propto t$$

and $\qquad \propto (\theta_2 - \theta_1)/l$ i.e. temperature gradient

Therefore, $Q = KS\,(\theta_2 - \theta_1)t/l$ or, $\dfrac{Q}{t} = KS\,(\theta_2 - \theta_1)/l$ and in differential

form, $\dfrac{dQ}{dt} = KS\dfrac{d\theta}{dl}$ the constant of proportionality $K$ which is the thermal

conductivity is related to the rate of flow of heat as:

$$K = \frac{dQ/dt}{S\dfrac{d\theta}{dl}}.$$

In this steady state the section within $C$ and $D$ does not absorb any quantity of heat and the entire heat entered through $C$ is transmitted through $D$ leading no change in the temperatures of $C(\theta_1{}^\circ C)$ and of $D(\theta_2{}^\circ C)$. Before this steady state is reached a portion of the heat entering this region will be absorbed by the material depending on the specific heat of the material and is manifested by the rise of temperatures. In this unsteady state the rate of increase of temperature is given by the ratio:

$$\frac{K}{\rho S} = \frac{\text{thermal conductivity}}{\text{thermal capacity per unit volume}}, \text{ where } \rho \text{ is the density and } S \text{ is}$$

the specific heat. This ratio is known as thermal diffusivity or thermometric conductivity.

## 4.3 LAWS OF THERMODYNAMICS

So far we have come across three thermodynamical parameters of a system pressure $(P)$, volume $(V)$ and temperature $(T)$. While pressure has its microscopic origin, temperature is defined unambiguously by thermodynamics from its own macroscopic domain. Let us start with the laws of thermodynamics.

### 4.3.1 Zeroth Law of Thermodynamics

Thermodynamic equilibrium of any system is said to be attained by the system when it is found that the thermodynamic parameters of it are not changing. "When two systems are found to remain in thermodynamic equilibrium with a third system separately or independently, then these two systems will also remain in thermodynamic equilibrium". This is known as Zeroth Law of Thermodynamics. "The parameter which determines thermal equilibrium and also determines the direction of flow of heat between two systems not in thermal equilibrium is known and defined by thermodynamics as temperature". Heat flows from a body at higher temperature to a body at lower temperature irrespective of the total content of heat in either of the systems.

**Conclusion:** Heat can only flow under normal condition from a body at higher temperature to another body at lower temperature when they are put in thermal contact.

### 4.3.2 First Law of Thermodynamics

The first law determines the relationship between the mechanical work and heat and vice versa. It is found that whenever two solids are rubbed together they become hot, also when a body is drilled tremendous heat is generated. This is related with the conversion of mechanical work in to heat. If $W$ is the amount of work spent for it and $Q$ is the quantity of heat generated, then:

$W \propto Q$ or, $W = JQ$. Where $J$ is known as Joule's constant. When 4.2 Joules of mechanical work is converted in to heat it produces one calorie of heat.

When heat is to be converted in to mechanical work, the heat is to be supplied to a system and the system is to transform the supplied heat in to mechanical work. The performance of work from a fixed quantity of heat is different for different given conditions or constraints. For better representation let us consider a gaseous system enclosed within a cylinder fitted with a piston.

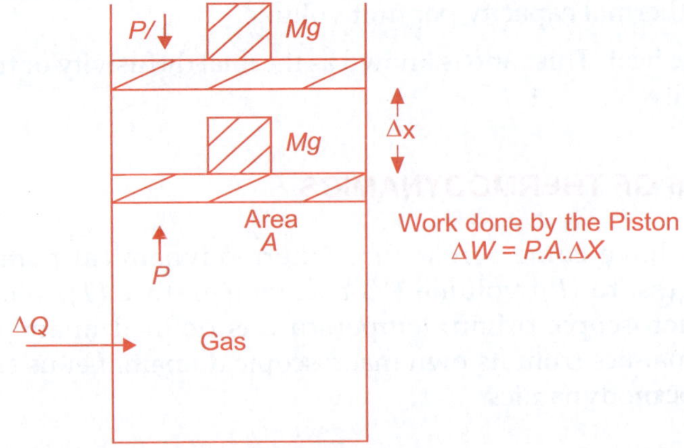


**Fig. 4.2** $\Delta Q$ is the heat supplied, work done by the piston $\Delta W = P.A. \Delta X = Mg. \Delta X$

When heat is supplied by the piston, the pressure within the gas increases to $P$ which becomes greater than the outside pressure $P'$ and as a result the gas forces the piston of mass $Mg$ to move up against gravity. In differential form: $dW = P \cdot A \cdot dX = P \cdot dV$, integrating from initial to final $W = \int_i^f P \cdot dv$. In general for any thermodynamic process from $i$ to $f$, the work done is $\int_i^f P \cdot dv$ and it is the area enclosed in the $P - V$ diagram known as Work diagram.

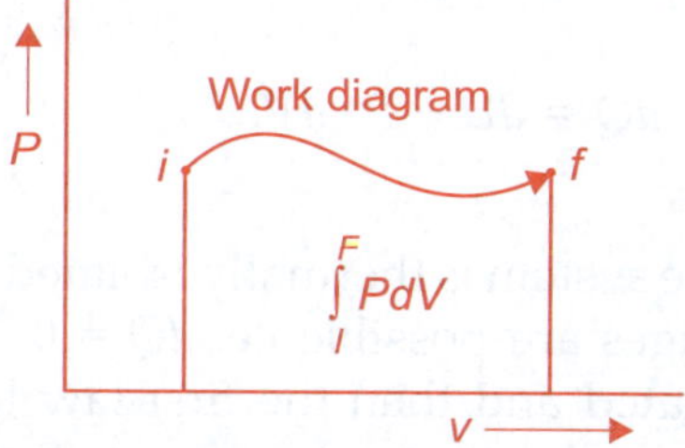


Now, the heat given to the system under general process is not totally converted to work, a certain portion of the supplied heat is retained by the system and as a result its "Internal Energy"($U$) increases. Therefore, the energy conversion equation is given by: $dQ = dU + PdV$. This conversion equation is also the statement of First Law of Thermodynamic like $W = JQ$. Now, this conversion or conservation of energy equation depends on the processes involved either on the mode of supply of heat or the constraints imposed on the system to perform work. This first law is essentially is the law of conservation of energy. Now, the different unique and fundamental processes possible are as under:

### 1. Isothermal Process

In this process the temperature must remain constant i.e. $dT = 0$. Now, it will be shown later that for an ideal system the internal energy, $U$ is a function of temperature only and so in this process $dU = 0$ and the first law then transforms in to: $dQ = P \cdot dV$.

The entire heat given to the system is fully converted in to work. We will see that this conclusion is not remain valid for any continuous process. This process however, gives the definition of mechanical work from thermodynamics. "The mechanical work is the equivalent heat given to a system isothermally".

### 2. Isochoric Process

In this process the volume of the working system remains constant i.e. $dV = 0$ and therefore as work done is equnl to $\int_i^f P \cdot dv$, $dW = 0$ and the first law transforms in to: $dQ = dU$. "The internal energy change is the equivalent of the heat energy given to the system isochorically" is then the definition of internal energy of the system.

Unlike work which is path dependent, internal energy, $U$ is a state dependent parameter. The change of this parameter depends only on the initial and final states of any process not on the process followed to bring the system from initial to final. $dU$ is then an exact differential.

### 3.  Isobaric Process

In this process the pressure should remain constant i.e. $dP = 0$ and the process is to be slow as that of isothermal process. The first law is then transformed in to:

$$dQ = dU + P \cdot dV.$$

### 4.  Adiabatic Process

In this process the entire system is thermally isolated from the surrounding so that no heat exchanges are possible i.e. $dQ = 0$. This is possible if the system is ideally insulated and then the first law leads to: $dU = -P \cdot dV = -dW$. As the work done by the system is positive and so an adiabatic work done by the system will decrease its internal energy. Conversely, as the work done on the system is negative, then for any adiabatic work done on the system will increase the internal energy. As internal energy for an ideal system is a function of temperature only, these increase or decrease of internal energy will be manifested by the increase or decrease of temperature.

These processes defined above may be combined partially or fully leading to in numerable thermodynamic processes. All these possible thermodynamic processes are then classified in to two categories which are intrinsically connected to the basic concept of thermodynamics and they are:

(A)  **Reversible Process**

(B)  **Irreversible Process.**

Reversible processes are those which after the completion of the cycle do not leave any change either within the system or its surrounding. This is an ideal process and cannot be observed or created. All natural processes are irreversible i.e. after the completion of any natural process some changes are left either in the system and its surrounding or in both. As thermodynamics discusses its principles considering the system and its immediate surroundings with which it may interact, there is nothing like reversible reactions or process. If there is no change left after the completion of the process there will definitely be some permanent changes in its surroundings. However, the reversible process is taken as standard and all the practical irreversible processes are compared with it to assess their efficiency and also feasibility. We have seen in isothermal process that complete conversion of heat in to mechanical work is possible. Now, let us raise a question can this complete conversion even by isothermal process be continued indefinitely and continuously?

If it is to be made possible then we require an infinite reservoir of heat so that any decrease of its temperature after supply of heat to the system can be neglected. Even if it can be assumed then the length of the working cylinder Fig. 4.2 must be infinitely long so that the piston can move up against gravity and perform work un interrupt. This is simply not possible. Therefore, to continue the work the piston has to come back to its initial position and in doing so it has to decrease its volume by releasing heat. Now this release of heat is not possible by returning back the heat to the reservoir from which the heat was taken during expansion as the temperature of the system must remain always same and less than the supplier heat reservoir and therefore, the process to be continuous a 'cold' reservoir having temperature less than the system temperature is essential. In short, the system can perform work continuously if it works between two reservoirs 'Hot' and 'Cold' compared to their temperatures with that of the system. This impossibility of working with only one reservoir can be illustrated by the impossibility of running a ship in ocean by extracting heat from the ocean bed and fully converting it in to work without requiring any 'cold' reservoir to release residual heat. This natural observation is formulated in thermodynamics as second law of thermodynamics.

**Conclusion:** Mechanical work can only be obtained from an equivalent quantity of heat energy and vice versa. One form of energy can only be produced by the conversion of other form of energy. This is an example of conservation of energy principle any violation of this principle is impossible and the hypothetical system which would violet this first law is known in thermodynamics as "Perpetual Motion of First kind".

### 4.3.3 Second Law of Thermodynamics

The second law of thermodynamics has two statements one is known as Kelvin-Planck Statement and the other Claussius Statement. Let us first introduce these two statements.

**Kelvin-Planck Statement**: It is impossible for an engine working in cycle to extract heat from a hot reservoir and convert it fully in to work, leaving no change either in the system or in the surrounding except the production of work.

**Claussius Statement:** It is impossible for a device working in cycle to transfer heat from a cold reservoir to a hot reservoir without leaving any change in the surrounding.

**Discussions on Kelvin-Planck Statement:** This statement means that an engine which is a device whose sole function is to convert heat energy in to mechanical energy, cannot convert the entire quantity of heat that it accepts from a hot reservoir in to work if it is to work in cycle. Certain amount of heat is to rejected after each cycle and amount of this rejected heat depends on whether the engine in question is a reversible engine or not.

For ideal performance assuming that there is no loss of heat:

$$Q_1 = Q_2 + W.$$

The Efficiency of the heat engine is given by

$$\eta = \frac{W}{Q_1} = \frac{Q_1 - Q_2}{Q_1} = 1 - \frac{Q_2}{Q_1}$$

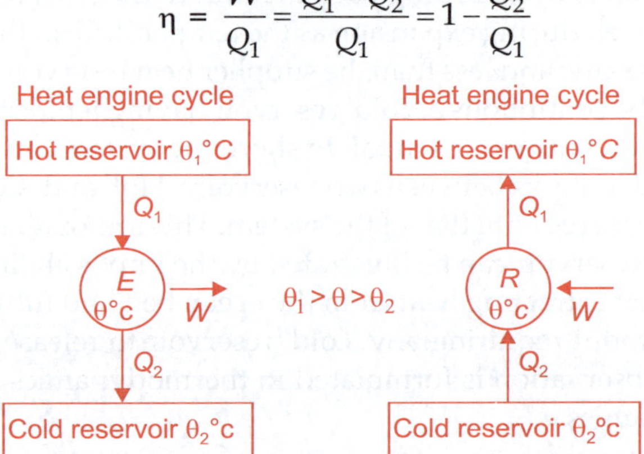


**Fig. 4.3** The Engine and Refrigerator cycle working between hot and cold reservoirs. Refrigerator cycle operates in the opposite direction as that of engine cycle

**Discussion on Claussius Statement:** The transfer heat from a cold reservoir to a hot reservoir is not a general or natural process as this is not permitted by zeorth law but to perform it it is essential that some quantity of work is to be done on the system. As this work is to be done by surrounding on the working system, the surrounding can deliver the required amount of work at the expense of its energy.

For an ideal performance, $Q_1 = Q_2 + W$

The Coefficient of Performance of the refrigerator $(COP)_{Ref.} = \dfrac{Q_2}{W} = \dfrac{Q_2}{Q_1 - Q_2}$. There is another device known as Heat pump whose sole function is to maintain the temperature of a body constant. This is necessary when a body is to be kept at a temperature higher than the surrounding temperature. The function of such Heat Pump can be explained by the following Figure.

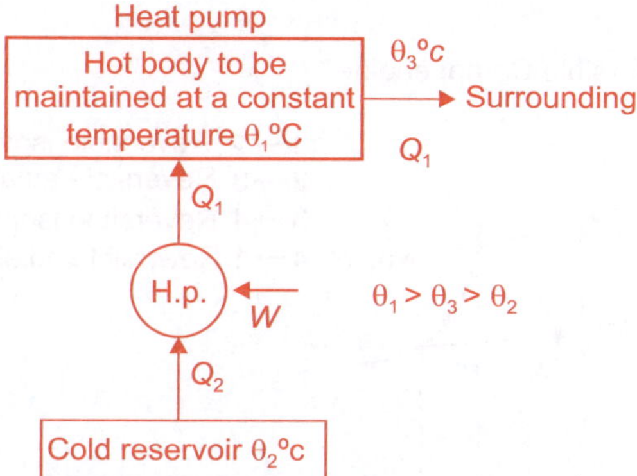


The coefficient of performance of Heat Pump: $(COP)_{H.P.} = \dfrac{Q_1}{W} = \dfrac{Q_1}{Q_1 - Q_2}$.

Therefore, $(COP)_{H.P} - (COP)_{Ref.} = \dfrac{Q_1}{Q_1 - Q_2} - \dfrac{Q_2}{Q_1 - Q_2} = 1$.

**Conclusion:** There is essential requirement of two reservoirs one having temperature higher than the working system of the engine known as 'Hot Reservoir' and the other having temperature lower than the working system known as 'Cold Reservoir'. Any system which violates this second law is known as "Perpetual Motion of Second Kind". If this impossible hypothetical system be found to exist then we would run a ship on ocean without spending any fuel. This hypothetical ship would then extract heat from ocean bed and convert it totally in to work by working with only one reservoir.

### 4.3.4 Ideal Reversible Carnot Cycle

As has been stated earlier the reversible cycle is an ideal cycle, let us now discuss about the ideal Carnot cycle. In the following diagram which is the work diagram of such Carnot cycle, let us consider that the working system be kept in thermal contact with a hot reservoir and extract heat $Q_1$ from it and expand isothermally at a temperature of $\theta_1$°C. The expansion of the cylinder from 1 to state 2 represent this isothermal expansion during which the temperature of the system remains constant at $\theta_1$°C. 2 to 3 represent adiabatic expansion during which the working system is detached from hot reservoir and insulated from the surrounding and during this process the temperature of the system decreases from $\theta_1$°C to $\theta_1$°C.

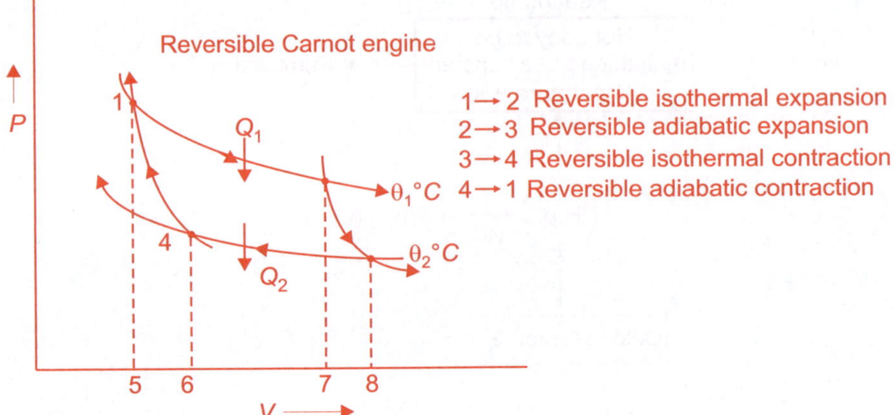


**Fig. 4.4** Reversible Carnot Cycle(Engine) from $1 \to 2 \to 3 \to 4 \to 1$ and work involved in each of such processes are $W_1 =$ Area 1275, $W_2 =$ Area 2387 (both positive) and $W_3 =$ Area 3468, $W_4 =$ Area 4156 (both negative). The network out put $W$(Area 1234) is then equivalent to $W = W_1 + W_2 - W_3 - W_4$

After 2 to 3, the work is then done on the system by external agency isothermally at temperature $\theta_2$°C. So that the volume decreases and pressure increases. In this isothermal process system is put in thermal contact with a cold reservoir, which accepts the released heat $Q_2$. The system after completion of this process is again put under insulating enclosure and work is performed on the system adiabatically by external agency so that it returns to initial state 1 by following 4 to 1 reversible adiabatic path.

$$W_1(\text{Area 1275}) = \int_1^2 PdV = R\int_1^2 dV/V = R(\ln[V_2/V_1]), \text{ where } R \text{ is constant}$$

and as for isothermal processes $PV = RT$. (for 1 gm mole)

$$W_2(\text{Area 2387}) = \int_2^3 PdV \text{ but as 2 to 3 is an adiabatic expansion,}$$

$$PV^3 = k \text{ (constant) and so, } W_2 = k\int_2^3 dV/V^\gamma = \frac{k}{1-\gamma}(V_3^{1-\gamma} - V_2^{1-\gamma}).$$

$$W_3(\text{Area 3468}) = \int_3^4 PdV \text{ and 3 to 4 is an isothermal compression i.e.}$$

$$PV = RT,$$

$$W_3 = R\int_3^4 dV/V = R(\ln[V_4/V_3]). \text{ For 4 to process which is an adiabatic}$$

compression i.e. $PV^3 = k$ (constant), $W_4 = k\int_4^1 dV/V^\gamma = \dfrac{k}{1-\gamma}(V_1^{1-\gamma} - V_4^{1-\gamma})$

and the network out from the entire cycle $1 \rightarrow 2 \rightarrow 3 \rightarrow 4 \rightarrow 1$ $W$ is equnl to: $W = W_1 + W_2 - W_3 - W_4$. In terms of heat $W = Q_1 - Q_2$ as this is an ideal reversible cycle and there is no loss of heat from the system or during its transport.

The efficiency of this Carnot engine, $\eta = W/Q_1 = \dfrac{Q_1 - Q_2}{Q_1}$. Which is always $< 1$.

When the cycle is reversed i.e. $1 \rightarrow 4 \rightarrow 3 \rightarrow 2 \rightarrow 1$, then the Carnot engine will work as a Carnot refrigerator and all the outcome of the processes will be reversed.

### 4.3.5 Equivalence of Kelvin-planck and Claussius Statements

The two statements of second law of thermodynamics (Kelvin-Planck and Claussius) as stated above seem to be two different statements but actually they are basically same and formulate the same phenomena but only in two different ways to express the two different aspects. The two statements are equivalent to each other. This equivalence between the two can be established on the principle of negation of negation i.e. first we will assume that even if one law is not necessarily be true other one can retain its validity and vice versa. This is the negation of the equivalence. Now, if it can be shown that this negation is not supported by thermodynamics then the two negative statements operated one over other lead to the positive conclusion of their equivalence. Let us now study the following Fig 4.5.

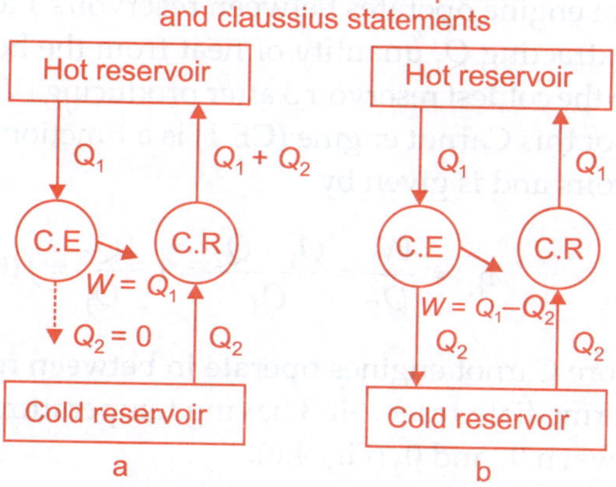


**Fig. 4.5** (a) Represents the negation of K-P statement and assuming the validity of claussius and (b) Represents the reverse

In Fig. 4.5 (a) let us assume that Kelvin-Planck statement is wrong but Claussius statement is correct. A Carnot engine operating between two reservoirs can then extract $Q_1$ quantity of heat from the hot reservoir and convert it fully into work, $W$ and becomes a perpetual motion of second kind. A Carnot refrigerator operating between same two reservoirs extracts $Q_2$ and $W$ be the work done on it and releases $W + Q_2$ amount of heat to the hot reservoir. Now, if we couple these two cycles, so that output work $W$ from Carnot engine be given to the refrigerator, then the total effect is extraction of heat from a reservoir and transfer heat to a hotter body without requiring any assistance in the form of work from the surrounding. This is an obvious violation of Claussius statement. Therefore, if Kelvin Plancks statement is wrong then the Claussius statement cannot remain valid.

Again in Fig. 4.5 b. as Kelvin-Planck statement is valid then let a Carnot engine operates between two reservoirs and extracts $Q_1$ heat performs work W and releases $Q_2$ heat to the cold reservoir. In between the same two reservoirs let a Carnot refrigerator operates and it needs not to obey Claussius statement as the statement is assumed to wrong. In the similar fashion if we now couple these two cycles, then the resultant cycle in effect needs $Q - Q$ heat from hot reservoir to convert it fully work violating the Kelvin-Planck statement. Therefore, considering these two aspects we can conclude that if one of the two statements is wrong the other one cannot remain correct. This is then the negation of negation and thus the equivalence of the two statements are established.

### 4.3.6 Kelvin's Absolute Scale of Temperature

Let us consider three reservoirs at temperatures in Celsius scale as $\theta_1$, $\theta_2$ and $\theta_3$ and let a Carnot engine operates between reservoirs 1 and 2 at $\theta_1$ and $\theta_2$ respectively by extracting $Q_1$ quantity of heat from the hottest reservoir 1 and releases $Q_2$ to the coldest reservoir 3 after producing $W$ amount of work. The efficiency $\eta_1$ of this Carnot engine (CE 1) is a function of temperatures of the two reservoirs and is given by

$$\eta_1 = \frac{W_1}{Q_1} = \frac{Q_1 - Q_2}{Q_1} = 1 - \frac{Q_2}{Q_1} = f(\theta_1, \theta_2). \qquad ...(4.1)$$

Now, two more Carnot engines operate in between reservoirs 1 and 2 with another intermediate reservoir 3 having temperature $\theta_3$ which is also intermediate between $\theta_1$ and $\theta_2$ (Fig. 4.6).

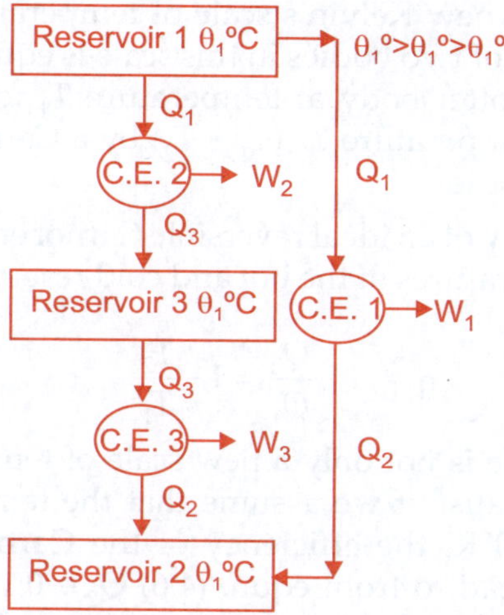


**Fig. 4.6** Carnot engine (CE1) operates between $\theta_1$ and $\theta_2$ and CE 2 and CE 3 operate between $\theta_1$, $\theta_3$ and $\theta_3$, $\theta_2$ giving work output respectively as $W_2$ and $W_3$

Now, if CE 2 and CE 3 respectively have their efficiencies as $\eta_2$ and $\eta_3$, then they are similarly related to:

$$\eta_2 = 1 - \frac{Q_3}{Q_1} = f(\theta_1, \theta_3) \qquad \ldots (4.2)$$

and

$$\eta_2 = 1 - \frac{Q_3}{Q_2} = f(\theta_3, \theta_2) \qquad \ldots (4.3)$$

Now, from equns. (4.1), (4.2) and (4.3) we can write:

$$\frac{Q_1}{Q_2} = \frac{1}{1 - f(\theta_1, \theta_2)} = \phi(\theta_1, \theta_2), \quad \frac{Q_1}{Q_3} = \phi(\theta_1, \theta_3) \text{ and } \frac{Q_3}{Q_2} = \phi(\theta_3, \theta_2)$$

But as $\dfrac{Q_1}{Q_3} = \dfrac{Q_1 / Q_2}{Q_3 / Q_2} = \dfrac{\phi(\theta_1, \theta_2)}{\phi(\theta_3, \theta_2)}$ and as the right hand side has $\theta_3$ both in the numerator and denominator and can be replaced by its relation with $\theta_1$ and $\theta_2$ then the above expression on simplification leads to the following equation where $\psi$ is a new function. Therefore,

$$\frac{Q_1}{Q_3} = \frac{\psi(\theta_1)}{\psi(\theta_3)} \qquad \ldots (4.4)$$

Now, this new function of temperature in Celsius scale leads to a different scale of temperature and denoting that by $T$ is known as Kelvin's scale of temperature so that in general we can write:

$$\frac{Q_1}{Q_2} = \frac{T_1}{T_2} \qquad \ldots (4.5)$$

**Conclusion:** This new Kelvin's scale of temperature is defined as the ratio of temperatures of two bodies in this scale is equnl to the ratio of heat extracted from the hotter body at temperature $T_1$ to the heat rejected to the colder body at temperature $T_2$ $(T_1 > T_2)$ by a Carnot engine operating between these two bodies.

Now, the efficiency of an ideal reversible Carnot engine can be modified in terms of the temperatures of the hot and cold reservoirs in Kelvin's scale as:

$$\eta = 1 - \frac{Q_2}{Q_1} = 1 - \frac{T_2}{T_1} \qquad \qquad ...(4.6)$$

This Kelvin's scale is not only a new scale of temperature but it is the "Absolute Scale" because if we assume that the temperature of the cold reservoir is at $T_2 = 0°K$, the efficiency of the Carnot engine would be $\eta = 1$ and this will lead to from equn. (4.6) $Q_2 = 0$ i.e. the Carnot engine would then require only one reservoir to operate with and thus violate the Kelvin-Planck statement of second law of thermodynamics. Therefore, $0°K$ is the absolute zero which can not be attained by anybody and taking this as the zero (absolute), the scale is known as **Kelvin's Absolute Scale of temperature.** It may be elaborated once more that as the temperature of a body approaches this absolute zero of Kelvin's scale, it becomes more and more difficult to decrease the temperature further and ultimately the time to reach will approach infinitely long. This practically non attainability of absolute zero which otherwise would lead to the violation of second law of thermodynamics is known as Third Law of Thermodynamics.

### 4.3.7 Carnot Theorem

This theorem establishes the difference between the ideal reversible process and the irreversible process. It states like this: "An irreversible engine working between two temperatures can not have its efficiency more than or equnl to the efficiency of a Carnot engine working between same two temperatures".

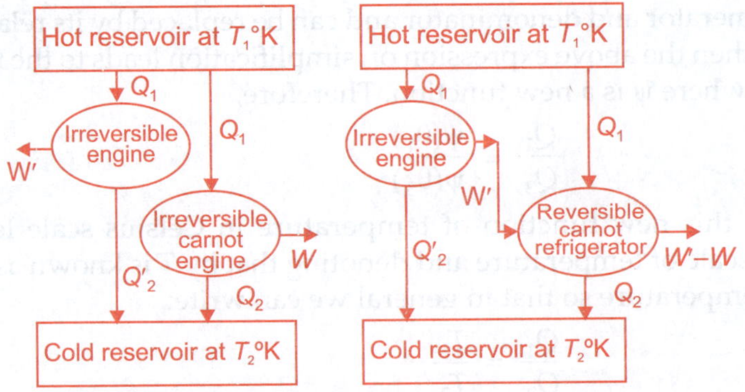


**Fig. 4.7**  The irreversible engine and reversible carnot engine and refrigerator work between same two reservoirs at $T_1$°K and $T_2$°K

To prove the Carnot theorem let us consider that the irreversible engine violates the principle and has more efficiency than the ideal Carnot engine i.e. $\eta_{IR} > \eta_{CE}$,

Therefore $\dfrac{W'}{Q_1} > \dfrac{W}{Q_1}$ , and so $Q_1 - Q'_2 > Q_1 - Q_2$ or, $Q_2 > Q'_2$.

Now, let us reverse the Carnot engine (right side of the Fig. 4.7) so that it can act as a Carnot refrigerator and to run it let the out put work from the irreversible engine $W'$ be given to the Carnot refrigerator. If we consider the net function of this irreversible engine and the Carnot refrigerator we can find that:

1. There is no change of the hot reservoir at $T_1°K$ and the coupled system extracts $Q_2 - Q'_2$ quantity of heat from the reservoir at $T_2°K$ and converts it into $W' - W$ amount of work.

2. The coupled system then violates the second law of thermodynamics and would then perform as a perpetual motion of second kind. Therefore, $\eta_{IR} > \eta_{CE}$ assumption is wrong.

The other possibility is that $\eta_{IR} > \eta_{CE}$, but under that condition the irreversible engine would then transform into a reversible engine and lose its practicaticality.

The only possibility which is then left is $\eta_{IR} > \eta_{CE}$ and then $W > W'$ and $Q_2 < Q'_2$.

The coupled system then receives $W - W'$ amount of work and delivers to the reservoir at $T_2°K$ an amount of heat equnl to $Q'_2 - Q_2$. Though there is no requirement of the reservoir at $T_1°K$, but the functioning of this coupled system is not forbidden by the second law as the process is simply the conversion of mechanical work into heat and such process is not restricted by the second law.

Therefore, it can be concluded that the efficiency of Carnot engine working between two temperatures is the maximum limit of efficiency by any engine working between same temperatures and it is higher than the efficiency of any irreversible engine working between same reservoirs i.e. $\eta_{IR} < \eta_{CE}$.

| Parameters | Irreversible Cycles | Reversible Cycle (e.g. Carnot) |
|---|---|---|
| Efficiency of an Engine | $\eta = 1 - \dfrac{Q_2}{Q_1}$ | $\eta = 1 - \dfrac{T_2}{T_1}$ |
| Coefficient of Performance (COP) of Refrigerator | $COP = \dfrac{Q_2}{Q_1 - Q_2}$ | $COP = \dfrac{T_2}{T_1 - T_2}$ |
| Coefficient of Performance (COP) of Heat Pump | $COP = \dfrac{Q_1}{Q_1 - Q_2}$ | $COP = \dfrac{T_1}{T_1 - T_2}$ |

**Note:** $\dfrac{Q_1}{Q_2} = \dfrac{T_1}{T_2}$ is only applicable for reversible ideal engines, refrigerators or heat pumps and the efficiency or COP calculated in terms of $T_1$ and $T_2$ in Kelvin's scale is the maximum limits of efficiency or COP, which no real irreversible engine, refrigerator or heat pumps can attain.

### 4.3.8 Claussius Theorem

It has been stated earlier that internal energy of an ideal system is a state dependent parameter and mechanical work is a path dependent parameter then what should be the relation between heat involved in a process and with the process itself?

Claussius theorem states that "In all the processes in between one initial and final state, the heat involved will be same if the work involved in such processes is also same".

In the following Fig. 4.8. $i \rightarrow f$ is an arbitrary reversible process and it is replaced within the same initial and final states by $i \rightarrow g$ (reversible adiabatic), $g \rightarrow h$ (reversible isothermal) and $h \rightarrow f$ (reversible adiabatic in such a way that the area subtended by $i \rightarrow f$ is same as sum totals of area subtended by series of processes i.e. $i \rightarrow g$, $g \rightarrow h$ and $h \rightarrow f$.

Therefore, the work involved in both of these processes will be same. Then from first law $Q_1 = U_f - U_i + W$ for original process and $Q_2 = U_f - U_i + W$. The difference of internal energy $U$ will remain same as it is only state dependent parameter.

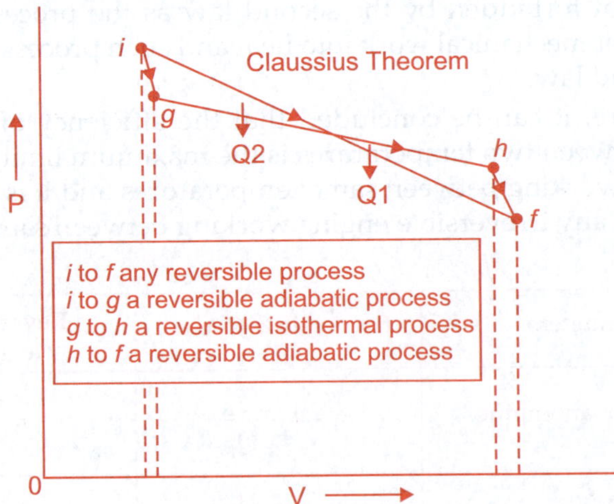


Fig. 4.8 Claussius theorem. $i$ and $f$ are initial and final stages. $g$ and $h$ are intermediate states

Therefore, $Q_1$, the heat absorbed in the original reversible process is same as $Q_2$, the heat involved in the substituted process. This Claussius theorem will be utilized in the introduction of one thermodynamic variable known as Entropy.

## 4.4 ENTROPY AND ENTROPY PRINCIPLE

So far we are introduced to thermodynamic parameters like Pressure, $P$; Volume, $V$; temperature, $T$ and Internal energy $U$. We will now introduce one of the most important parameter known as Entropy and will see and realize during the course of progress of this Chapter the importance of it. Let us consider an arbitrary reversible process is followed by an ideal system and from the following work diagram the area enclosed will give the net work output during the forward cycle. Now, if we sectioned the entire cycle area by a series of reversible adiabatic and join the ends with reversible isotherm then the entire area can be divided in to a series of Carnot cycles. In the following (Fig. 4.9) such series of Carnot cycles are 1, 2, 3, 4, 5, 6, 7, 8 etc. Now, let $Q_1$ and $Q_2$ be the heat absorbed and heat rejected at temperatures $T_1$ and $T_2$ by the Carnot cycle (1, 2, 3, 4) then we can write:

$$\frac{Q_1}{Q_2} = \frac{T_1}{T_2} \quad \text{or,} \quad \frac{Q_1}{T_1} = \frac{Q_2}{T_2}.$$

And similarly for the second Carnot cycle (5, 6, 7, 8)

$$\frac{Q_3}{T_3} = \frac{Q_4}{T_4}$$

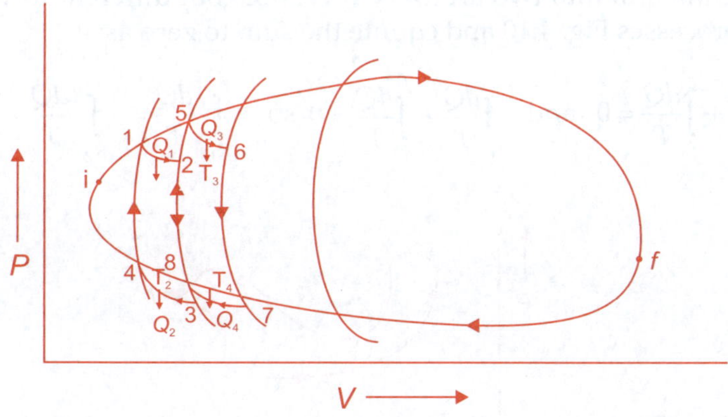


**Fig. 4.9** The work diagram of an arbitrary reversible cycle and Carnot cycle sections

Now, as heat absorbed by a system is considered positive and heat rejected by the system is negative according to the sign convention of thermodynamics, we get from the above expressions after assuming $Q_1$, $Q_3$ ... as positive and $Q_2$, $Q_4$ ... as negative

$$\frac{Q_1}{T_1} + \frac{Q_2}{T_2} = 0 \quad \text{and} \quad \frac{Q_3}{T_3} + \frac{Q_4}{T_4} = 0$$

Considering the total effect of all such possible Carnot cycles in to which the entire area can be divided we get

$\sum\limits_{n} \dfrac{Q_n}{T_n} = 0$ where $n$ is the all such number of Carnot cycle sections. Now, if we bring these Carnot cycles infinitesimally close to each other then we can assume that;

(i) The heat involved is also infinitesimally small and can the be replaced as differential.

(ii) The temperature difference between each cycle may be ignored and

(iii) The summation sign can be replaced by integral.

(iv) The integrated areas of such all possible infinitesimal Carnot cycles can be the equnted to the entire area of the original arbitrary reversible cycle.

Then from the Claussius Theorem, we can write for the entire reversible cycle as:

$$\oint_{\text{Reversible Cycle}} \dfrac{dQ}{T} = 0 \qquad \qquad ...(4.7)$$

As the above equation is valid for any reversible cycle we can then divide the cyclic integral into two arbitrary reversible but different forward and reverse processes Fig. 4.10 and equnte the sum to zero as:

$$\int_i^f \dfrac{dQ}{T} + \int_f^i \dfrac{dQ}{T} = 0 \quad \text{and} \quad \int_i^f \dfrac{dQ}{T} - \int_i^f \dfrac{dQ}{T} = 0 \text{ so, } \int_{i(R_1)}^f \dfrac{dQ}{T} = \int_{i(R_2)}^f \dfrac{dQ}{T} \qquad ...(4.8)$$

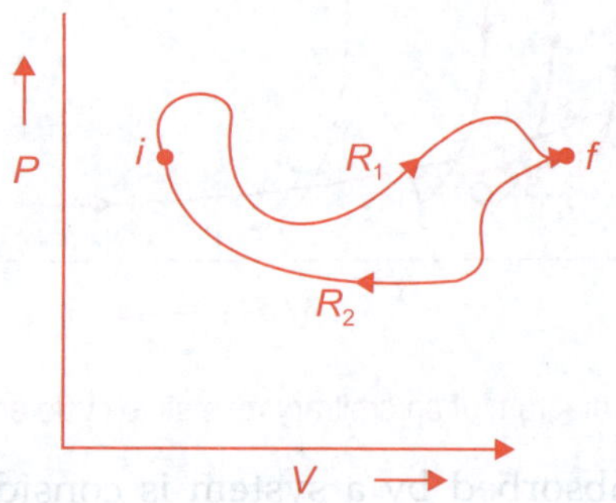


**Fig. 4.10** $R_1$ and $R_2$ are two arbitrary reversible processes between same initial and final states

As $R_1$ and $R_2$ are any two arbitrary reversible processes and as per equn. (4.8), we can write that over any reversible path say $R$ between initial $i$ and final $f$ states;

$$\int_{\substack{i \\ 1(R)}}^{f} \frac{dQ}{T} = S_f - S_i ,$$ Where $S$ is a state dependent parameter and is

independent of the path. This state dependent parameter $S$ is introduced in thermodynamics as Entropy and the change of which between initial and final state is given by $\int_i^f \frac{dQ}{T}$ when the integration is carried over any arbitrary reversible path between $i$ to $f$.

**Conclusion:** Entropy like Internal Energy is a state dependent thermodynamic parameter and also like internal energy its changes between initial and final states can only be measured and not its absolute value. The difference between internal energy and entropy is that while internal energy of a system can increase or decrease, the entropy of a system together with its immediate surrounding can only increase. This is what is known as Entropy Principle and discussed in the following section. Therefore, to summarize the observations made so far:

$$\oint \frac{dQ}{T} = 0 ,$$ For over any reversible complete cycle and, $$\int_i^f \frac{dQ}{T} = S_f - S_i$$ i.e.

the change of entropy between the initial and final states for any reversible path.

### 4.4.1 Entropy Principle

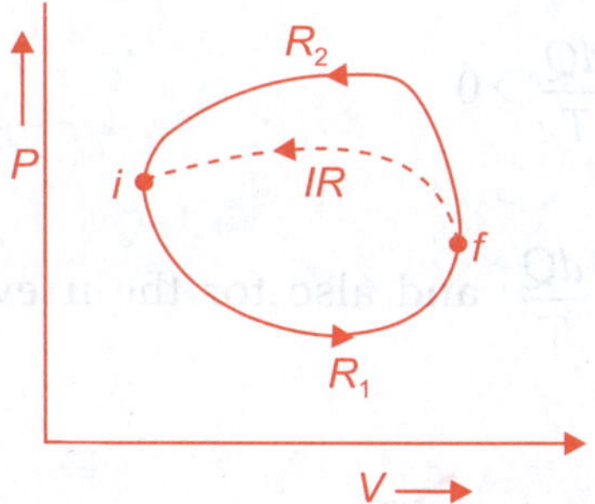


Fig. 4.11 An irreversible cycle, having two steps $i \to f$ by reversible, $R_1$ path and $f \to i$ by an irreversible arbitrary path (IR). The complete cycle comprising

$R_1$ and $IR$ is then irreversible. Another reversible return path $R_2$ is also shown. $R_1$ forward and $R_2$ return constitute then a reversible cycle.

Let us recall the Carnot theorem which concludes that in between two fixed reservoirs $\eta_{IR} < \eta_{CE}$ therefore referring to the Fig. 4.7.

$$1 - \frac{Q'_2}{Q_1} < 1 - \frac{Q_2}{Q_1} \quad \text{and} \quad 1 - \frac{Q'_2}{Q_1} < 1 - \frac{T_2}{T_1} \quad \text{or} \quad \frac{Q'_2}{Q_1} > \frac{T_2}{T_1}$$

And therefore, $\dfrac{Q'_2}{T_2} > \dfrac{Q_1}{T_1}$ and in differential form $\displaystyle\int_{IR} \frac{dQ}{T} > \int_{R_2} \frac{dQ}{T}$

More explicitly: $\displaystyle\int_{f \; IR}^{i} \frac{dQ}{T} > \int_{f \, R_2}^{i} \frac{dQ}{T}$ then $\displaystyle\int_{f \; IR}^{i} \frac{dQ}{T} - \int_{f \, R_2}^{i} \frac{dQ}{T} > 0$

or, $\displaystyle\int_{f \, R_2}^{i} \frac{dQ}{T} - \int_{f \, IR}^{i} \frac{dQ}{T} < 0$ ...(4.9)

but we know for the reversible cycle

$$\oint_{R} \frac{dQ}{T} = \int_{iR_1}^{f} \frac{dQ}{T} + \int_{fR_2}^{i} \frac{dQ}{T} = 0$$

...(4.10)

Now, subtracting (4.10) from (4.9), we get:

$$\int_{fR_2}^{i} \frac{dQ}{T} - \int_{fIR}^{i} \frac{dQ}{T} - \int_{iR_1}^{f} \frac{dQ}{T} - \int_{fR_2}^{i} \frac{dQ}{T} < 0$$

or, $\displaystyle\int_{fIR}^{i} \frac{dQ}{T} + \int_{iR_1}^{f} \frac{dQ}{T} > 0$ and therefore, $\displaystyle\int_{fIR}^{i} \frac{dQ}{T} - \int_{fR_1}^{i} \frac{dQ}{T} > 0$ ...(4.11)

As $R_1$ is an arbitrary reversible path and so, for any reversible path,

$$\int_{f \; IR}^{i} \frac{dQ}{T} - \int_{f \; R}^{i} \frac{dQ}{T} > 0$$

or, $\displaystyle\int_{f \; IR}^{i} \frac{dQ}{T} > \int_{f \; R}^{i} \frac{dQ}{T}$ and also for the irreversible and reversible

cyclic processes:

$$\oint_{IR} \frac{dQ}{T} > \oint_{R} \frac{dQ}{T} \quad \text{but as} \quad \oint_{R} \frac{dQ}{T} = 0$$

Therefore, $\displaystyle\oint_{IR} \frac{dQ}{T} > 0$ and this is valid for any irreversible cyclic

process.

For any process in general, the change in entropy, $dS = \oint \dfrac{dQ}{T} \geq 0$ ...(4.12)

Where, the equality sign holds good for reversible process and inequality sign for an irreversible process. Let us now consider a practical irreversible transfer or exchange of heat between a system and its surrounding and let the temperatures of the system and its surrounding are respectively as $T_1$ and $T_2$ and $T_2 > T_1$. There will be then a flow of heat say $dQ$ from the surrounding to the system. The change of entropy of the universe comprising the system and its surrounding is then $dS_{universe}$ and it is then:

$dS_{universe} = dS_{system} + dS_{surrounding}$ and for the flow of heat dQ from the surrounding to the system:

$dS_{system} = \dfrac{dQ}{T_1}$ and $dS_{surrounding} = -\dfrac{dQ}{T_2}$ and therefore, $dS_{universe}$

$= \dfrac{dQ}{T_1} - \dfrac{dQ}{T_2} > 0$ or, $dS_{universe} > 0$ and for an isolated system for no such exchange of heat energy:

$\Delta S \geq 0$, the equnlity holds for reversible processes and inequnlity holds for the presence of any irreversibility in the process. This leads to the conclusion which is known as "Entropy Principle" and that is: Irreversible processes which are spontaneous and can only occur if that lead to the increase of the entropy of the universe or that of the system if it is isolated. As all practical and natural processes are irreversible, entropy of the universe or that of an isolated system will always increase. Therefore, for an isolated system:

$\Delta S > 0$ for all irreversible processes

$\Delta S = 0$ for ideal reversible processes and

$\Delta S < 0$ is an impossible process.

### 4.4.2 Consequence of Entropy Principle

This entropy principle and its result showed above have some very important consequence. Let us consider a natural process of conduction of heat through a solid having its two ends at different temperature. Heat will then automatically flow from the end at higher temperature to the end at lower temperature and this process will continue until the temperatures of the two ends equalize.

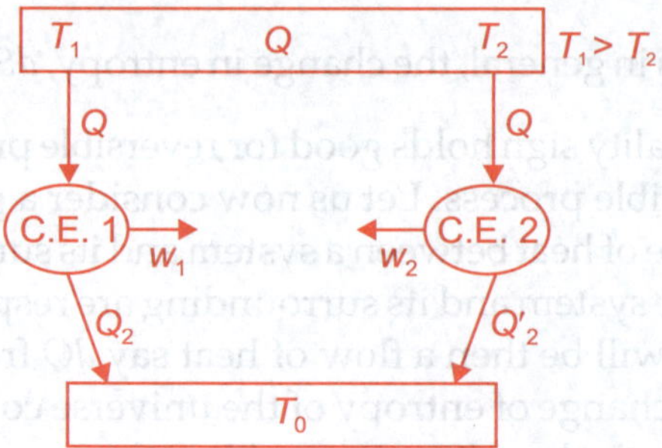



**Fig. 4.12** Conduction of heat from higher temperature end to lower temperature end, a natural irreversible process. Two carnot engines 1 and 2 operate between two ends of the rod and a heat sink

For Carnot engine, CE 1 (Efficiency $\eta_1$) operating at end $T_1$ extracts heat $Q$ before it is transferred to the lower temperature end, releases $Q_2$ to the heat sink at temperature $T_0$ and delivers $W_1$ amount of work.

Then
$$\eta_1 = \frac{W_1}{Q} = 1 - \frac{Q_2}{Q} = 1 - \frac{T_0}{T_1} \qquad \ldots(4.13)$$

Now, let a second Carnot engine (Efficiency $\eta_2$) which operates in between the lower temperature end $T_2$ and the common heat sink at $T_0$ extracts the same quantity of heat immediately after it reaches the second end i.e. immediately after the natural process of heat conduction is completed. Then

$$\eta_1 = \frac{W_2}{Q} = 1 - \frac{Q'_2}{Q} = 1 - \frac{T_0}{T_2} \qquad \ldots(4.14)$$

Now, as $T_1 > T_2$, $\eta_1 > \eta_2$ and so $W_1 > W_2$. Therefore, the work output from the same quantity of heat decreases after the natural process of heat conduction is completed.

This will lead to a conclusion that after each general natural process the amount of mechanical energy in to which the heat energy can be converted, decreases and as natural processes are to continue in this universe for its existence, the quantity of derivable mechanical energy converted from heat energy and intended to be utilized for human benefit and for the survival, decreases continuously. As time moves only in the forward direction and with this, natural irreversible processes continue unabated, entropy of this universe always increases and as a consequence efficiency with which the heat energy can be converted into mechanical work decreases. This can never be avoided or reversed. This can be stated in slightly different way that is the flow of time is directed towards the direction of increase of entropy i.e. direction of increase of disorder in all systems together and that is the only possible direction we find the time to flow as we can never find any incidence occurring in our nature that will set any system from less ordered state to more ordered state.

The above observations can be demonstrated as below:

→ Direction of flow of time

→ Direction of change of entropy

← Direction of change of efficiency of conversion of heat into work

We know that for reversible process: $S_f - S_i = \int\limits_{i(R)}^{f} \dfrac{dQ}{T}$ and $dQ = TdS$.

So, at constant volume $\left[\dfrac{dQ}{dT}\right]_V = C_V = T\left[\dfrac{\partial S}{\partial T}\right]_V$ and

at constant pressure $\left[\dfrac{dQ}{dT}\right]_P = C_P = T\left[\dfrac{\partial S}{\partial T}\right]_P$.

Therefore, knowing $C_V$ and $C_P$

At isochoric process: $S_f - S_I = \int\limits_{i}^{f} \dfrac{C_V}{T} dT$ and at isobaric: $S_f - S_i = \int\limits_{i}^{f} \dfrac{C_P}{T} dT$.

The temperature, $T$ vs. Entropy $S$ diagram is commonly known as $T$-$S$ diagram is also important and the area under initial and final state in $T$-$S$ diagram gives the total heat interaction in the concerned process.

Now, from first law of thermodynamics, $dQ = dU + PdV$ and recalling

$$\left[\dfrac{dQ}{dT}\right]_V = C_V = \left[\dfrac{dU}{dT}\right]_V \qquad \text{so,} \qquad dU = C_V dT$$

and similarly, $\left[\dfrac{dQ}{dT}\right]_P = C_P = \left[\dfrac{dU}{dT}\right]_P + P\left[\dfrac{dV}{dT}\right]_P$ as for ideal gas $PV =$

$nRT$, $P\left[\dfrac{dV}{dT}\right]_P = nR$ and as $U$ is only state dependent parameter $\left[\dfrac{dU}{dT}\right]_P$

$= \left[\dfrac{dU}{dT}\right]_V = C_V$.

Therefore, $C_P - C_V = nR$, Again from the first law, we can get

$\dfrac{dQ}{T} = \dfrac{C_V}{T} dT + \dfrac{P}{T} dV$ and as from gas law $\dfrac{P}{T} = \dfrac{nR}{V}$, we get

$$\int\limits_{i}^{f} \dfrac{dQ}{T} = S_f - S_i = C_V \int\limits_{i}^{f} \dfrac{dT}{T} + nR \int\limits_{i}^{f} \dfrac{dV}{V}$$

and similarly, $\int\limits_{i}^{f} \dfrac{dQ}{T} = S_f - S_i = C_P \int\limits_{i}^{f} \dfrac{dT}{T} - nR \int\limits_{i}^{f} \dfrac{dP}{P}$.

## 4.5 THERMODYNAMIC FUNCTIONS

**Enthalpy:** Enthalpy is defined as:

$$H = U + PV \text{ and therefore, in differential form:}$$

$$dH = dU + P\,dV + V\,dP \text{ and from first law}$$

$$dH = dQ + V\,dP \qquad \qquad \ldots(4.15)$$

and so, $\dfrac{dH}{dT} = \dfrac{dQ}{dT} + V\dfrac{dP}{dT}$ for reversible isobaric process we get:

$$\left[\frac{dH}{dT}\right]_P = \left[\frac{dQ}{dT}\right]_P = C_P, \; H_f - H_i = \int_i^f C_P dT \text{ and } H_f - H_i = Q.$$

Therefore, Enthalpy is a thermodynamic function, so the change of which in an isobaric process is equnl to the heat that is transferred.

Now as $dQ = T\,dS$ so the equn. (4.15) can be written as:

$$dH = T\,dS + V\,dP \qquad \qquad \ldots(4.16)$$

Considering that $H$ is a function of entropy, $S$ and pressure, $P$

$$dH = \left[\frac{\partial H}{\partial S}\right]_P dS + \left[\frac{\partial H}{\partial P}\right]_S dP$$

and comparing with equn. (4.16), We get:

$\left[\dfrac{\partial H}{\partial S}\right]_P = T$ and $\left[\dfrac{\partial H}{\partial P}\right]_S = V$. Now, from the first relation we get that

from the slope of the isobar drawn on $H$-$S$ diagram of a system gives the temperature of the system in Kelvin scale. This $H$ vs. $S$ values of a system are known as Mollier Chart.

**Helmholtz Function:** The Helmholtz function is defined as:

$$A = U - TS \text{ and in differential form}$$

$$dA = dU - T\,dS - S\,dT, \text{ but from first law } T$$

$$dS = dU + P\,dV$$

Therefore, $\qquad dA = -P\,dV - S\,dT \qquad \qquad \ldots(4.17)$

Now, for reversible isothermal process:

$$A_f - A_i = -\int_i^f P dV ,$$

and for reversible isothermal and isochoric process, $A = $ constant. The equation of state for a system is, then can be obtained from the following relationships:

$$P = -\left[\frac{\partial A}{\partial V}\right]_T \quad \text{and} \quad S = -\left[\frac{\partial A}{\partial T}\right]_V$$

**Gibbs Function:** The Gibbs function is defined as:

$$G = H - TS \text{ and in differential form:}$$

$$dG = dH - T\,dS - S\,dT \text{ and from equn. (4.16)}$$

$$dG = V\,dP - S\,dT \qquad\qquad \ldots(4.18)$$

for reversible isobaric and isothermal process,

$$dG = 0, G = \text{constant.}$$

## Maxwell's Thermodynamic Equations

For a chemical system of constant mass the equilibrium states are defined by three thermodynamic coordinates $P$, $V$, and $T$. In describing the behaviour of such system, we are already introduced to four other thermodynamic functions like Internal energy $U$, Enthalpy $H$, Helmholtz function $A$ and also Gibbs function G. These functions are defined already by first law, equns. (4.15), (4.17) and (4.18). Now, as $U$, $H$, $A$ and $G$ are actual functions, their differentials are exact differential of the type:

$dz = M\,dx + N\,dy$ where $z$, $M$ and $N$ are functions of $x$ and $y$ and

therefore, $\left(\dfrac{\partial M}{\partial y}\right)_x = \left(\dfrac{\partial N}{\partial x}\right)_y$. $\qquad\qquad \ldots(4.19)$

Applying this result to the four exact differentials $dU$, $dH$, $dA$ and $dG$ we get:

$$dU = TdS - P\,dV; \text{ using (4.19)} \quad \left(\frac{\partial T}{\partial V}\right)_S = -\left(\frac{\partial P}{\partial S}\right)_V \qquad \ldots \text{(i)}$$

$$dH = TdS + V\,dP; \text{ using (4.19)} \quad \left(\frac{\partial T}{\partial P}\right)_S = \left(\frac{\partial V}{\partial S}\right)_P \qquad \ldots \text{(ii)}$$

$$dA = -PdV - S\,dT; \text{ using (4.19)} \quad \left(\frac{\partial P}{\partial T}\right)_V = \left(\frac{\partial S}{\partial V}\right)_T \qquad \ldots \text{(iii)}$$

$$dG = VdP - S\,dT; \text{ using (4.19)} \quad \left(\frac{\partial V}{\partial T}\right)_P = -\left(\frac{\partial S}{\partial P}\right)_T \qquad \ldots\text{(iv)}(4.20)$$

These four equns. (i),(ii),(iii) and (iv) which relate the four thermodynamic functions of a chemical system are known as Maxwell's Equations.

### TdS Equations:

There are two $TdS$ equations; the first one is derived considering that the entropy $S$ is a function of temperature $T$ and volume $V$ of a chemical system and the second one when $S$ is considered as a function of $T$ and pressure $P$.

**First *TdS* equation:**

$$S = f(T, V) \text{ so,}$$

$dS = \left(\dfrac{\partial S}{\partial T}\right)_V dT + \left(\dfrac{\partial S}{\partial V}\right)_T dV$. Now multiplying both sides by $T$, we get:

$TdS = T\left(\dfrac{\partial S}{\partial T}\right)_V dT + T\left(\dfrac{\partial S}{\partial V}\right)_T dV$. Now, $T\left(\dfrac{\partial S}{\partial T}\right)_V = \left(\dfrac{\partial Q}{\partial T}\right)_V = C_V$ and

From Maxwell's equn. (4.20 iii):

$\left(\dfrac{\partial S}{\partial V}\right)_T = \left(\dfrac{\partial P}{\partial T}\right)_V$ . Replacing these relations in the above *TdS* equation:

$$TdS = C_V dT + T\left(\dfrac{\partial P}{\partial T}\right)_V dV \qquad \qquad ...(4.21)$$

**Application:** Heat transferred in reversible isothermal expansion in a van der Waals gas.

Let us consider 1 mole of a van der Waals gas, undergoing a reversible isothermal expansion from volume $v_i$ to volume $v_f$. The equn. (4.21) is then be written as:

$Tds = c_V dT + T\left(\dfrac{\partial P}{\partial T}\right)_V dv$ and using van der Waals equation of state as:

$P = \dfrac{RT}{v-b} - \dfrac{a}{v^2}$ $\left(\dfrac{\partial P}{\partial T}\right)_V = \dfrac{R}{v-b}$ and so, $Tds = c_V dT + RT\dfrac{dv}{v-b}$.

Now, as the process is isothermal $dT = 0$ and as the process is reversible

$q = \int Tds$ . Therefore, $q = RT\displaystyle\int_{vi}^{vf} \dfrac{dv}{v-b}$ which is $q = RT\ln\dfrac{v_f - b}{v_i - b}$ .

### 4.5.1 Application: Energy Equation

In a chemical system for an infinitesimal reversible process the change of internal energy, $U$ is: $dU = TdS - PdV$ from first law, and from First *TdS* equn. (4.21) we get:

$TdS = C_V dT + T\left(\dfrac{\partial P}{\partial T}\right)_V dV$ . Now, combining these two equations:

$dU = C_V dT + \left[T\left(\dfrac{\partial P}{\partial T}\right)_V - P\right]dV$ . Now again as $U$ is a function of $T$ and

$V$ we get: $dU = \left(\dfrac{\partial U}{\partial T}\right)_V dT + \left(\dfrac{\partial U}{\partial V}\right)_T dV$ and equating the coefficient of $dV$.

We get:

$$\left(\frac{\partial U}{\partial V}\right)_T = T\left(\frac{\partial P}{\partial T}\right)_V - P. \qquad \ldots(4.22)$$

This is 'Energy equation'.

Now, for an ideal gas i.e. $P = \dfrac{nRT}{V}$ we get:

$$\left(\frac{\partial P}{\partial T}\right)_V = \frac{nR}{V} \text{ and so, } \left(\frac{\partial U}{\partial V}\right)_T = \frac{TnR}{V} - P = 0.$$

Therefore, for an ideal gas system internal energy is a function of temperature only and not dependent on volume.

Second $TdS$ equation:

$$S = f(T, P)$$

$dS = \left(\dfrac{\partial S}{\partial T}\right)_P dT + \left(\dfrac{\partial S}{\partial P}\right)_T dP.$ Now, multiplying both sides by $T$

$TdS = T\left(\dfrac{\partial S}{\partial T}\right)_P dT + T\left(\dfrac{\partial S}{\partial P}\right)_T dP.$ Now as $T\left(\dfrac{\partial S}{\partial T}\right)_P = C_P$

And from Maxwell's equn. (4.20 iv)

$\left(\dfrac{\partial V}{\partial T}\right)_P = -\left(\dfrac{\partial S}{\partial P}\right)_T$, we get the $TdS$ equation as:

$$TdS = C_P dT - T\left(\frac{\partial V}{\partial T}\right)_P dP. \qquad \ldots(4.23)$$

### 4.5.2 Application: Reversible Adiabatic Change of Pressure

For reversible adiabatic Entropy, $S$ remains constant so the equn. (4.23) can be written as: $dT = \dfrac{T}{C_P}\left(\dfrac{\partial V}{\partial T}\right)_P dP$ and introducing the volume expansion coefficient of solid or liquid system $\beta = \left(\dfrac{\partial V}{V\partial T}\right)_P$, we get,

$dT = \dfrac{TV\beta}{C_P} dP$. Therefore, as for solid or liquid the change of $C_P$ is negligible over a wide range of pressure: $\Delta T = \dfrac{TV\beta}{C_P}(P_f - P_i)$. So, for an adiabatic increase of pressure there will be increase of temperature of the system.

### 4.5.3 Throttling Process: Liquefaction of Gas

The throttling process, thermodynamically known as Joule-Kelvin expansion is an important phenomenon for the development of cryogenic, the low temperature generation and its use in low temperature Physics and technology.

Let us first start with the experiment as below.

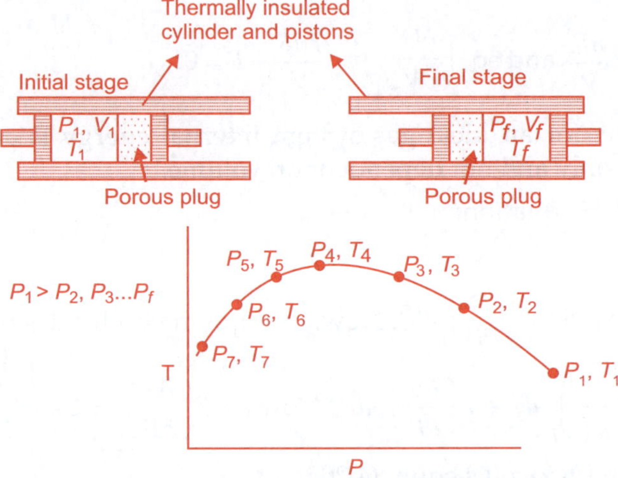


**Fig. 4.13** Joule-Kelvin expansion experiment and result

In the left side diagram at the top a gas is taken at an initial pressure say $P_1$ in the left side chamber, the left side piston is pushed right slowly towards the porous plug and right side piston is slowly withdrawn so that the pressure in the right side chamber is always maintained at a predetermined pressure say $P_2$, so that $P_2$ is constant and always less than $P_1$. The temperature in the second chamber is measured when final stage is reached and suppose that temperature is found to be $T_2$. The experiment is repeated with the same value of initial pressure in the left chamber at $P_1$ and the process is repeated with new predetermined pressure of the right chamber say $P_3$ but still keeping the condition that $P_1$ is greater than $P_1$. The value of temperature say $T_3$ is noted. The experiment is continued to be repeated with new sets of predetermined pressures as $P_4, P_5, P_6$ ...etc. which are always less than initial pressure $P_1$ and their corresponding temperatures $T_4, T_5, T_6$ ... are measured and plotted in $T - P$ graph. The plot shows an important feature that is it shows a maxima and slope reverses on both sides of it. As the process is continued in insulated surrounding, the enthalpy remains constant. This expansion which is not free is called Joule-Kelvin expansion the slope m given by following is known as Joule-Kelvin coefficient.

$$\mu = \left(\frac{\partial T}{\partial P}\right)_h \qquad \qquad \text{...(4.24)}$$

The reversal of slope $m$ on both sides of the maxima indicates that the gas undergoing throttling process goes heated i.e. temperature increases if the predetermined pressure lies to the right of the maxima and the slope $m$ becomes negative whereas the gas is cooled if the predetermined pressure is at the left side of the maxima and the slope is positive. Now, if the experiment described above is repeated with new initial pressure in the left chamber and throttling is continued with predetermined new sets of values of pressures which are always kept less than initial pressure in the left chamber then we get the following variation.

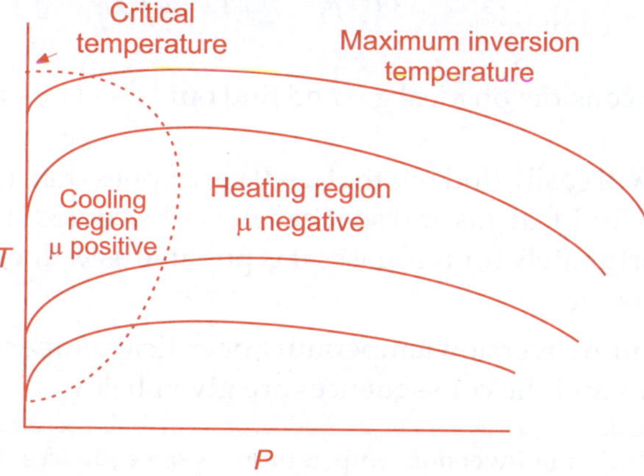


**Fig. 4.14** $T - P$ curves for various initial pressures and resultant temperature, the region within dotted curve shows cooling region due to throttling and critical temperature

It should be noted that below critical temperature the gas throttled through the porous plug will be cooled and outside this region it will be heated. Different real gases will have different critical temperatures and cooling region. Now, recalling the above expression (4.24):

$$\mu = \left(\frac{\partial T}{\partial P}\right)_h$$

But $\qquad\qquad h = u + Pv$ and so,

$$dh = du + pdv + vdP$$

and as $\qquad\qquad dq = du + Pdv$ and $dq = Tds$

we get $dh = Tds + vdP$ now, using second $Tds$ equn. (4.23)

$$dh = c_P\,dT - \left[T\left(\frac{\partial v}{\partial T}\right)_P - v\right]dP$$

or

$$dT = \frac{1}{c_P}\left[T\left(\frac{\partial v}{\partial T}\right) - v\right]dP + \frac{1}{c_P}dh$$

Writing $T$ as a function of $P$ and $h$

$$dT = \left(\frac{\partial T}{\partial P}\right)_h dP + \left(\frac{\partial T}{\partial h}\right)_P dh$$

Now, comparing coefficients of $dP$ we get the Joule-Kelvin coefficient, $m$ as:

$$\mu = \left(\frac{\partial T}{\partial P}\right)_P = \frac{1}{c_P}\left[T\left(\frac{\partial v}{\partial T}\right)_P - v\right] \qquad \ldots(4.25)$$

Now, if we consider an ideal gas and find out $\left(\dfrac{\partial v}{\partial T}\right)_P$ using gas equation

$Pv = nRT$. We will easily find out that $\mu = 0$ i.e. implies that, there is no slope for $T - P$ curve for ideal gas and so it can never be cooled due to throttling process. But fortunately for real gases it is possible as $m$ does vanish to zero for van der Waal gas.

The maximum inversion temperature or critical temperatures for some important gases and the consequences are given below

| Gas | Maximum Inversion temperature Critical Temperature in °K | Consequences |
| --- | --- | --- |
| Carbon dioxide | ~ 1500 | The gas can be Liquefied and also solidified by application of pressure alone. |
| Nitrogen | 621 | Can be easily liquefied if temperature is kept below. |
| Hydrogen | 202 | Pre-cooling is to be done Either by liquefied Nitrogen or air. |
| Helium | ~ 25 to 60 | Pre-cooling is essential by liquid Hydrogen. A liquid helium plant must be supported by liquid Nitrogen followed by liquid Hydrogen. |

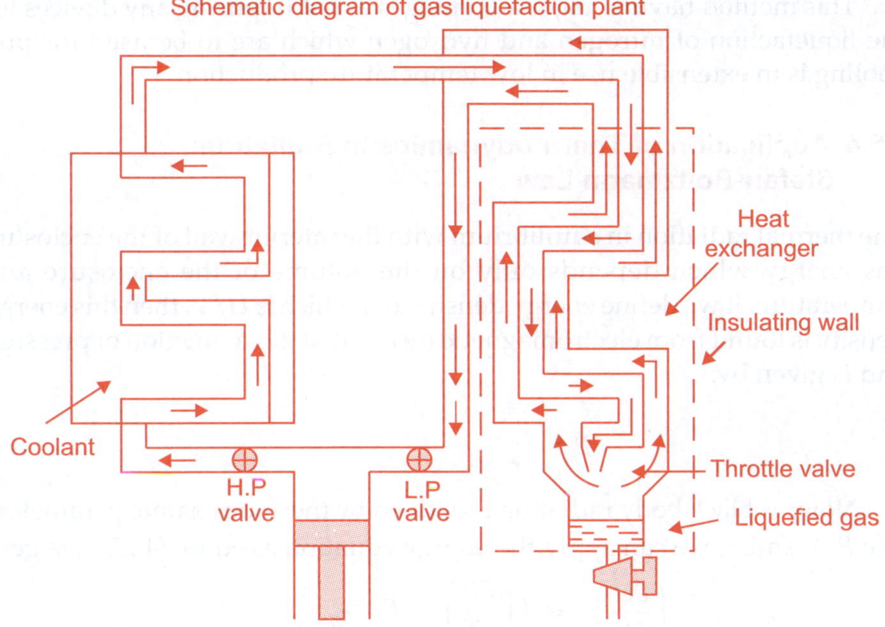


**Fig. 4.15** Gas liquefaction plant using Joule-Kelvin process. Pre-cooling is done by liquid as given in the table above

The liquid gas which is cable of cooling down the temperature of the compressed gas below the critical temperature is taken. The cooled gas passes through the inner tube and is further cooled after suffering expansion through the 'Throttle valve'. The throttled gas is sucked through outer tube in the cylinder and in passing through it further cooled the incoming gas through 'Heat exchanger'. The cooled gas is again compressed. As this process is repeated in cycle, the gas is finally liquefied. However, it should be mentioned here that the use of Joule-Kelvin effect to produce liquefaction of gases has two advantages: (1) As there are no moving parts working at low temperature, so there is no difficulty in lubrication (2) The lower the temperature, the larger will be the drop in temperature for the same difference of pressure. But for the liquefaction of gases like hydrogen and helium large amount of pre cooling is necessary and this requires as mentioned before liquid helium plant has to be a gigantic plant if it works on throttling process. The latest development in the field of gas liquefaction is the 'Collins helium liquefier' in which helium undergoes adiabatic expansion in a reciprocating engine. The expanded gas is then used to cool the incoming gas in the usual countercurrent heat exchanger. When the temperature of the gas is low enough, the gas is passed through throttling valve so that Joule-Kelvin cooling is used finally for the liquefaction of helium.

This method having major advantages of not requiring any devices for the liquefaction of nitrogen and hydrogen which are to be used for pre-cooling is in extensible use in low temperature production.

### 4.5.4 Application of Thermodynamics in Radiation: Stefan-Boltzmann Law

The thermal radiation in equilibrium with the interior wall of the enclosure has energy which depends only on the volume of the enclosure and temperature. If we define energy density as $u$ which is $U/V$, then this energy density is found from electromagnetic theory that it is a function of pressure and is given by:

$$P = \frac{u}{3}$$

Now, as black body radiation is defined by thermodynamic parameters like $P$, $V$ and $T$, we can apply the energy equation as equn. (4.22), we get:

$$\left(\frac{\partial U}{\partial V}\right)_T = T\left(\frac{\partial P}{\partial T}\right)_V - P.$$

Since $u = U/V$ and $P = u/3$, where $u$ is a function of $T$ only, we get:

$$u = \frac{T}{3}\frac{du}{dT} - \frac{u}{3}$$

and $\dfrac{du}{u} = 4\dfrac{dT}{T}$ Integrating this equation we get:

$$\ln u = \ln T^4 + \ln b$$

Therefore, $\qquad\qquad u = bT^4$ $\qquad\qquad\qquad\qquad\qquad$ ...(4.26)

Here, $b$ is a constant. This equn. (4.26) is known as Stefan-Boltzmann Law.

### 4.5.5 Application of Thermodynamics in First Order Transition of Phase: Clapeyron's Equation

We have introduced Gibb's function G in section 4.5 as

$$G = H - T S \text{ and so,}$$
$$dG = VdP - S dT$$

Now, in the familiar phase transition like melting, vaporization, sublimation and also the change of structure of one crystal phase to other, the temperature and pressure remain constant. Therefore, such phase transformation the Gibb's function remain constant. Therefore, for unit mass, $g^{(i)} = g^{(f)}$.

For a phase change at $T + dT$ and $P + dP$

$$g^{(i)} + dg^{(i)} = g^{(f)} + dg^{(f)} \text{ and so}$$

$$dg^{(i)} = dg^{(f)}$$

or, $\qquad v^{(i)}\, dP - s^{(i)}\, dT = v^{(f)}\, dP - s^{(f)} dT$

and so, $\qquad \dfrac{dP}{dT} = \dfrac{s^{(f)} - s^{(i)}}{v^{(f)} - v^{(i)}}.$

But, for the reversible phase transition the heat required is Latent heat and is related to specific entropy $l$ as:

$$l = T(s^{(f)} - s^{(i)}) \text{ and finally we get}$$

$$\dfrac{dP}{dT} = \dfrac{l}{T[v^{(f)} - v^{(i)}]} \qquad \qquad \text{...(4.27)}$$

This is known as Clapeyron's equation for the first order change of phase.

## 4.6 STEAM AND STEAM CYCLE

When water is heated to boil, it first starts evaporating from the surface with an enhanced rate and with increase of temperature. There will also be a small increase of volume. After some time depending on the quantity of water and the rate of heat supply, bubbles are formed on the inner wall of the vessel. These bubbles are due to the release of dissolved air in water. Soon signs of internal activity appear. Small bubbles are formed on and near the heating surface; they rise a little through water and collapse. These are steam bubbles and this collapse of bubbles is the reason for 'singing' of kettles. The temperature rises with the supply of heat and ultimately the steam bubbles are able to reach the top surface of water. This is a state of turbulence and is known as boiling or ebullition. As soon as this boiling starts the temperature of water attains a constant value known as 'boiling point'. As long as there is water present, it is impossible to increase the temperature even if the supply of heat continues. This boiling point temperature is called 'saturation temperature'.

As steam is produced when water present is under turbulence, the steam also carries with it water molecules and the steam is then called 'Wet steam' and it can be seen to comeout from the peak of the kettle. As long as water is present in the container, steam remains to be wet and temperature also remains constant at the saturation temperature. On continued supply of heat, the water droplets present in the wet steam vapourize and when no more water is present in the container and when all the water droplets are converted into vapour, the existing stem is then called 'Dry saturated steam'. The temperature then again starts increasing and this dry steam is then completely transparent i.e. cannot be seen. As there is no scattering of light by water droplets so long present in wet steam.

## Schematical Representation of the Formation of Super heated Steam

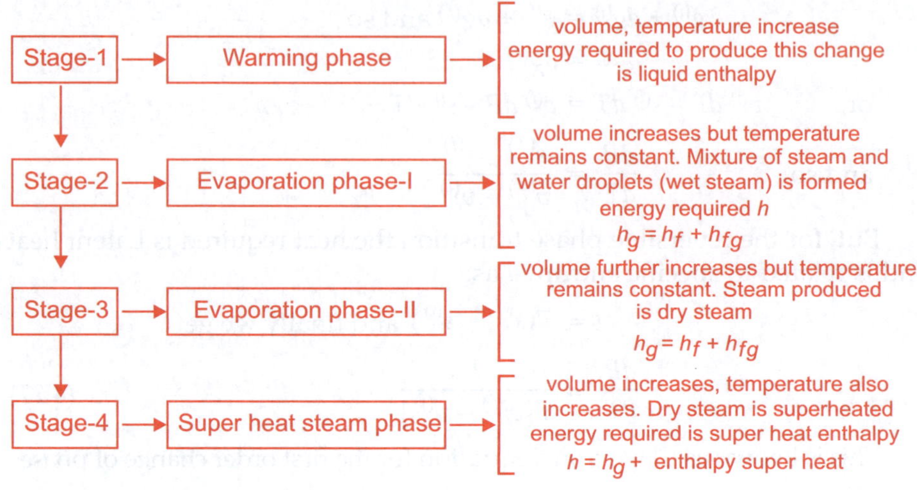



Specific Liquid Enthaply $= h_f$

Specific Enthalpy for Dry Saturated Vapour $= h_g$

Specifi Evaporation Enthalpy $h_{fg}$.

If we study the variation of temperature with specific enthalpy increase for water when the supply of heat continues, we would get the curve like the Fig. 4.16. As the pressure above water surface is increased, we get different curves of same nature and the higher pressure observation will be placed above the lower one. ABEH line indicates the boundary between water (liquid) and vapor and is called saturated liquid line. The plateau which immediately follows is the constant temperature line and represents the evaporation. After completion of this process, temperature increases rapidly with the supply of heat and the region beyond this line represents the super heat steam phase.

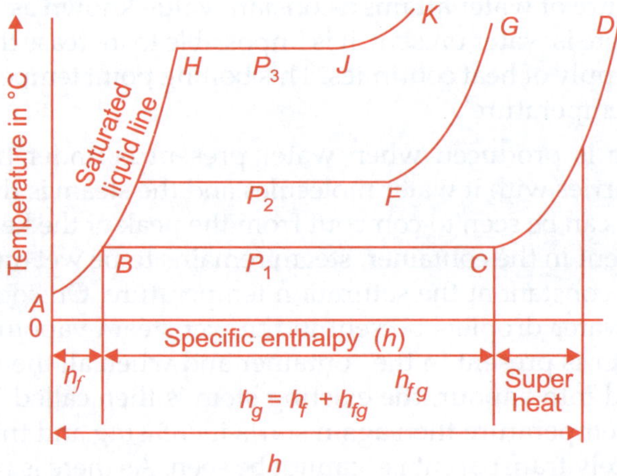



**Fig. 4.16** Temperature-Enthalpy diagram for water at pressures $P_1 < P_2 < P_3$

If we now join the points like *A, B, E, H* etc. and *J, F, C* etc., we get a curve like that represented in the following Fig. 4.17. The important criteria which determine the utility of steam is called 'Dryness Fraction'. This parameter is defined as:

$$\text{Dryness Fraction} = \frac{\text{Mass of Dry steam in a certain volume}}{\substack{\text{Mass of Wet steam containig the} \\ \text{Dry steam in the same volume}}}$$

$$x = \frac{M}{M + m}.$$

There are different methods to determine this dryness fraction, the description of which are beyond the scope of this book. However, it should be emphasized here that more is the value of *x* the dryness fraction, the amount of energy required for producing the same quantity of steam into dry saturated steam becomes less. The maximum of the curve indicates the boundary between gas phase and vapor phase. Above this maximum the steam is super saturated, invisible and behaves like a perfect gas obeying gas laws. Whereas below this line it behaves like vapour, which does not follow the gas laws. The steam above this temperature can not be liquefied by the application of pressure alone and temperature has to be brought down by cooling the steam in order to condense it. However, the vapour can be liquefied by the application of pressure alone. Due to content of high heat energy and large kinetic energy of the water molecules, the dry saturated steam above this critical temperature can be used to run turbines. The successful production of this super saturated dry steam is industrially done in 'Boilers'. There are two different types of boilers (i) Fire tube boilers and (ii) Water tube boilers. The detailed discussions on these boilers are however beyond the scope of this book.

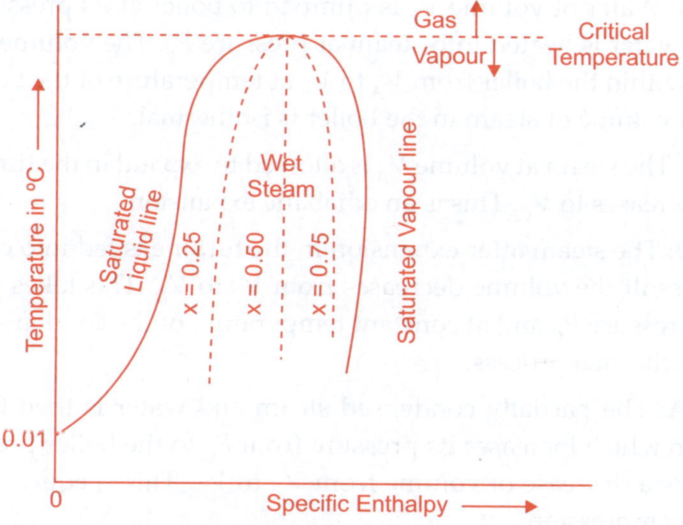


**Fig. 4.17** Boundary line between gas and vapour phases of steam

## 4.6.1 Carnot Steam Cycle

Please refer to the section (4.3.4) on ideal reversible Carnot cycle. This cycle is a reversible and perfectly ideal cycle dealing with ideal gas. We have seen that the efficiency of this ideal Carnot cycle is maximum and no real engine or cycle can attain such efficiency. If we now apply this Carnot cycle on steam, then let us see how far that will be applicable. The Fig. 4.18 shows such hypothetical steam cycle.

Here the left side figure is the $P$–$V$ and right side the $T$–$S$ diagrams. $AB$ and $CD$. Represent the isotherms and as natural in Carnot cycle, $BC$ and $DA$ the two pairs of adiabatic. The areas $ABCD$ for the left figure and $a\,b\,c\,d$ for the right respectively represent the work output and heat involved Let us now briefly explain the actions of each step.

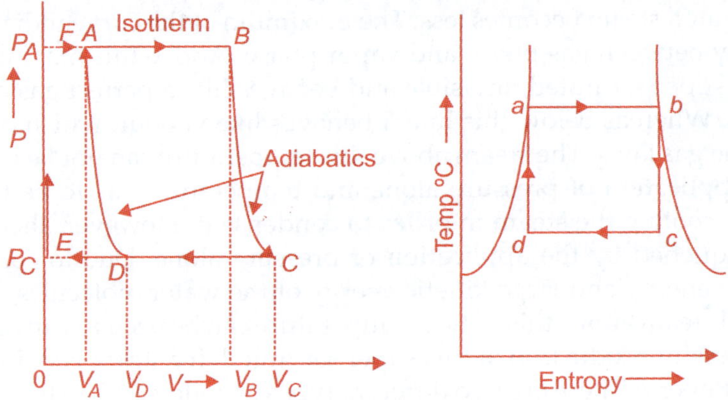


**Fig. 4.18** Carnot steam cycle. Left shows the $P$–$V$ diagram and right the $T$–$S$ diagram

**A to B**: Water of volume $V_A$ is pumped to boiler at its pressure $P_A$. In the boiler water is heated into steam at pressure $P_B$. The volume of steam increases within the boiler from $V_A$ to $V_B$ at temperature of the boiler. This expansion volume of steam in the boiler is isothermal.

**B to C**: The steam at volume $V_B$ is allowed to expand in the turbine. The volume increases to $V_C$. This is an adiabatic expansion.

**C to D**: The steam after expansion in the turbine is fed into condenser and as a result the volume decreases from $V_C$ to $V_D$. This takes place at a constant pressure $P_C$ and at constant temperature of the condenser. This is then an isothermal process.

**D to A**: The partially condensed steam and water is then fed to the feed pump which increases its pressure from $P_C$ to the boiler pressure $P_A$. This results a decrease of volume from $V_D$ to $V_A$. This is equivalent to an adiabatic compression.

**Observation**: The cycle diagram representing turbine is represented by two isotherms and two adiabatic and is represented by area *FBCE*. The feed pump diagram represented by area *EDAF* is the work input. The net work output is then represented by *ABCD*. The isothermal expansion of steam in the boiler and adiabatic expansion of steam in the turbine are reasonable but condensation of wet steam at C fully into water and wet steam is not possible and it is also not reasonable practically that feed pump can deal with successfully both wet steam and condensed water to increase the pressure of the mixture to the boiler pressure.

Therefore, Carnot steam cycle is not feasible both theoretically and practically and more reasonable cycle for steam as system is Rankine Cycle. This is discussed in the next section.

## 4.6.2 Rankine Cycle

Rankine cycle is a modification over the Carnot steam cycle just discussed. The modification concerns the full condensation of wet steam into water so that the feed pump can effectively pump the condensed water into boiler at boiler pressure. The condensation is continued up to saturated liquid line to *d* Fig. 4.19 (b). There will be slight increase of temperature due to this increase of pressure and ultimately this increase continues until it equnls the boiler temperature *a* shown by *d'* to a step.

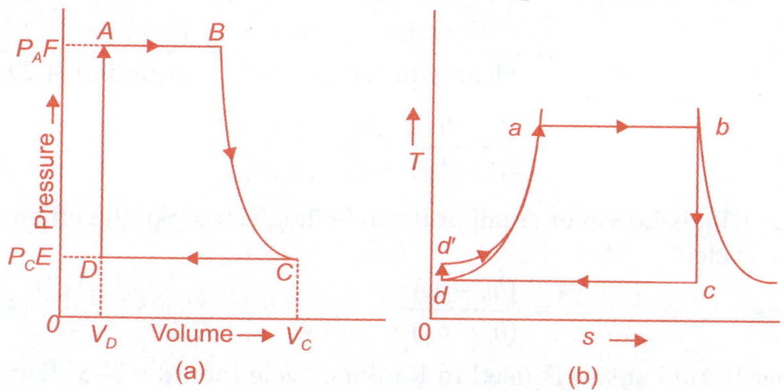

Fig. 4.19  Rankine Cycle. a its P-V diagram and b the T-S diagrams

The complete Rankine cycle is then represented by *a b c d d' a*. In *P-V* diagram (Fig. 4.19 a), the work done in the turbine is represented by area *F B C E* which includes feed pump work *E D A F* and hence, the

Work done per cycle = Area *ABCD*

1. During Adiabatic expansion *B* to *C*, $q = 0$ and so, the energy equation from steady flow equation $h_1 = h_2 + w$ i.e. $w = h_1 - h_2$

This relation can however be shown as follows:

$$\left\{\begin{array}{l} h = u + pv \\ dh = du + pdv + vdp \quad \text{for adiabatic expansion } B \to C, dq = 0 \\ dh = v\,dp; \; h_2 - h_1 = v(p_2 - p_1) = -w \text{ as } pv = \text{const. and so,} \\ v\,dp + p\,dv = 0 \text{ and so, } \int pdv = -\int v\,dp = -w \end{array}\right.$$

$$w = h_b - h_c \qquad \text{(applicable for Fig. 4.16 b)}$$

or, $$= \text{area } FBCE \qquad \text{(applicable for Fig. 4.16 a)}$$

Feed pump work/unit mass = area $EDAF$

$$= (P_B - P_C)V_D$$

Net work done per cycle $= (h_b - h_c) - (P_B - P_C)V_D$     ... (4.28)

The heat transfer required in the boiler to convert the water at $d'$ to steam at $b = h_b - h'_d$.

Total energy of water entering the boiler at $d'$ = liquid enthalpy at $d$ + Feed pump work.

or, $$h'_d = h_d + (P_B - P_C)V_D$$

Therefore, heat transfer required in the boiler to convert it into saturated steam

$$= h_b - h_d' = h_b - h_d + (P_B - P_C)V_D \qquad ...(4.29)$$

The thermal efficiency of the cycle:

$$= \frac{\text{Work done per cycle}}{\text{Heat required per cycle}} = \frac{\text{Equation (4.28)}}{\text{Equation (4.29)}}$$

$$= \frac{(h_b - h_c) - (P_B - P_C)V_D}{(h_b - h_d) - (P_B - P_C)V_D}$$

$(P_B - P_C)V_D$ is however small and can be neglected. So, the efficiency of Rankine cycle:

$$= \frac{(h_b - h_c)}{(h_b - h_d)} \qquad ...(4.30)$$

If super heated steam is used in Rankine cycle then the $T$–$S$ diagram will be slightly modified as follows:

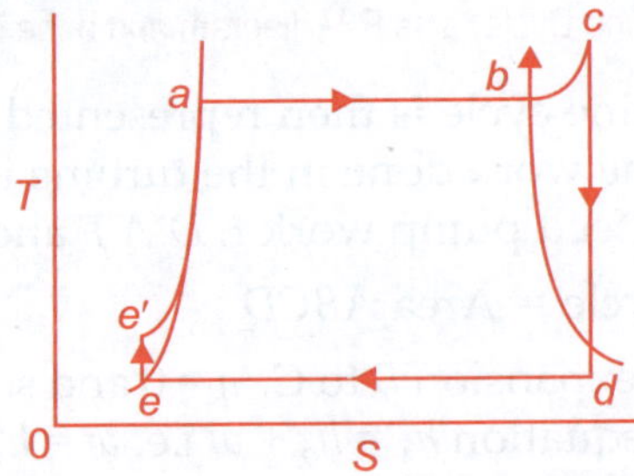


**Fig. 4.20** Rankine cycle with super heated steam

Thermal efficiency of this modified Rankine cycle is given in the same form as:

$$\eta = \frac{h_c - h_d}{h_c - h_e} \qquad \ldots (4.31)$$

Therefore, there will be a slight increase of the efficiency if instead of saturated steam super saturated steam is used in Rankine cycle.

## 4.7   OTTO AND DIESEL CYCLES

Otto cycle is air standard cycle of the spark ignition engine. In most of the spark ignition cycles the piston executes four complete strokes within the cylinder and crankshaft completes two complete revolutions. These engines are called 'Four stroke internal combustion engines. The engine devised on this cycle was proposed by N.A. Otto, a German engineer in 1876 and is known after him. The four strokes constitute the following actions:

1. **Compression Stroke:** During this stroke piston moves upwards and compresses the air-fuel mixture. At the end of this stroke, the spark plug fires and ignites the fuel mixture, pressure inside increases and the piston is forced to move down.

2. **Expansion Stroke:** During the downward movement of the piston the crankshaft rotates and produces the useful work. At the end of this stroke the piston moves to its lowest position.

3. **Exhaust Stroke:** During this stroke, the piston moves up and the burnt fuel is ejected out

4. **Intake Stroke:** During this stroke the piston again moves down and fresh air and fuel is sucked inside the cylinder.

The execution of this four stroke air standard cycle or Otto cycle is given below with the description of the actions of each step as above.

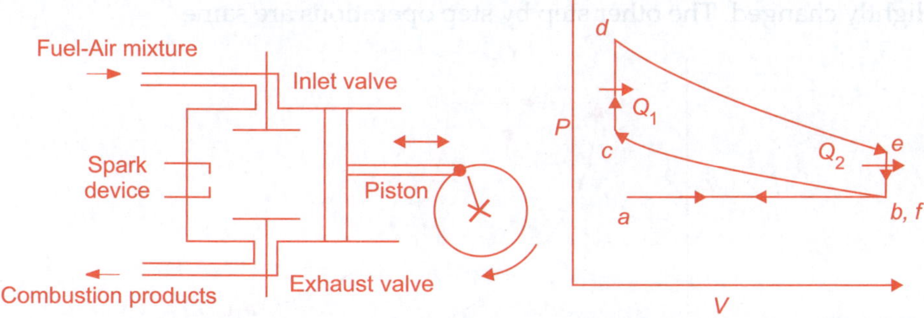


**Fig. 4.21**  Four stroke otto engine and the $P - V$ diagram of otto cycle

***a* to *b*** = Intake stage, Inlet valve opens and the mixture of air and fuel is inducted in the cylinder at constant pressure.

*b* **to** *c* = Compression stroke, the both valves are closed and the mixture of fuel and air is compressed within the cylinder.

*c* **to** *d* = Combustion stage, the mixture is fired by spark plug, the pressure of burnt mixture within the cylinder increases with piston at the same position.

*d* **to** *e* = Expansion stroke, due sudden increase of pressure the piston moves right. This is the stage of production of work. Pressure and temperature decrease.

*e* **to** *f* = The blow-down stage, exhaust valve opens and burnt mixture returns to decreased pressure.

*f* **to** *a* = Exhaust stroke, the piston moves left and ejects out the used fuel-air mixture.

Heat supplied

$$Q_1 = Q_{b \to c} = mc_v(T_c - T_b)$$

Heat rejected

$$Q_2 = Q_{d \to a} = mc_v(T_d - T_a)$$

Efficiency:  $\eta = 1 - \dfrac{Q_2}{Q_1} = 1 - \dfrac{mc_v(T_c - T_b)}{mc_v(T_d - T_a)}$

$$= 1 - \frac{T_c - T_b}{T_d - T_a}. \qquad \qquad \dots(4.32)$$

## Diesel Cycle

The only difference between the Otto cycle and Diesel cycle so far as their $P - V$ diagrams are concerned is that while heat is supplied for Otto cycle through the constant volume step, it is done through constant pressure step and instead of injecting the mixture of fuel and air, two are injected separately and mixing is done in the cylinder itself. Therefore, the $P - V$ diagram is slightly changed. The other step by step operations are same.

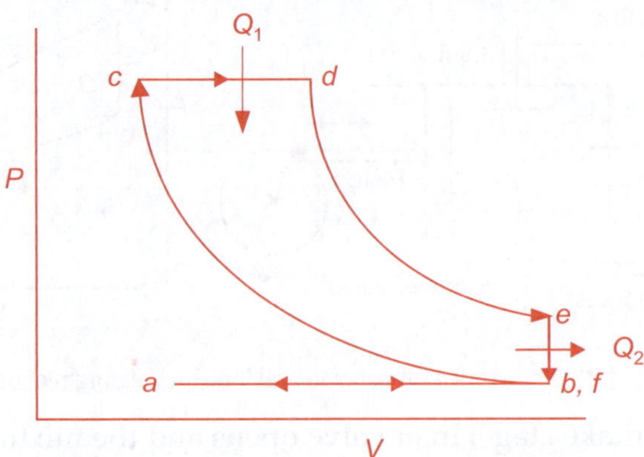


**Fig. 4.22** The diesel cycle: *P–V* diagram

The efficiency expression is then modified for Diesel cycle as $Q_1$ is applicable for constant pressure and $c_p$ is used. The expression (4.27) is modified as using $\dfrac{c_p}{c_v} = \gamma$.

$$\eta = 1 - \frac{(T_c - T_b)}{\gamma(T_d - T_a)}. \qquad \dots (4.33)$$

## REVIEW QUESTIONS

1. Explain why and when any amount of mechanical work can be converted fully and indefinitely into heat the reverse is not correct?

2. Explain the different fundamental thermodynamical processes to convert heat into mechanical work and show that the different processes result different quantity of work.

3. Explain the second law of thermodynamics and establish the equivalence between its two statements.

4. Define reversible and irreversible processes and show by stating examples that all natural processes are irreversible.

5. Introducing Carnot cycle, establish the Carnot theorem.

6. Explain and prove the Entropy principle and also show its outcome in natural processes.

7. Derive the first and the second $TdS$ equations and state some of their applications.

8. Explain Joule-Kelvin throttling process. Prove that this is an isenthalpic process.

9. Explain the critical temperature and state its importance in explaining the difference between gas and vapour. Why super saturated dry steam is used as working substance in steam cycles, than wet steam?

10. Explain the working diagram of Otto and Diesel cycles.

## REFERENCES


1. Heat and Thermodynamics: M. W. Zemansky, McGraw–Hill Co. Inc., NY.1957.

2. Thermodynamics: J. F. Lee and F. W. Sears, Addison-Wesley, Cambridge, Mass, 1962.

3. Engineering Thermodynamics: P. K. Nag, Tata McGraw-Hill Pub. Co., New Delhi, 2005.

# Electrostatics

## 5.1 INTRODUCTION

When the charges are not moving then the characteristics that they demonstrate are known as "Electrostatics". An atom is electrically neutral having equal amount of positively charged core and negatively charged electrons in its surrounding. These electrons may be removed by applying frictional force by rubbing two materials. In the electrostatic series given below if two materials of the series is rubbed then the material in the top of the series develops positive electricity and the other which occurs later in the series develops negative electricity. However, these charges will remain for sometime on the surface of the materials if they are insulated and otherwise the charges will move to earth. For example, if Glass is rubbed with Flannel then glass will develop negative electricity and flannel positive and if it is rubbed with silk cloth then it develops positive and silk negative. There is however no creation of charge but simply there is a transfer of charge from one to other making the earlier one deficit of charge that is transferred and the latter one excess of the charge transferred. The demonstration of static electricity can best be observed in insulators as these materials can not conduct it to the other end or to earth easily.

**Electrostatic Series:**

(1) Fur (2) Flannel (3) Sealing Wax (4) Glass (5) Paper (6) Silk (7) Wood (8) Metals (9) India-rubber (10) Sulphur (11) Ebonite (12) Gutta-percha.

The charges can also be stationed in insulator and conductors by a process known as Induction. When a charged body, say positive is brought close to another then on the nearest end of the body an opposite charge that is negative in this case will develop and at the furthest end of the body similar charge that is positive will develop. This is because of the shift of centre of mass of the positively charged core and that of orbital negatively charged electrons of the atoms of the body where in the charges are induced. The opposite charge developed due to this induction process at the nearest end is known as bound charge as the charge does not move as long as the inducing charged body is kept close to this end. But however the similar charge developed at the furthest end is know as free charge as it has freedom of moving away if and only if it gets a scope. It has been found that opposite charges attract each other and same charges repel each other. Thus, there is a force between two charges and the magnitude and direction of this force between two charges say $q_1$ and $q_2$ is a central force and act along their line of centres.

This is known as Coulomb's inverse square law as the force between these two charges is directly proportional to the product of the charges and inversely proportional to the square of the distance between them.

## 5.2 ELECTROSTATICS AND ELECTRIC FIELD

The force $F$ between two point charges $q_1$ and $q_2$ at distance $r$ is given by Coulomb's Law as:

$$F = \frac{1}{4\pi\varepsilon_0} \frac{q_1 q_2}{r^2} \, n_r = f(r) \cdot n_r \qquad \ldots(5.1)$$

Here $\varepsilon_0$ is the permittivity of free space and $n_r$ is the unit vector giving the direction between the charges.

If there are a number of charges then the force on a particular charge will be the vector sum of forces between this particular charge pairing with each individual charges. The ratio of this force on a charge $q$ to the charge $q$ i.e. force per unit charge is defined as the intensity of the electro-static field and it is given as:

$$E = F/q = \frac{f(r)}{q} \, n_r \qquad \ldots(5.2)$$

When there are number of charges $q_1, q_2, q_3 \ldots$ etc., then the field at a point $P$ is given by:

$$E_P = \frac{1}{4\pi\varepsilon_0} \sum \frac{q_i}{r_i^3} \, r_i$$

Now, if there is a continuous distribution of charge giving rise to a volume charge density $\rho$, then the intensity at a point $P$ is given as:

$$E_P = \frac{1}{4\pi\varepsilon_0} \int_V \frac{\rho}{r^3} \, r \, dV.$$

Where $r$ is the distance between the volume element $dV$ and the point $P$.

This electric intensity at a point say, $P$ can also be understood as the total electric flux passing through an unit area at $P$.

### 5.2.1  Gauss's Law of Electrostatics

Let an arbitrary closed surface contains a charge $q$ and charge $q'$ lays outside it.

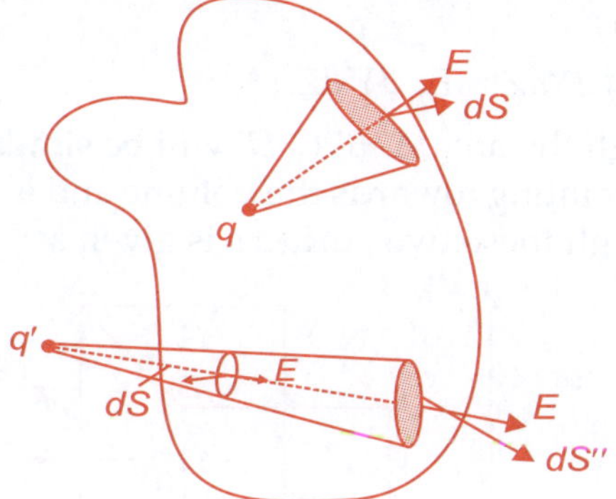


**Fig. 5.1** Gauss's law $E$ is the intensity (flux density) of electric field on the surface

The flux of $E$ across an element of area $dS$ on $S$ is $E \cdot dS$ and so the total flux of the electric field through the entire surface is given by:

$$\oint_S E \cdot dS = \oint_S E\cos\theta\, dS = \frac{q}{4\pi\varepsilon_0}\oint_S \frac{dS\cos\theta}{r^2} = \frac{q}{4\pi\varepsilon_0}\oint_S d\Omega \qquad \ldots(5.3)$$

where $\theta$ is the angle between $E$, the intensity and the normal to $dS$ and $d\Omega$ is the solid angle that area $dS$ subtends at the charge $q$. Now as $\oint_S d\Omega = 4\pi$ and so we get

$$\oint_S E \cdot dS = \frac{q}{\varepsilon_0}.$$ Now, if the charge, say $q'$ stays outside the surface then the flux that enters the surface $dS'$ will be the same that will pass out of $dS''$ on the other side of the area and so the net flux inside the surface will be zero. Now, if there are a number of such charges inside the closed area then:

$$\oint_S E \cdot dS = \frac{1}{\varepsilon_0}\sum q_i$$ and similarly, if there is uniform charge distribution

resulting in to a charge density $\rho$ inside the surface then as $\sum q_i = \int_V \rho\, dV$ we get:

$$\oint_S E \cdot dS = \frac{1}{\varepsilon_0}\int_V \rho\, dV \qquad \ldots(5.4)$$

The equn. (5.4) is known as Integral Form of Gauss's Law and it states that: "The total flux of electric intensity $E$ across a closed surface equals the total charge inside the surface divided by $\varepsilon_0$".

**Differential Form of Gauss's Law:**

$dV = dX \cdot dY \cdot dZ$ and $ABCD$ and $A'\,B'\,C'\,D'$ are the two faces each of area $dYdZ$ and respectively at positions $X + dX$ and at $X$.

Let us take one element of volume $dV = dX \cdot dY \cdot dZ$ and whose edges are parallel to $XYZ$ axes. If **E** is the field intensity on the surface $ABCD$, then the flux through it is:

$$\mathbf{E} \cdot \mathbf{dS} = (E \cos\theta)\, dYdZ = E_X\, dY\, dZ.$$

The flux through the area $A'\,B'\,C'\,D'$ will be similar but with negative sign as the field is pointing towards the volume and is given as $- E'_X\, dY\, dZ$. The total flux through these two surfaces is given as:

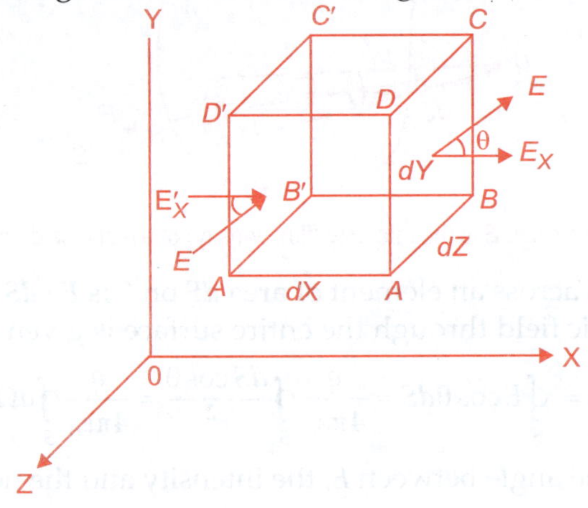


**Fig. 5.2**  Differential form of gauss's law. In cartesian system the volume element

$E_X\, dY\, dZ - E'_X\, dY\, dZ = (E_X - E'_X)\, dY\, dZ$. Now as the distance between these two surfaces is very small and equal to $dX$ then $(E_X - E'_X) = dE_X = \dfrac{\partial E_X}{\partial X} dX$ and the total flux in X direction is:

$\dfrac{\partial E_X}{\partial X} dX\, dY\, dZ = \dfrac{\partial E_X}{\partial X} dV$. Using the similar results for other four faces the total flux through the volume element is;

$$\phi_E = \frac{\partial E_X}{\partial X} dV + \frac{\partial E_Y}{\partial Y} dV + \frac{\partial E_Z}{\partial Z} dV = \left( \frac{\partial E_X}{\partial X} + \frac{\partial E_Y}{\partial Y} + \frac{\partial E_Z}{\partial Z} \right) dV.$$

But from Gauss's Law if $dq$ is the charge contained in the volume element $dV$:

$$\left( \frac{\partial E_X}{\partial X} + \frac{\partial E_Y}{\partial Y} + \frac{\partial E_Z}{\partial Z} \right) dV = \frac{dq}{\varepsilon_0} \text{ and since } dq = \rho dV, \text{ we get}$$

$$\left( \frac{\partial E_X}{\partial X} + \frac{\partial E_Y}{\partial Y} + \frac{\partial E_Z}{\partial Z} \right) = \frac{\rho}{\varepsilon_0} \qquad \text{... (5.5)}$$

This is Gauss's Law in differential form. Now left hand side (L. H. S.) of the above equation is Divergence of *E*, then equn. (5.5) can also be written as:

$$\text{div } E = \frac{\rho}{\varepsilon_0} \quad \text{or,} \quad \nabla \cdot E = \frac{\rho}{\varepsilon_0} \qquad \ldots (5.6)$$

Now, equating the equns. (5.4) and (5.6) we get the total flux of the electric field across a closed surface can be equated to the volume integral of the divergence of *E* over the volume:

$$\oint_S E \cdot dS = \int_V \text{div } E \, dV = \int_V \nabla \cdot E \, dV = \frac{1}{\varepsilon_0} \int_V \rho dV \quad \ldots (5.7)$$

The physical meaning of Gauss's Law is that the electric charges are sources of electric field and their distribution and magnitude determine the electric field at each point of the space.

Like other central forces the electrostatic force and the field is also be conservative. Then the total work done in moving one unit charge around the uniform field in a closed path will be zero.

In the following Fig. 5.3 let *E* be the uniform field intensity due a charge *q* and integrating over the closed path

$$\oint_L E . dl = 0$$

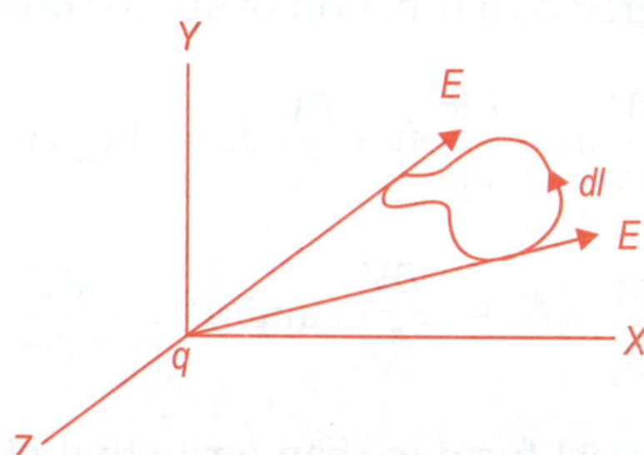


**Fig. 5.3** A closed path in an electric field

Now, applying Stoke's Theorem which is for a vector field of intensity *A* states that the value of curl of *A* at a given point is equal to the line integral around a closed path per unit surface:

$$\nabla \times A = \lim_{S \to 0} \frac{C}{S} = \frac{dC}{dS} \quad \text{and where the line integral of } A \text{ along an elementary}$$

path *dl* is given by: $dC = A \cdot dl$

Therefore, $\int_S (\nabla \times A) \cdot dS = \oint_L A \cdot dl$. Now, using this Stoke's theorem

we get

$$\oint_L E \cdot dl = \int_S \text{curl } E \cdot dS = \int_S (\nabla \times E) \cdot dS = 0, \text{ is true for any surface and so}$$

$$\nabla \times E = 0.$$

This Gauss's Law either in integral or differential form is called as the axioms of the theory of electrostatics.

## 5.2.2 Electrostatic Potential

As has been stated that for an uniform electrostatic field the line integral of $E$ depends only on the initial and final positions and not on the path, we can define the electrostatic scalar potential as:

$$V(r) = -\int_i^f E \cdot dr \text{ or } dV = -E \cdot dr$$

The negative sign shows that $E$ points towards the decrease in potential. This electrostatic potential may also be seen as the electrostatic energy per unit charge. We can however measure only the change of the potential within a field and zero potential may only be taken as a potential at a point at infinite distance from a charge distribution. In Cartesian coordinate system this $V$ may be considered as a function of space coordinates $x$, $y$ and $z$.

Therefore, $dV = \dfrac{\partial V}{\partial x} dx + \dfrac{\partial V}{\partial y} dy + \dfrac{\partial V}{\partial z} dz = -\left( E_x dx + E_y dy + E_z dz \right)$ so that

$$E_x = -\frac{\partial V}{\partial x}, \ E_y = -\frac{\partial V}{\partial y}, \text{ and } E_z = -\frac{\partial V}{\partial z}.$$

And in more compact form we can write that electric field $E$ is minus the gradient of the electrostatic potential

$$E = -\text{ grad } V = -\nabla V. \qquad \qquad \ldots(5.8)$$

## 5.2.3 Field Due to an Electric Dipole

An electric dipole is two equal and oppositely charged point charges situated at a distance which is very small. In the following figure let two charges of $+q$ and $-q$ be separated by a small distance say $d$ which is measured from negative charge to the positive charge. $P$ is a point in the vicinity of the dipole where the separate effect from both of the charges can be felt. The dipole moment $p = qd$.

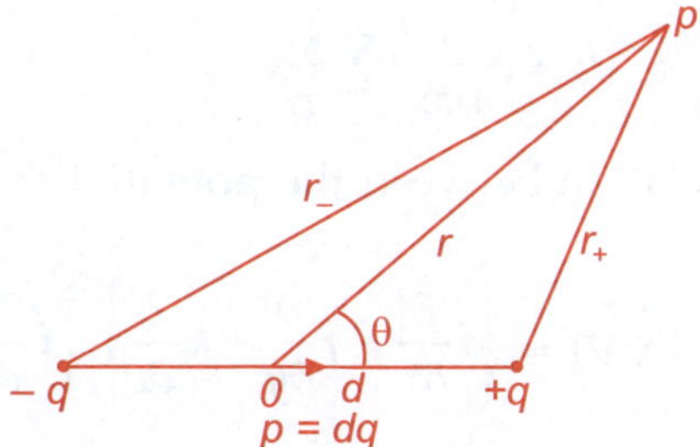

**Fig. 5.4** Potential and field due to an electric dipole

The potential at the point $P$, $V(r) = \dfrac{1}{4\pi\varepsilon_0}\left(\dfrac{q}{r_+} - \dfrac{q}{r_-}\right) = \dfrac{1}{4\pi\varepsilon_0}\dfrac{q(r_- - r_+)}{r_+ r_-}$.

Now as distance $d$ compared with $r$ then we can safely approximate as: $r - r_1 = d\cos\theta$ and $r_- r_+ = r^2$. This results in to:

$$V(r) = \frac{qd\cos\theta}{4\pi\varepsilon_0 r^2} = \frac{p\cos\theta}{4\pi\varepsilon_0 r^2}$$

$$= \frac{\boldsymbol{p}\cdot\boldsymbol{r}}{4\pi\varepsilon_0 r^3}. \qquad\qquad \ldots(5.9)$$

Now, the electric field due to this dipole at the point $P$ is from equn. (5.8) as:

$$\boldsymbol{E} = -\nabla V(r) = -\frac{1}{4\pi\varepsilon_0}\nabla\{(\boldsymbol{p}\cdot\boldsymbol{r})/r^3\}$$

$$= \frac{-1}{4\pi\varepsilon_0}\left\{\frac{1}{r^3}\nabla(\boldsymbol{p}\cdot\boldsymbol{r}) + (\boldsymbol{p}\cdot\boldsymbol{r})\nabla\left(\frac{1}{r^3}\right)\right\}$$

But as: $\nabla(\boldsymbol{p}\cdot\boldsymbol{r}) = \boldsymbol{p}$ as $\boldsymbol{r} = \boldsymbol{i}x + \boldsymbol{j}y + \boldsymbol{k}z$, and $r = (x^2 + y^2 + z^2)^{1/2}$

So, $\dfrac{\partial r}{\partial x} = \dfrac{\partial}{\partial x}(x^2 + y^2 + z^2)^{1/2} = \dfrac{x}{r}$, and similarly $\dfrac{\partial r}{\partial y} = \dfrac{y}{r}$ and $\dfrac{\partial r}{\partial z} = \dfrac{z}{r}$.

Therefore, $\nabla r = 1$ and $\nabla\left(\dfrac{1}{r^3}\right) = -\dfrac{3}{r^4}\left\{\dfrac{x}{r}\boldsymbol{i} + \dfrac{y}{r}\boldsymbol{j} + \dfrac{z}{r}\boldsymbol{k}\right\} = \dfrac{-3}{r^5}\boldsymbol{r}$

Hence,

$$\boldsymbol{E} = \frac{1}{4\pi\varepsilon_0}\{3(\boldsymbol{p}\cdot\boldsymbol{r})/r^5 - \boldsymbol{p}/r^3\}$$

For a group of point charges $q_i$ at distances $r_1$ from point $P$ the electrostatic potential at point $P$ is

$$V_P = \frac{1}{4\pi\varepsilon_0} \sum_i \frac{q_i}{r_i}$$

Now, to find a relation between the potential and charge density, let us write

$$\nabla \cdot (-\nabla V) = \left\{ i\frac{\partial}{\partial x} + j\frac{\partial}{\partial y} + k\frac{\partial}{\partial z} \right\} \cdot \left\{ i\frac{\partial V}{\partial x} + j\frac{\partial V}{\partial y} + k\frac{\partial V}{\partial z} \right\}$$

$$= \frac{\partial^2 V}{\partial x^2} + \frac{\partial^2 V}{\partial y^2} + \frac{\partial^2 V}{\partial z^2} = \frac{1}{\varepsilon_0}\rho$$

i.e. 
$$\nabla^2 V = \frac{1}{\varepsilon_0}\rho. \qquad \qquad \text{...(5.10)}$$

This equation is known as Poisson's equation. It allows us to find the potential function $V$ given the distribution of charge density $\rho$ at every point. When the region does not contain any charge density, the expression (5.10) reduces to

$$\nabla^2 V = 0 \quad \text{or,} \quad \frac{\partial^2 V}{\partial x^2} + \frac{\partial^2 V}{\partial y^2} + \frac{\partial^2 V}{\partial z^2} = 0 \qquad \text{...(5.11)}$$

This is known as Laplace's equation.

### 5.2.4 Some Applications of Poisson's and Laplace's Equations

**A :** To verify that the potential of a point charge satisfies Laplace's equations at all points except at the point where the charge is situated.

Let the point charge be situated at the origin of a Cartesian coordinate system and a point $P$ is at a distance $r$ which is given by:

$r^2 = x^2 + y^2 + z^2$ and so, the derivative relative to $x$ is $\dfrac{\partial r}{\partial x} = \dfrac{x}{r}$

Again $\dfrac{\partial}{\partial x}\left(\dfrac{1}{r}\right) = -\dfrac{1}{r^2}\dfrac{\partial r}{\partial x} = -\dfrac{x}{r^3}$ and $\dfrac{\partial^2}{\partial x^2}\left(\dfrac{1}{r}\right) = \dfrac{\partial}{\partial x}\left(-\dfrac{x}{r^3}\right) = -\dfrac{1}{r^3} + \dfrac{3x}{r^4}\dfrac{\partial r}{\partial x}$

$= -\dfrac{1}{r^3} + \dfrac{3x^2}{r^5}$ . Then similarly after finding the derivatives with respect to $y$ and $z$ and then adding we get:

$$\frac{\partial^2}{\partial x^2}\left(\frac{1}{r}\right) + \frac{\partial^2}{\partial y^2}\left(\frac{1}{r}\right) + \frac{\partial^2}{\partial z^2}\left(\frac{1}{r}\right) = -\frac{3}{r^3} + \frac{3(x^2 + y^2 + z^2)}{r^5} = 0$$

Now, multiplying the both sides of the above equation by $\dfrac{q}{4\pi\varepsilon_0}$ we get as

$$V = \frac{q}{4\pi\varepsilon_0 \, r}$$

$\dfrac{\partial^2 V}{\partial x^2} + \dfrac{\partial^2 V}{\partial y^2} + \dfrac{\partial^2 V}{\partial z^2} = 0$, this conclusion is however is not valid when

$r = 0$ as $1/r$ goes to infinity and at that point which is the position of charge and which is here taken as origin the Laplace's equation is not applicable.

**B :** Using Laplace's equation obtain the electric potential and the electric field in the empty region between two parallel plates at potentials $V_1$ and $V_2$.

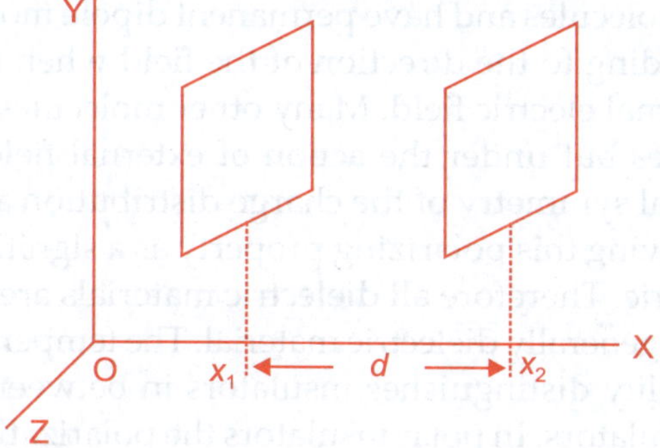

**Fig. 5.5** Two parallel plates at potentials $V_1$ and $V_2$ and at places $x_1$ and $x_2$ respectively

The field as per the problem depends only on $x$-coordinates and as there are no charges in the space between the plates, from Laplace's equation we get $\dfrac{d^2 V}{dx^2} = 0$ without using the partial derivative as the potential $V$ is only $x$ dependent. Now, integrating $\dfrac{d^2 V}{dx^2} = 0$, we get $\dfrac{dV}{dx} = $ constant But

the electric field $E = -\dfrac{dV}{dx}$ and so we can conclude that the field in between

the plates is constant. Now, we can write

$$\int_{V1}^{V2} dV = -E \int_{x1}^{x2} dx \quad \text{and} \quad E = -\frac{V_2 - V_1}{x_2 - x_1} = -\frac{V_2 - V_1}{d}$$

## 5.3 POLARIZATION OF DIELECTRIC

In this section we will discuss the effect of electric field on a substance when it is placed within the field. If the substance is a conducting material having free electrons then the free electrons will move in the direction of field and this movement of free electrons then constitute current. If the substance is an insulator or even a semiconductor and if the band gap between the valence band and conduction band is large then the field may not be able to transport the electrons from the valence band to conduction band and the substance does not show any current. But there is generally some effect of the electric field on the molecules of the substance. The atoms have spherical symmetrical distribution of charge and they are electrically neutral but some molecules do not possess such spherical symmetry and the separation of the centre of mass of the positive and negative charges renders the molecules as electric dipoles. Such molecules are called polar molecules and have permanent dipole moment. They orient themselves according to the direction of the field when the molecules are subjected to external electric field. Many other molecules are not normally as polar molecules but under the action of external field such molecules loose the spherical symmetry of the charge distribution and are polarized. The insulators having this polarizing property as a significant property are known as dielectric. Therefore all dielectric materials are insulators but all insulators are not generally dielectric material. The temperature dependence of this polarizability distinguishes insulators in between polar insulators and non polar insulators. In polar insulators the polarization is temperature dependent where as in non polar it is not.

As a consequence of this induced dipole moments, layers of positive and negative charges develop on the two surfaces of the dielectric normal to the direction of the external applied field $E$. The elementary charge induced in a layer of thickness $r$ and area $dS$ can be expressed in terms of "Polarized charge density" $\sigma_P$ such that $\sigma_P = \dfrac{dq}{dS}$ an also: $\sigma_P \, dS = \left(\dfrac{1}{V}\sum q\right) r \, dS$

$$= \left(\dfrac{1}{V}\sum qr\right) dS = \left(\dfrac{1}{V}\sum p\right) dS = P \, dS = \mathbf{P} \cdot d\mathbf{S}$$

Therefore, $\qquad\qquad P = \dfrac{1}{V}\sum p \qquad\qquad\qquad \dots (5.12)$

This $P$ is called polarization or dipole moment per unit volume of the element of dielectric.

We may also integrate the surface charge density over the closed surface of a given surface of a volume $V$ to obtain.

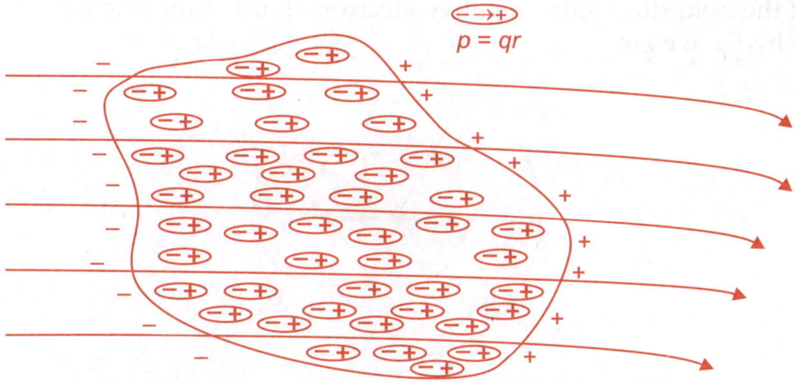


**Fig. 5.6** Polarization of the molecules and the dielectric in an electric field

$$\oint_S \sigma_P \, d\boldsymbol{S} = \oint_S \boldsymbol{P} \cdot d\boldsymbol{S} = \int_V \nabla \cdot \boldsymbol{P} \, dV \text{ using Gauss's divergence theorem.}$$

Now, since each element of dielectric remain neutral under an applied electric field we must introduce volume polarization charge of density $\rho_P = \dfrac{dq}{dV}$ such that total charge is zero

$$\oint_S \sigma_P \, d\boldsymbol{S} + \int_V \rho_P \, dV = 0 \text{ and } \int_V \rho_P \, dV = -\oint_S \sigma_P \, d\boldsymbol{S} = -\int_V \nabla \cdot \boldsymbol{P} \, dV$$

Therefore, the charge density resulting from polarization of the bulk is :

$$\rho_P = -\,div\, \boldsymbol{P} = -\nabla \cdot \boldsymbol{P} \qquad\qquad \text{... (5.13)}$$

Therefore, wherever the divergence of polarization is not zero, there exists volume polarization charge density distributed throughout the bulk of the dielectric. However this should be retained in the mind that both $\sigma_P$ and $\sigma_P$ are bound charge densities arising from the bound electrons and nuclei in the dielectric.

The polarization is determined by the charge separation in the individual atoms or molecules so that a functional relationship between $\boldsymbol{P}$ and the electric field applied, $\boldsymbol{E}$ may be derived.

### 5.3.1 Dipole Moment of an Atom

Consider an individual atom of radius $R$ having the charge density of the electron cloud $\rho$ given by : $\rho = -\dfrac{q}{\dfrac{4}{3}\pi R^3}$. Now if we can draw a sphere of radius $\boldsymbol{r}$ about the centre of the electron cloud after it has been displaced by $\boldsymbol{r}$, under the action of an electric field $E$. Then applying the Gauss's Law to get the field due to the negative electron cloud at the nucleus which is given by $E_C$, we get:

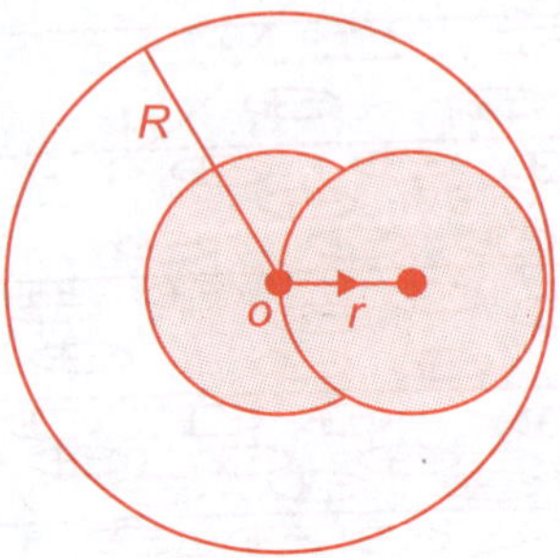

**Fig. 5.7** The charge distribution in an electron cloud

$$E_C \, 4\,\pi r^2 = \frac{1}{\varepsilon_0}\left(-\frac{3q}{4\pi R^3}\right)\frac{4\pi}{3}r^3$$

Which is $E_C = -\dfrac{q\,r}{4\pi\varepsilon_0\,R^3} = -\dfrac{p}{4\pi\varepsilon_0\,R^3}$, Now to maintain equilibrium of the nucleus the field at the nucleus $E_C$ is to be equal and opposite to the external field $E$.

$$p = 4\,\pi\varepsilon_0 R^3 E \qquad\qquad \ldots(5.14)$$

This expression shows that dipole moment $p$ is proportional to the external applied field $E$. This is the condition of "linear" dielectric where in as per the equn. (5.12) the polarization $P$ is also directly proportional to external field. If in addition the electrical properties are same in all directions for an isotropic dielectric we can write:

$$P = \chi_e\,\varepsilon_0 E \qquad\qquad \ldots(5.15)$$

Where $\chi_e$ is a scalar called "Electric susceptibility". In anisotropic dielectric $P$ and $E$ may have different directions and the electric susceptibility becomes a second rank tensor relating their components.

Now, recalling the Gauss's law in differential form Equn. 5.5 and 5.6 and applying the same for the dielectric we get

$$\nabla \cdot E = \frac{1}{\varepsilon_0}(\rho + \rho_P) = \frac{1}{\varepsilon_0}(\rho - \nabla \cdot P) \qquad\qquad \ldots(5.16)$$

where $\rho_P$ is the polarization charge density and $\rho$ is the free charge density and also using equn. (5.13). Now from the above equation separating the distribution of free charges from polarization charges we get :

$\nabla \cdot (\varepsilon_0 E + P) = \rho$. We can now introduce a vector field $D$ and can write the Gauss's law for a dielectric by equating the divergence of the field $D$ with the free charge density.

$$\nabla \cdot D = \operatorname{div} D = \rho \qquad \ldots (5.17)$$

Where the field $D$ is called "Electric displacement" and must always be associated with a particular dielectric :

$$D = \varepsilon_0 E + P \qquad \ldots (5.18)$$

While polarization vector $P$ is defined in presence of a dielectric, the electric field $E$ is defined irrespective of the environment. Using equn. (5.15) for a linear isotropic dielectric, the displacement $D$ becomes:

$$D = \varepsilon_0 E + \chi_e \varepsilon_0 E = \varepsilon_0 (1 + \chi_e) E = \varepsilon_0 \varepsilon_r E \qquad \ldots (5.19)$$

Which is often called the constitutive relation of electrostatics, where: $\varepsilon_r = 1 + \chi_e$ is known as "Relative permittivity". $\mu_r$ is a dielectric constant except for anisotropic dielectrics where the electric susceptibility and so $\mu_r$ is represented by a second rank tensor.

### 5.3.2 Electrostatic Energy

When we consider a system of point charges under mutual electrostatic internal forces, the potential energy can be expressed as:

$U = \dfrac{1}{2} \sum_i q_i \sum_{j \neq i} V_{ij}$. This expression shows that each charge experiences

the potential due to all other charges and then for a continuous distribution of charges we get:

$U = \dfrac{1}{2} \int_V \rho(r) V(r) dV$. Now in presence of a dielectric using equn. (5.17)

we get

$$U_E = \frac{1}{2} \int_V (\nabla \cdot D) V(r) dV = \frac{1}{2} \int_V \{ \nabla \cdot [VD] - D \cdot \nabla V \} dV.$$

But according to Gauss's Law the first term on the right hand side vanishes as:

$\displaystyle \int_V \nabla \cdot (VD)\, dV = \oint_S VD \cdot dS = 0$ as the surface $S$ is chosen to be an

infinite sphere. It follows then:

$U_E = \dfrac{1}{2} \int_V E \cdot D\, dV$ since $E = -\nabla V$. Therefore the electrostatic energy

density stored in a charge distribution is given as:

$$U_E = \frac{1}{2} E \cdot D \qquad \ldots (5.20)$$

### 5.3.3 Claussius-Mossotti Equations

In dielectric the dipole moment of each atom or molecule is determined by an effective field. This effective field consists of the applied field $E$ and the total of all other dipoles, added together. This effective field may also be called local field $E_{eff}$ can be calculated in a linear isotropic dielectric by considering a spherical region within the dielectric having radius $R$ and centre at a given molecule.

The field resulted due to the uniform distribution of molecules inside the sphere averages to zero so that the contribution of all other molecules to $\mathbf{E}_{eff}$ reduces to the field associated with the polarization surface charges on the sphere. In the arrangement illustrated in the following figure the surface charge density at angle $\theta$ to the direction of the field is equal to the normal component of polarization $P$ with negative sign. Where $P$ is the polarization field outside the spherical region of the dielectric

$$\rho = -\mathbf{P} \cdot \mathbf{n} = -P \cos \theta$$

The surface area of the annular shaded ring $R \sin \theta$ and of width $Rd\theta$ is given as $2\rho R \sin\theta \cdot Rd\theta$ and the charge on it is given by :

$$dq = (-P \cos \theta)\, 2\pi R \sin\theta \cdot Rd\theta.$$

This produces an electric field at the origin parallel to the applied field

$$dE_P = \frac{dq \cos \theta}{4\pi \varepsilon_0\, R^2} = \frac{-2\pi R^2\, P \cos^2 \theta \sin \theta\, d\theta}{4\pi\varepsilon_0\, R^2} = -\frac{P}{2\varepsilon_0} \cos^2 \theta\, d\,(\cos \theta)$$

Integrating over $\theta$ from $0$ to $\pi$ to take the entire sphere we get the contribution of the polarization charges to the local field as

$$E_P = P/3\varepsilon_0 \qquad\qquad \dots(5.21)$$

Now, as
$$E_{eff.} = E + E_P$$
$$E_{eff.} = E + P/3\varepsilon_0$$

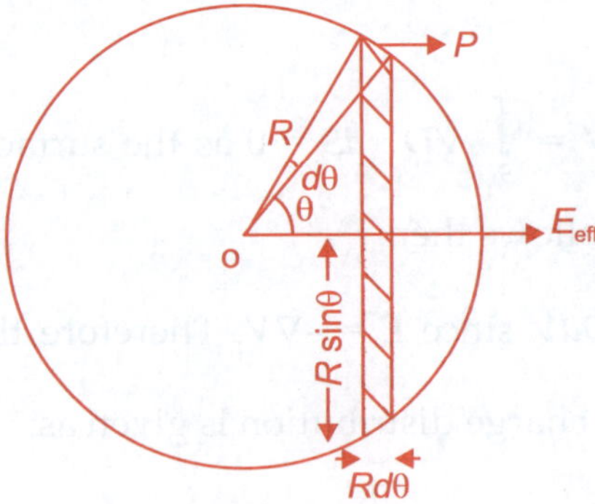


**Fig. 5.8** The electric field due to the polarization surface charge density on a spherical region of the dielectric

Now, assuming the dipole moment of a molecule $p$ is proportional to the effective field $E_{eff}$, we can write after introducing $\alpha$ as "polarizability" of the molecule

$$p = \alpha\varepsilon_0\, E_{eff.} = \alpha\varepsilon_0\, [E + P/3\varepsilon_0]$$

The polarization P for N molecules per unit volume is given by

$$P = Np = N\alpha\varepsilon_0\, [E + P/3\varepsilon_0]$$

i.e.
$$P = \frac{N\alpha}{1-\dfrac{N\alpha}{3}}\,\varepsilon_0 E \qquad \qquad \dots(5.22)$$

Now using equn. (5.19) which introduces the dielectric constant $\varepsilon_r$, we get the relation between the dielectric constant $\varepsilon_r$ in terms of polarizability $\alpha$ and number of moleculies per unit volume $N$ as :

$$\varepsilon_r = \frac{1 + 2N\alpha/3}{1 - N\alpha/3}$$

or,

$$\frac{\varepsilon_r - 1}{\varepsilon_r + 2} = \frac{N\alpha}{3\varepsilon_0} \qquad \qquad \dots(5.23)$$

This equation is known as "Claussius-Mossotti" equation.

## REVIEW QUESTIONS

1. Two point charges are placed on the $x$-axis, one of $+1\,e\,s\,u$ at $x = 2$ cm and the other of $-4\,e\,s\,u$ at $x = -2$ cm.. Calculate the magnitude and direction of the electric field at the point $(0, 3.0)$. Is there any point where the field is zero?

2. A water droplet of $10^{-2}$ cm in diameter carries a negative charge and due to this the field on the surface is 20 stat volts/cm. How strong a vertical field is required to hold the drop from falling down?

3. An infinite plane has a uniform surface charge distribution s on its surface. Right next to it is an infinite parallel layer of charge of volume density $r$. Find the field E everywhere.

4. Considering a spherical charge distribution of density $r$ from $r = 0$ to $r = a$. Find the field due to this charge distribution for values of $r$ both less than and greater than $a$.

5. A thin rod along $z$-axis is from $z = -a$ to $z = +a$. The rod carries a charge density per unit length as l. Find the potential at all points along the $x$-axis from $x > 0$.

6. Let us consider two different processes while inserting a dielectric slab between two parallel plates of a capacitance, one when the plates of the capacitor is connected to battery and in the other when they are not. Find and compare the energy stored in the capacitor and force on the slab.

## REFERENCES


1. Berkeley Physics course (Vol. II), McGraw-Hill book Co., New York.

2. R. A. Serway and J. W. Jewett – Principles of Physics, Thomson, 3rd Edn., New York, 2006.

3. S. B. Palmer, M. S. Rogalski- Advanced University Physics, Gordon Breach Pub.,1996.

4. M. Alonso, E. J. Finn- Fundamental University Physics, Addison-Wesley Pub. Co., 1968.

# Magnetism - I

## 6.1 INTRODUCTION

Long before Christ, it was observed that certain naturally available material known as "Lodestone" has the potentiality to attract iron, cobalt and also nickel. This specific property is only found to exist in the particular iron ore and is not found in other general bodies. Moreover, this property of attraction is apparently not related to electrical interaction as it does not attract paper pieces or cork and also not related to gravitational attraction. It is also found to be concentrated at some spots in the mineral ore. As this, it, is a different interaction named as magnetic interaction and the phenomena is said to belong a new natural property known as "Magnetism". The specific materials showing this property are known as "Magnet" and the spot on the material where the property is found to be concentrated is known as "Magnetic poles". The interaction between two such magnets was also found interesting. Ends of the two cylindrical shaped magnets either attract or repel. The poles of the two bars which attract each other are called unlike poles and which repel each other are known as like poles. If the rod is freely suspended then it has been found that a particular end always point toward north and south. The pole at the end of the magnetic bar pointing towards north is known as north seeking pole or "North pole", (N) and the other pole pointing south is "South pole", (S). This also demonstrated the fact that earth is also a huge magnet as it was seen that a magnet is influenced only by another magnet. These magnetic poles are basically different from electric charges. The electric charges though are of two kinds positive and negative, it has been found that they can have their separate existence. A body can be positively charged and also of negative charge of different quantity but magnetic bodies always show the poles in pairs, equal and opposite. Fundamental particles can be either positive or negative or even neutral and for charged fundamental particles the charges can be of different quantity but we cannot find a fundamental particle having only one kind of magnetism.

However, the electric and magnetic interactions are very closely related and in fact they are the two different aspects of the one property of the matter. When an electric charge is stationary, it shows only electric field but when it moves it demonstrate magnetic field and so magnetism is the manifestation of electric charges in motion and so electric and magnetic interactions should be considered together under more general name "Electro-magnetic interactions". Therefore, it is necessary first to introduce charge flow before the magnetic effects and interactions.

## 6.2 MAGNETIC INDUCTION

It has been mentioned that the electric and magnetic interactions are actually two different aspects of the same property of the matter. In order to establish this it is necessary to introduce first the charge flow constituting current and then the effect of this current in the magnetic induction.

### 6.2.1 Electric Current, Current Density and Ohm's Law

In dielectrics the charge densities are bound and the electrostatic behaviour of these materials are described by a vector field $D$, which is proportional to the applied electric field $E$. In conductors the charges are free and they move under the action of an applied electric field $E$. This steady charge flow under the influence of the field constitutes current. A steady electron flow in conductors requires the condition that at one end of the conductor the surface charges balancing the field $E$ must move out to the source of the field and at the other the charges are supplied back from the source and thus maintaining the conductor as a part of the complete circuit.

Now, let $m_e$ be the mass and $-e$ be the charge of the electron and $v$ be the "drift velocity" of the electron in the direction opposite the direction of the field. Then

$$m_e \frac{d}{dt} v = -eE - \alpha v \qquad \qquad \text{... (6.1)}$$

where $-\alpha v$ represents the damping force against the direction of velocity. This damping force results due to the collision between the moving electrons with the ions in the conductor. Now if there are $N$ number of free electrons in volume $V$ then free charge density is given by: $\rho = -\dfrac{Ne}{V} = -ne$. Now, as

all these electrons are moving with same velocity $v$, the magnitude of the electric current is defined as the time rate of charge through the normal cross section of area $S$ of the conductor. Then

$$I = \frac{dq}{dt} = -neSv = \rho Sv \qquad\qquad \ldots(6.2)$$

and from equn. (6.1)

$$\frac{dI}{dt} = \frac{ne^2 SE}{m_e} - \frac{\alpha}{m_e}I$$

or, $\qquad \dfrac{dI}{I - ne^2 SE/\alpha} = -\dfrac{\alpha}{m_e}dt$ and integrating we get

$$\ln\left(I - \frac{ne^2 SE}{\alpha}\right) = -\frac{\alpha}{m_e}t + C \quad \text{but from the initial condition the current } I = 0$$

at $t = 0$, we get

$$I = \frac{ne^2 SE}{\alpha}\left(1 - \exp\left[-\frac{\alpha t}{m_e}\right]\right), \text{ which shows that the current}$$

exponentially increases to its steady value:

$$I = \frac{ne^2 SE}{\alpha} \qquad \text{or} \qquad I = \frac{ne^2 S}{\alpha}\frac{V}{l} \qquad\qquad \ldots(6.3)$$

where $V$ is the potential difference at the two ends of the conductor and $l$ is the length of the conductor as $E = \dfrac{V}{l}$.

Now, introducing $\dfrac{\alpha}{ne^2} = \dfrac{1}{\sigma}$ and writing: $R = \dfrac{l}{\sigma S}$ we get from equn.(6.2)

$$V = RI \qquad\qquad \ldots(6.4)$$

This is known as Ohm's Law, where $R$ is called as Resistance and $\sigma$ is the conductivity.

Now, the charge flow is always tangential to the electric field lines and so we can define a current density field $j$, the magnitude of it is equal to the current flow per unit normal area. So that from equn. (6.2) $j = \dfrac{dI}{dS} = \rho v$

or, $j = \sigma v$ or, $I = \int_S j \cdot dS$ and from equn. (6.3) and expression of conductivity

$\sigma$ we get after equating: $j \cdot dS = \sigma E \cdot dS$ and therefore, the current density $j = \sigma E$ where the conductivity $\sigma$ acts here as the constant of proportionality.

Now, considering the surface $S$ to enclose a volume $V$, the total current flowing out of the surface $S$ is equal to the time rate of decrease of charge inside $S$. Therefore,

$$\oint_{S} j \cdot dS = -\frac{\partial}{\partial t}\int_{V}\rho\, dV = -\int_{V}\frac{\partial\rho}{\partial t}\, dV$$

$$\oint_{S} j \cdot dS + \int_{V}\frac{\partial\rho}{\partial t}\, dV = 0$$

Applying Gauss's divergence theorem:

$$\int_{V}\nabla \cdot j\, dV + \int_{V}\frac{\partial\rho}{\partial t}\, dV = 0$$

Which is equivalent to: $\nabla \cdot j\ \dfrac{\partial\rho}{\partial t} = 0.$ This is the equation of continuity

of current.

For a steady state flow of charge within the conductor producing steady current $\dfrac{\partial\rho}{\partial t} = 0$, indicating no accumulation of charge inside the surface S, we get:

$$\nabla \cdot j = 0 \qquad\qquad\qquad …(6.5)$$

There are considerable similarities between the equations of current flow expressed in terms of current density field $j$ in conductor and the equations of electrostatic interactions in dielectric expressed in terms of electric displacement field $D$. This can be seen as:

For dielectric for conductors

$$D = \varepsilon_0\,\varepsilon_r\,E \qquad \text{and} \qquad j = \sigma E$$
$$\nabla \cdot D = \rho \qquad\qquad\qquad\qquad \nabla \cdot j = 0$$
$$\nabla \times E = 0\nabla \times E \qquad\qquad\qquad\qquad = 0 \qquad …(6.6)$$

As stated before the electric and magnetic interactions are two different aspects of the same property of matter. The magnetic effects on macroscopic and microscopic scale are explained in terms of current flow as this effect finds its origin in the flow of charge. Although this assumption does not allow us to ascribe the source of magnetic fields to the magnetic poles even though it is a similar in concept of electric charge, the theory of magnetostatics can be developed accepting its similarity with electrostatics. It can be once again emphasized that a stationary charge positive or negative shows only electrostatic field where as a moving charge constituting current shows magnetic field.

## 6.2.2 Magnetic Effects

Let a vector field $B$, called magnetic induction associated with the flow of a charge $dq$ with velocity $v$, $dB$ is given by: $dB = k\,\dfrac{dq}{r^3}\,(v \times r)$. That is the reason for the deflection of magnetic needle when placed near the current carrying wire. The direction of this resultant magnetic field is perpendicular over the plane containing $v$ giving the direction of current and the distance vector $r$ of the point of observation from the current due to the flow of elementary charge $dq$.

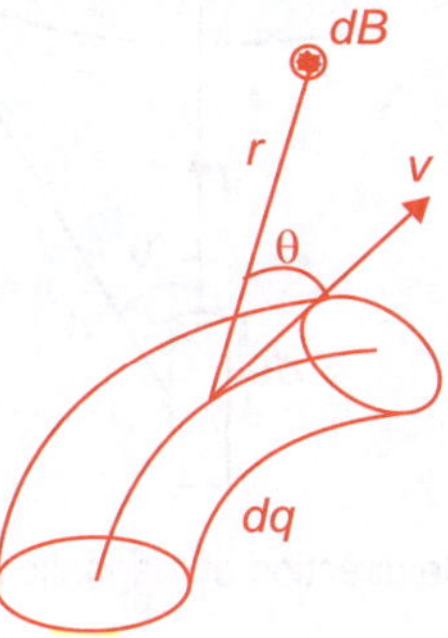


**Fig. 6.1** Magnetic induction due to charge flow

This magnetic field $B$ is an inverse square dependence on position of observation. The constant of proportionality $k$ is equal to $\dfrac{\mu_0}{4\pi}$ when all the parameters are measured in $SI$ units and $\mu_0$ is the permeability of free space.

$$d\mathbf{B} = \frac{\mu_0}{4\pi} \frac{dq}{r^3} (v \times r) \qquad \ldots (6.7)$$

Let us now express this magnetic induction in terms of steady current I flowing in an infinitesimal element of conductor of length $dl$.

Now as $dq = \rho\, dV = \rho\, Sdl$ where $\rho$ is the volume charge density and S is the cross section of the element of conductor of length $dl$ the equn. (6.7) can be written as

$$d\mathbf{B} = \frac{\mu_0}{4\pi} \frac{\rho Sdl}{r^3} (v \times r) = \frac{\mu_0}{4\pi} \frac{Sdl}{r^3} (\rho v \times r) = \frac{\mu_0}{4\pi} \frac{Sdl}{r^3} (j \times r)$$

Where as introduced before $j = \rho v$ and now as $j$, the current density and $dl$, the element of length of the conductor are parallel, so we can replace $jdl$ by $\mathbf{j}dl$ and so the above expression can be written as:

$$d\mathbf{B} = \frac{\mu_0}{4\pi} \frac{Sj}{r^3} (dl \times r) = \frac{\mu_0}{4\pi} \frac{I}{r^3} (dl \times r) \qquad \ldots (6.8)$$

This is known as Biot-Savart Law. In a more convenient form, the Biot-Savart law (equn. 6.8) can be written in scalar form as:

$$dB = \frac{\mu_0 I}{4\pi} \frac{dl \sin\theta}{r^2} \qquad \ldots (6.9)$$

This expression of Biot-Savart Law can be demonstrated by the following Fig. 6.2.

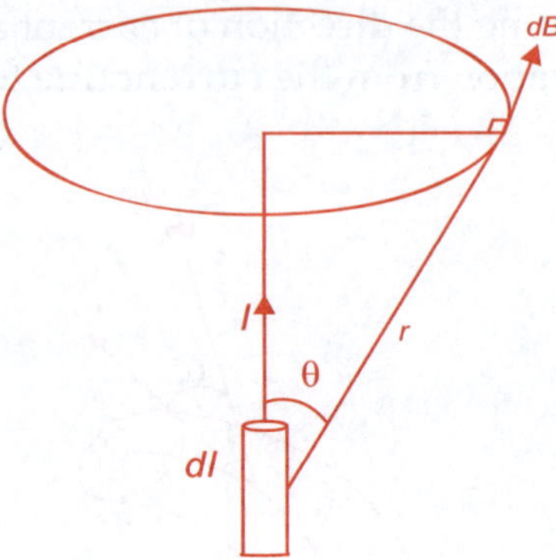

**Fig. 6.2** Biot-Savart law and the direction of magnetic field due to the current element

Now, recalling equn.(6.7) and modifying it for a total charge $q$ moving with velocity $v$ the total magnetic field $B$ at a point $P$ which is at a distance $r$ from the charge at an instant, we get

$$B = \frac{\mu_0}{4\pi}\frac{q}{r^2}\,(v \times n) \qquad \qquad \ldots(6.10\ a)$$

where $n$ is an unit vector which defines the direction of $r$.

Now, the electric field $E$ due to this charge at that instant and at the point $r$ is given by

$$E = \frac{q}{4\pi \varepsilon_0\, r^2}\, n \qquad \qquad \ldots(6.10\ b)$$

Now, comparing equns. (6.10 a ) and (6.10b) we can establish a relation between $B$ and $E$ as

$$B = \mu_0\varepsilon_0 v \times E \quad \text{Now, writing } c = \frac{1}{\sqrt{\mu_0\,\varepsilon_0}},$$

we get

$$B = \frac{1}{c^2}\,v \times E. \qquad \qquad \ldots(\ 6.10\ c)$$

This equation establishes the relation between the electric field and magnetic field of induction produced by a moving charge. The relation between the directions of $B$ and $E$ is shown in the following diagram. (6.3).

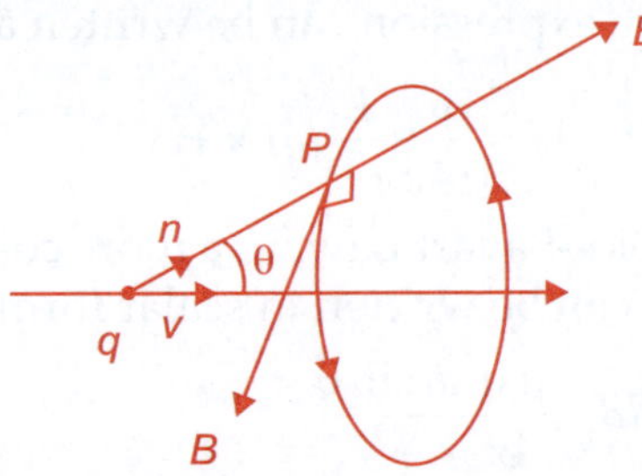

**Fig. 6.3** The magnetic field produced due to a moving charge

### 6.2.3 Magnetic Induction in Simple Common Circuits

**(A)  Due to long straight wire of very long length (infinite)**

Let us take a wire of an infinite length carrying a constant current I through it. We are to find the magnetic induction field at a point $P$ which is at a normal distance of $R$.

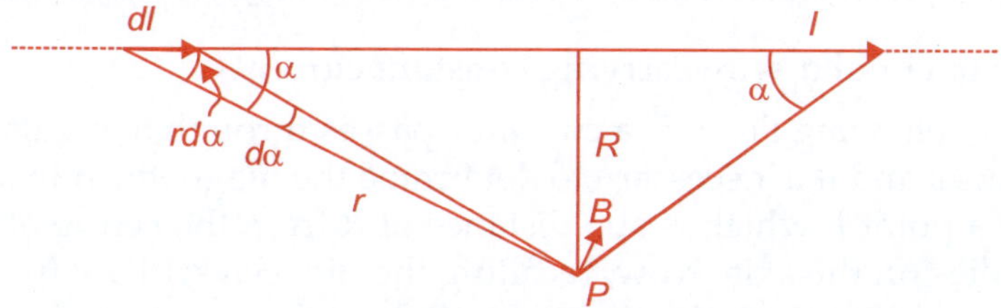


**Fig. 6.4** The long straight wire carrying current

Now, consider a current element $dl$ at a distance of $r$ from the point of observation and $\pm$ is the angle that this end that point $P$ makes. Then

$$r = \frac{R}{\sin\alpha} \quad \text{and the element of wire of length } dl = \frac{r\,d\alpha}{\sin\alpha} = \frac{R\,d\alpha}{\sin^2\alpha}$$

Now, recalling the equn. (6.9) we get as $\theta = \alpha$ here

$$dB = \frac{\mu_0}{4\pi}\frac{I\,R\,d\alpha\sin\alpha\sin^2\alpha}{R^2\sin^2\alpha} = \frac{\mu_0}{4\pi}\frac{I}{R}\sin\alpha\,d\alpha .$$

Now, if we consider another such element at the other end and if the element of length $dl$ makes angle $b$ and as these two elements make the magnetic induction field in the same direction (perpendicular to the plane containing both $dl$ and $r$) then we can add these two contributions and get

$$dB = \frac{\mu_0}{4\pi}\frac{I}{R}(\sin\alpha\,d\alpha + \sin\beta\,d\beta)$$

and then integrating we get

$$B = \frac{\mu_0}{4\pi}\frac{I}{R}(\cos\alpha + \cos\beta)$$

Now, for a very long wire both $\alpha$ and $\beta$ become zero and then the total magnetic induction field $B$ at point $P$ at a normal distance of $R$ from the straight wire of infinite length carrying current $I$ in scalar form as:

$$B = \frac{\mu_0}{2\pi}\frac{I}{R} . \qquad\qquad …(6.11)$$

The direction of the field at every of such points $P$ on the plane of the figure is perpendicular to the plane. The pictorial presentation of this field is given as below:

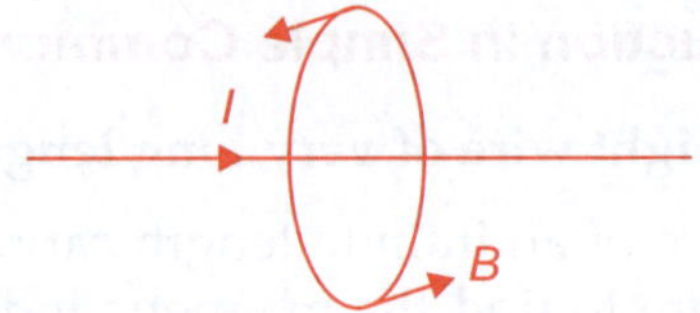

**Fig. 6.5** The direction of the circular field due to current *I* through the straight conductor

## (B) Due to circular wire carrying constant current

In the following Fig. 6.5. a current *I* passes through a circular loop of radius a and it is necessary to determine the magnetic induction field *B* at a point *P* which is at a distance of *R* from the centre of the loop and lies on the axis. Now, recalling the Biot-Savart Law from equns. (6.8) and (6.9) the induction field *dB* due to an element of wire *dl*

$$dB = \frac{\mu_0}{4\pi} \frac{I}{r^3} (dl \times r)$$

As direction of **B** is perpendicular to both current element *dl* and distance vector *r*, it is expressed as cross product between *dl* and *r* and so the other form of the law is: $dB = \frac{\mu_0}{4\pi} \frac{I \, dl \sin\theta}{r^2}$. Now as per the figure below θ being 90°.

The expression reduces to

$$dB = \frac{\mu_0}{4\pi} \frac{I \, dl}{r^2}. \qquad\qquad \ldots(6.12)$$

Now, as the direction of *dB* is perpendicular to both *dl* and *r* i.e. normal on the plane *ABCD*, the *dB* can be resolved in to *dB* perpendicular and parallel as noted in the Fig. 6.6.

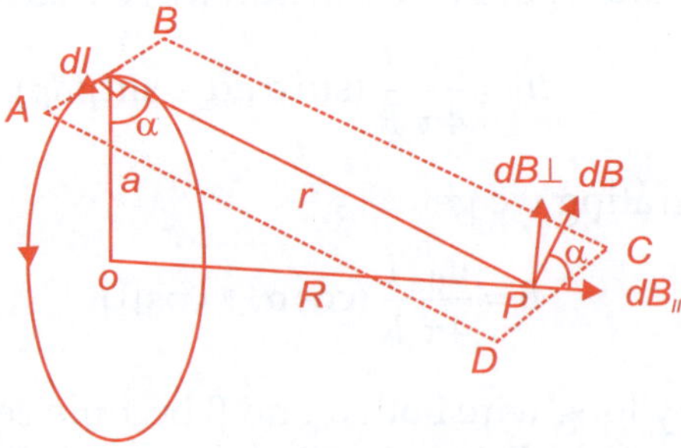


**Fig. 6.6** Circular loop of wire and the magnetic field of induction along the axis

The perpendicular components of magnetic induction field *dB* for the wire element *dl* taken at the two diametrically opposite direction cancel each other and the only component that will remain is $dB_{\parallel}$ for all positions of the wire element along the circumference of the loop.

Therefore,
$$dB_{\parallel} = dB \cos \alpha = \frac{a}{r} dB = \frac{\mu_0 I a}{4\pi r^3} dl$$

Then
$$B = \oint dB_{\parallel} = \frac{\mu_0 I a^2}{4\pi r^3} \oint dl = \frac{\mu_0 I a^2}{2 r^3}$$

But as: $r = (a^2 + R^2)^{1/2}$ and using this we get the expression for the magnetic induction field for the complete loop

$$B = \frac{\mu_0 I a^2}{2(a^2 + R^2)^{3/2}}. \qquad \qquad …(6.13)$$

The magnetic flux round a circular loop carrying current is shown below. The view from the right of the loop appears that this end where from the flow of current is anticlockwise acts like North pole as the magnetic flux emerge from this side and the opposite side acts like South pole. Therefore, this loop develops a magnetic moment and acts like a magnetic dipole provided the loop is very small. The magnetic moment $M = I(\pi a^2)$ and replacing this in equn. (6.12) we get

$$B = \frac{\mu_0 M}{2\pi (a^2 + R^2)^{3/2}}.$$

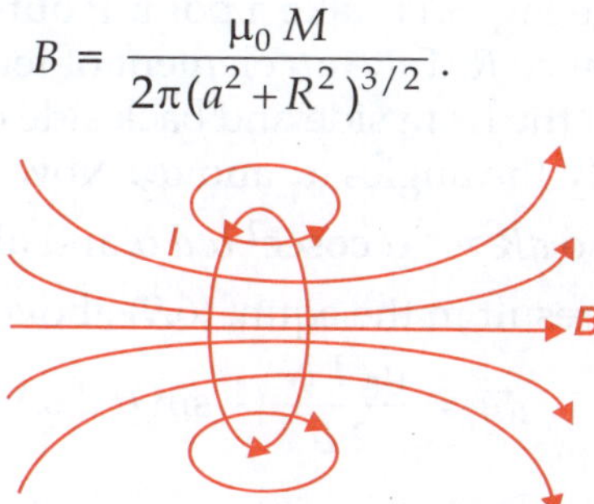


**Fig. 6.7** The magnetic induction lines of forces emerging through the loop of wire

## (C)  Due to solenoidal current

Now, if we have a large number of such loops close to each other and if the current flows through each of such loops in the same direction then the induction field produced either at the centre of the solenoid or at its ends will be dependent on the total number of such loops or turns (the insulated wire is wound on a cylinder in the same mode and the current passes through each loop of the cylinder in the same direction).

Recalling equn. (6.12) and writing in modified form

$$dB = \left[ \frac{\mu_0 I a^2}{2(a^2 + R^2)^{3/2}} \right] \frac{N}{L} dR \qquad …(6.14)$$

where $N$ is the total number of such turns and $L$ is the length of the solenoid.

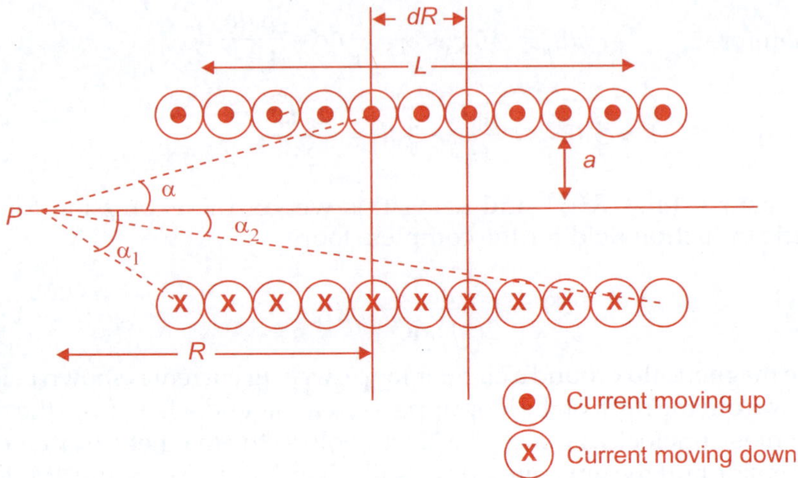



**Fig. 6.8** Solenoid of length $L$ having total turns $N \cdot dR$ is the element of length of the solenoid $\alpha$'s are the angles subtended as shown

Now, in the above Fig. 3.11, take a point $P$ outside the solenoid but on the axis and at a distance $R$. Take an element of length $dR$ which subtends an angle $\pm$ at $P$ and let the front side and back side of the solenoid of length L subtend respectively the angles $\alpha_1$ and $\alpha_2$. Now we can find that

$R = \alpha \cot \alpha$  and so $dR = - \alpha \, \text{cosec}^2 \, \alpha \, d \alpha$ and also $\alpha^2 + R^2 = a^2 \, \text{cosec}^2 \alpha$.

Substituting this result in the equn. (6.9) above we get

$$dB = \frac{\mu_0 \, I \, N}{2 \, L} (- \sin \alpha \, d \alpha).$$

The resultant field can be found by integrating from one appropriate limit to other so that it can cover the solenoid.

Therefore, the resultant field is

$$B = \frac{\mu_o IN}{2 \, L} \int_{\alpha_1}^{\alpha_1} - \sin \alpha \, d\alpha = \frac{m_o IN}{2 \, L} (\cos \alpha_2 = \cos \alpha_1).$$

Now, putting the appropriate values of $\alpha$'s and assuming the solenoid is very long, then at the centre of the solenoid we can assume $\alpha_1 = p$ and

$\alpha_2 = 0$ then the field at the center of the solenoid $B = \dfrac{\mu_o IN}{L}$ $\qquad$ ...(6.15 a)

Similarly, at one end of the solenoid (long) we can assume

$\alpha_1 = \dfrac{\pi}{2}$ and $\alpha_1 = 0$ or for the other end $\alpha_1 = \pi$ and $\alpha_2 = \dfrac{\pi}{2}$ and for either

case we can write $B = \dfrac{\mu_{0 \, IN}}{2 \, L}$ $\qquad\qquad$ ...(6.15 b)

## 6.2.4 Ampere's Law

Recalling Biot-Savart Law of equn. (6.8) as: $\mathbf{B} = \dfrac{\mu_0 I}{4\pi} \oint \dfrac{1}{r^3} (d\mathbf{l} \times \mathbf{r})$ Now, writing this form in terms of line integral in the form of volume integral and introducing $I d\mathbf{l} = jS\, dl = j\, dV$ we get $\mathbf{B} = \dfrac{\mu_0}{4\pi} \int_V \dfrac{1}{r^3} (\mathbf{j} \times \mathbf{r}) dV.$

Now taking the divergence of the magnetic induction $\mathbf{B}$

$$\nabla \cdot \mathbf{B} = \frac{\mu_0}{4\pi} \int_V \nabla \frac{1}{r^3} (\mathbf{j} \times \mathbf{r}) dV = \frac{\mu_0}{4\pi} \int_V \left[ \frac{1}{r^3} \mathbf{r} \cdot (\Delta \times \mathbf{j}) - \mathbf{j} \cdot \left( \Delta \times \mathbf{r} \frac{1}{r^2} \right) \right] dV$$

Now, $\nabla \times \mathbf{j} = 0$ and we have

$$\nabla \cdot \mathbf{B} = \frac{\mu_0}{4\mu} \int_V \mathbf{j} \cdot \left[ \mathbf{r} \times \nabla \left( \frac{1}{r^3} \right) - \left( \frac{1}{r^3} \right) (\nabla \times \mathbf{r}) \right] dV.$$

Now, $\nabla \left( \dfrac{1}{r^3} \right) = -\dfrac{1}{r^5} \mathbf{r}$ and so the cross product in the first term vanishes to zero and also $\nabla \times \mathbf{r} = 0$ and putting these results we get

$$\nabla \cdot \mathbf{B} = 0. \qquad \qquad \text{...(6.15)}$$

This result of the divergence of the magnetic induction indicates that the strength of source field at a point is zero and therefore, the magnetic induction $\mathbf{B}$ which is a solenoidal field has no source as the field lines have no beginning or end.

Now, using Gauss's divergence theorem the equn. (6.15) gives $\oint_S \mathbf{B} \cdot d\mathbf{S}$

$= \int_V \nabla \cdot \mathbf{B}\, dV = 0$, which shows that the flux of magnetic induction through any closed surface is zero and $\mathbf{B}$ is often referred as magnetic flux density.

Now, the line integral of $\mathbf{B}$ around any closed path $dL$ will be given after using equn. (6.11) and the following Fig. 6.8 as:

$$\oint_L \mathbf{B} \cdot d\mathbf{L} = \oint_L BRd\alpha = \frac{\mu_0}{2\pi} \frac{I}{R} \oint_\alpha Rd\alpha = m_0 I \qquad \text{...(6.16 a)}$$

This result will only be added up if in place of a single wire carrying current $I$ we have a large number of straight wires each carrying currents of different magnitude so that:

$$\oint_L \mathbf{B} \cdot d\mathbf{L} = \mu_0 \int_S \mathbf{j} \cdot d\mathbf{S} \qquad \text{...(6.16 b)}$$

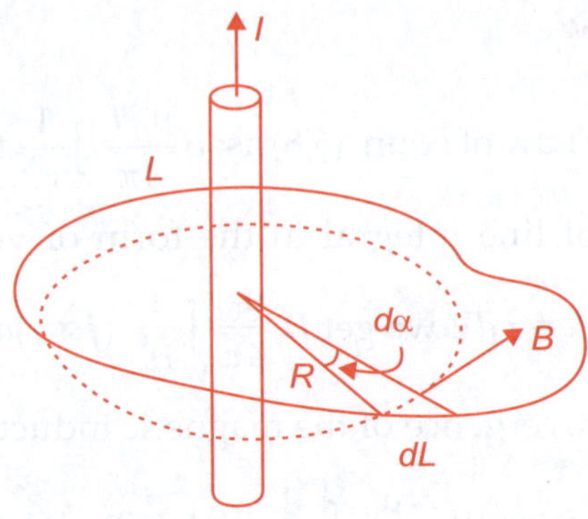

**Fig. 6.9** Current path for the ampere's circulation law

This equn. (6.16) is known as Ampere's Circuital Law. Which implies that the line integral of $B$ around any closed path is equal to $\mu_0$ times the total current I enclosed by the closed path or crossing any surface bounded by the path.

As for arbitrary distribution of current given by current density $j$, the equn. (6.16) can be written as:

$$\oint_L B \cdot dL = \mu_0 \int_S j \cdot dS \qquad \ldots(6.16\ c)$$

Using Stoke's theorem , the equn. (6.16 c) can be written as :

$$\int_S (\nabla \cdot B) \cdot dS = \mu_0 \int_S j \cdot dS$$ which is valid irrespective of the smallness of the surface element and so, the Ampere's Law can be written as:

$$\nabla \times B = \mu_0 j \qquad \ldots(6.16\ d)$$

This may also be called as Ampere's Law in differential form. It establishes a relation between magnetic field $B$ at a point and the current density $j$ at the same point. It is similar to the Gauss's Law which relates the electric field and the charges at the same point of space. As the charge distribution is the source of electric field the electric current is then the source of magnetic field.

## 6.3 MAGNETIC VECTOR POTENTIAL

The Ampere's Law which is stated in equn. (6. 16) shows that magnetic field $B$ is conservative in the region of space which does not enclose any current and in the regions where it is conservative, it can be written as: $B = -\nabla\varphi$, where $\varphi$ is the magneto static scalar potential. This equation has exact similarity with the equn. (5.8). for the conservative electric field.

Now, recalling the equn. (6.15) which states that the divergence of $B$ representing the strength of the source field at a point vanishes:

$$\nabla \cdot B = 0$$

Now, from this relation indicates that a vector field $A$ known as vector potential can be defined as

$$B = \nabla \times A.$$

Recalling Biot-Savart Law equn.(6.8) as

$$B = \frac{\mu_0 I}{4\pi} \frac{1}{r^3} (dl \times r)$$

Where both $dl$ and $r$ can be written in Cartesian coordinates as:

$$dl = idl_1 + jdl_2 + kdl_3 \quad \text{and} \quad r = ix + jy + kz$$

then substituting $dl$ and $r$

$$B = \frac{\mu_0 I}{4\pi} \frac{1}{(x^2 + y^2 + z^2)^{3/2}} [(idl_1 + jdl_2 + kdl_3) \times (ix + jy + kz)]$$

$$= \frac{\mu_0 I}{4\pi(x^2 + y^2 + z^2)^{3/2}} [kdl_1\, y - jdl_1\, z - kdl_2 x + idl_2\, z + jdl_3\, x - idl_3 y]$$

$$= \frac{\mu_0 I}{4\pi(x^2 + y^2 + z^2)^{3/2}} [i(dl_2 z - dl_3 y) + j(dl_3 x - dl_1 z) + k(dl_1 y - dl_2 x)]$$

Now, as $B = iB_x + jB_y + kB_z$ comparing the two expressions of $B$ we get:

$$B_x = \frac{\mu_0 I}{4\pi(x^2 + y^2 + z^2)^{3/2}} (dl_2 z - dl_3\, y) \qquad \ldots(6.17)$$

Now, let that vector potential $A$ be defined as:

$$A = \frac{\mu_0 I}{4\pi(x^2 + y^2 + z^2)^{1/2}} dl. \qquad \ldots(6.18)$$

Now, taking curl of $A$

$$\nabla \times A = \frac{\mu_0 I}{4\pi(x^2 + y^2 + z^2)^{1/2}} \nabla \times dl], \text{ writing}$$

$$dl = id\,l_1 + jdl_2 + kdl_3$$

$$= \frac{\mu_0 I}{4\pi\left(x^2 + y^2 + z^2\right)^{1/2}} \left[\left(i\frac{\partial}{\partial y} + j\frac{\partial}{\partial y} + k\frac{\partial}{\partial y}\right) \times (idl_1 + jdl + kdl_3)\right]$$

$$= \frac{\mu_0 I}{4\pi(x^2 + y^2 + z^2)^{1/2}} \left[k\frac{\partial}{\partial x}dl_2 - j\frac{\partial}{\partial x}dl_3 - k\frac{\partial}{\partial x}dl_1 + i\frac{\partial}{\partial x}dl_3 + j\frac{\partial}{\partial x}dl_1 - \frac{\partial}{\partial x}dl_2\right]$$

$$\frac{\mu_0 I}{4\pi}\left[ i\,\frac{\left\{\dfrac{\partial}{\partial y}dl_3 - \dfrac{\partial}{\partial z}dl_2\right\}}{(x^2+y^2+z^2)^{1/2}} + j\,\frac{\left\{\dfrac{\partial}{\partial z}dl_1 - \dfrac{\partial}{\partial x}dl_3\right\}}{(x^2+y^2+z^2)^{1/2}} + k\,\frac{\left\{\dfrac{\partial}{\partial x}dl_2 - \dfrac{\partial}{\partial y}dl_1\right\}}{(x^2+y^2+z^2)^{1/2}} \right]$$

Now, considering the $x$ component only we get

$$[\nabla \times A]_X\,\frac{\mu_0 I}{4\pi}\,\frac{\left\{\dfrac{\partial}{\partial y}dl_3 - \dfrac{\partial}{\partial z}dl_2\right\}}{(x^2+y^2+z^2)^{1/2}}$$

$$= \frac{\mu_0 I}{4\pi}\left[ dl_3\,\frac{-(1/2)2y}{(x^2+y^2+z^2)^{3/2}} - dl_2\,\frac{-(1/2)2z}{(x^2+y^2+z^2)^{3/2}} \right]$$

$$[\nabla \times A]_X = \frac{\mu_0 I}{4\pi(x^2+y^2+z^2)^{3/2}}\,[zdl_2 - ydl_3]$$

Now, comparing this with equn. (6.17)

We can conclude that $B_x = [\nabla \times A]_X$ and therefore:

$[B = \nabla \times A$, where the vector potential $A$ which is introduced and defined and is given as:

$$|A| = \frac{\mu_0 I}{4\pi}\oint \frac{dl}{r}. \qquad\qquad \dots (6.19)$$

### 6.3.1 Magnetic Field and Magnetizaton

So long what we have discussed is termed as Magnetic Induction which is the fact that a moving charge not only show the electric field but also a magnetic field associated with this moving charge. This resultant magnetic field will cease to exist as soon as the charge concerned ceases to move. We have seen in earlier sections that a current carrying wire produces a magnetic field around it without it self getting magnetized. If we pass a current through a copper wire and if this wire is dipped through some iron powders, the powders do not cling to the wire but instead they are arranged in circular path surrounding the wire. This shows a dependence of induced magnetic field with existing current and also vice versa. We can however draw this conclusion as the electrons in the atom also constitute a closed current. Therefore this small current electron orbit constitutes a magnetic dipole. Atoms however may or may not exhibit a net magnetic dipole moment depending on the symmetry or on the relative orientation of their electronic orbits. Since most of the molecules are not spherically symmetric, they therefore exhibit magnetic dipole moment because of special orientation of the electron orbit. But the matter in bulk (exception ferromagnetic material) fails to exhibit a net magnetic dipole moment because of the random orientation of molecules. However, the presence of an external magnetic field distorts the electronic motion, giving rise to a net magnetic 'polarization' or magnetization of the material.

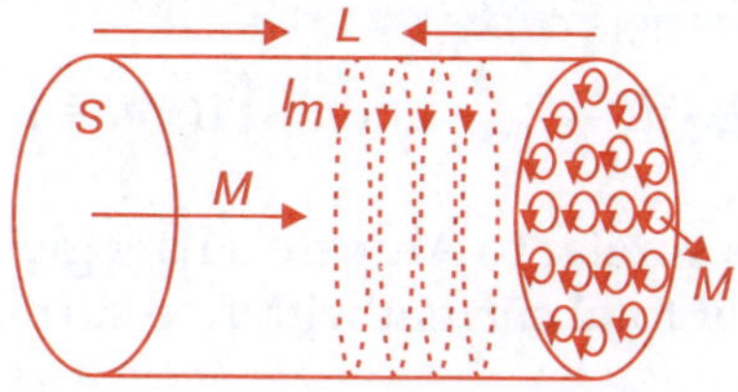


**Fig. 6.10** Magnetization surface current $I_M$ and elementary currents due to electron orbits in a magnetized cylinder

Now, if this cylindrical matter is put in a solenoidal current I in the same direction as $I_M$ and if the solenoid has $n$ turns per unit length, this magnetized cylinder is then equivalent to a cylinder carrying a current per unit length as nI + M. This effective solenoidal current gives rise to a resultant magnetic field $B$ parallel to the axis of the cylinder. Then

$$B = m_0 (nI + M) \quad \text{or,} \quad \frac{1}{\mu_0} B - M = nI. \qquad \qquad …( 6.20)$$

This expression relates the free currents per unit length, $nI$ on the surface of the cylinder in terms of the magnetic field $B$ in the medium and the magnetization $M$ of the medium.

Now, as both $B$ and $M$ are vectors in the same direction we can introduce a new vector field called magnetizing field $H$ defined as

$$H = \frac{1}{\mu_0} B - M \qquad \qquad …(6.21)$$

If we consider the cylinder of length L then we can write:

$$HL = LnI = I_{\text{free}}$$

$$\text{or,} \quad \oint_L H \cdot dl = I_{\text{free}} \qquad \qquad …(6.22)$$

Therefore, it may be stated that "The circulation of the magnetizing field $H$ along a closed line is equal to the total free current $I_{\text{free}}$ through the path". From equn. (6.21), we get

$$B = \mu_0(H + M)$$

Now, introducing a physical parameter $\chi_m$ as Magnetic susceptibility of the material by writing: $M = \chi_m H$ we get:

$$B = \mu_0 ( 0 + \chi_m H) = \mu_0(1 + \chi_m) H = \mu H \qquad \qquad …(6.23)$$

where, $\mu \dfrac{B}{H} = \mu_0(1 + \chi_m)$ is known as "permeability" of the medium.

The relative permeability is defined as

$$\mu_r = \frac{\mu}{\mu_0} = (1 + \chi_m).$$ The equn. (6.23) when used in (6.22)

We get

$$\oint_L \frac{1}{\mu}\, \boldsymbol{B} \cdot d\boldsymbol{l} = I_{\text{free}} \quad \text{and} \quad \oint_L \boldsymbol{B} \cdot d\boldsymbol{l} = \mu\, I_{\text{free}}$$

The above result is similar to Ampere's Law given in equn. (6.16) only with the replacement of total current with free current and $\mu_0$ by $\mu$.

## 6.4 MAGNETIC FORCE ON A MOVING CHARGE

When an electric charge moves in a magnetic field it experiences a force which is never observed on a stationary charge and this force is different from that of gravitational or the electric interactions. The force $\boldsymbol{F}$ exerted by a magnetic field is proportional to the charge $q$, intensity of magnetic field $\boldsymbol{B}$ and the velocity of the moving charge v within the magnetic field. This is expressed in mathematical form as: $\boldsymbol{F} = q\, v \times \boldsymbol{B}$ and when the charged particle moves in field where there are both magnetic field and electric field, the force $\boldsymbol{F}$ is expressed as:

$$\boldsymbol{F} = q(\boldsymbol{E} + v \times \boldsymbol{B}) \qquad \qquad \dots(6.24)$$

This expression is known as Lorentz Force.

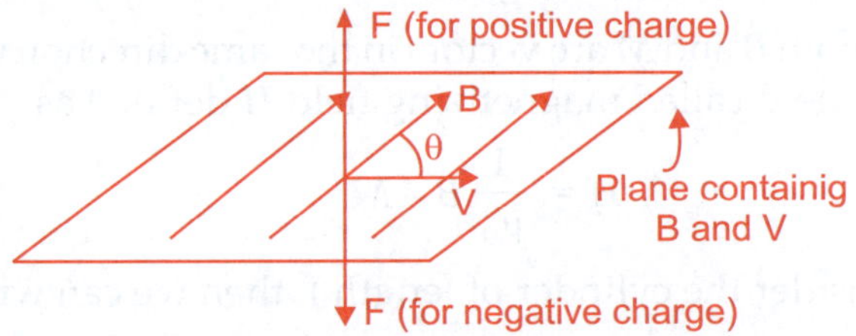


**Fig. 6.11** Vector relation between magnetic force, magnetic intensity and velocity

The scalar form of this magnetic force (Lorentz force) in the absence of electric field is given by: $\boldsymbol{F} = qvB \sin\theta$ as this force is always perpendicular to the direction of $v$ as long as the charge remains within the magnetic field, it acts as a centripetal force and the path of the moving charge will be circular for $\theta = 90°$ and then $F = qvB$, again as this force $\boldsymbol{F}$ is perpendicular to both $\boldsymbol{B}$ and $v$ therefore it is a "No work force" and it cannot increase the kinetic energy of the charge particle.

### 6.4.1 Motion of a Moving Charge in a Magnetic Field

As stated before this magnetic force $F$ will be perpendicular and for angle between $v$ and $\boldsymbol{B}$ as 90° this will have effect on the direction of the moving charge and equating with acceleration $v^2/r$, we get : $\dfrac{mv^2}{r} = qvB$ and the radius of the resulting circular path of the charge particle is given by:

$$r = \frac{mv}{qB} \qquad \qquad \dots (6.25)$$

or, $\qquad \qquad \omega = \dfrac{q}{m}B$

Now, as the acceleration of a uniform circular motion $a = \omega \times v$, therefore writing the equation of motion as $F = ma$ $m\omega \times v = qv \times B$ and now if we reverse the vector product as $B \times v$.

$$\omega \times v = -\left(\frac{q}{m}\right)B \times v$$

Therefore, $\qquad \qquad \omega = -\left(\dfrac{q}{m}\right)B. \qquad \dots (6.26)$

This expression gives the magnitude and the direction of $\omega$. The negative sign indicates that $\omega$ is opposite in direction of $B$ for a positive charge and for a negative charge the negative sign is changed to positive and then $\omega$ is in the same direction as that of $B$. This $\omega$ is known as cyclotron frequency as this expression is used in discussing the motion of charge particle in cyclotron.

This may be concluded that when a charged particle moves in a magnetic field of uniform intensity in a direction same as the magnetic field of induction then there will be no force acting on the charged particle i.e. $\theta = 0$ and the path will be circular if the direction of motion is perpendicular to the magnetic field. Now, if the angle between $v$ and $B$ i.e. $\theta$ is in between $0°$ and $90°$ then $v$ can be resolved parallel to $B$ and perpendicular to $v$, which are respectively as $v_{II}$ and $v_\perp$. The path of moving charge will then be helical of constant pitch as long as the charged particle remains within magnetic field of constant intensity. This movement of charged particle in a magnetic field is demonstrated in the following Fig. 6.12.

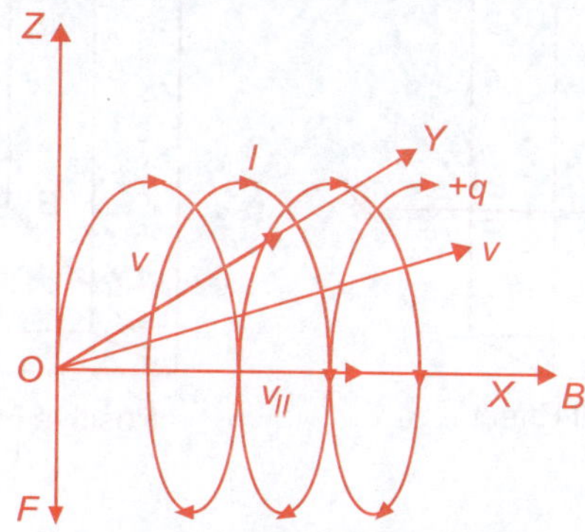


**Fig. 6.12** The helical path of a moving positive charge +q in an uniform magnetic field of intensity **B**

This may be interesting to note that if the magnetic field instead of remaining constant increases continuously then the helical path's pitch will decrease and the cyclotron frequency (6.26) will increase. This may also be mentioned here that the earth being a huge magnet, the lines of force of its magnetic field are perpendicular (normal) on the poles and horizontal at the equator. As a result, the fast charged particles from outer space (Cosmic rays) while entering earth's field near equatorial regions will execute helical path of decreasing pitch as the radius of the helix is to decrease with the increase of earth's magnetic field equn. 6.25 and finally suffer a reflection back (magnetic reflection) and can not reach the earth surface where as the particle approaching earth in the polar region can enter earth's atmosphere and create the most interesting polar incidence of "Aurora" in the night's sky.

### 6.4.2 Hall Effect

One of the most important applications of Lorentz force is the "Hall Effect". When a field is placed at the two ends of a semiconductor crystal, current is constituted in the direction of the field due to majority carrier which are electrons in $n$-type semiconductors and positive charge "holes" in the $p$-type semiconductors. The details of this will however be discussed later in the second part of the book. Now, if this semiconductor crystal is then subjected to a magnetic field in a direction perpendicular to the direction of majority current, a potential difference results in a direction transverse to the direction of original potential difference. This important phenomenon is known as Hall Effect. On the basis of the direction of the transverse potential difference, Hall Effect is classified in to two categories as Positive Hall effect and Negative Hall effect. This has immense importance in semiconductor physics and is used in identifying $n$–type and $p$-type semiconductors. In the following Fig. 6.13 the Hall Effect is explained.

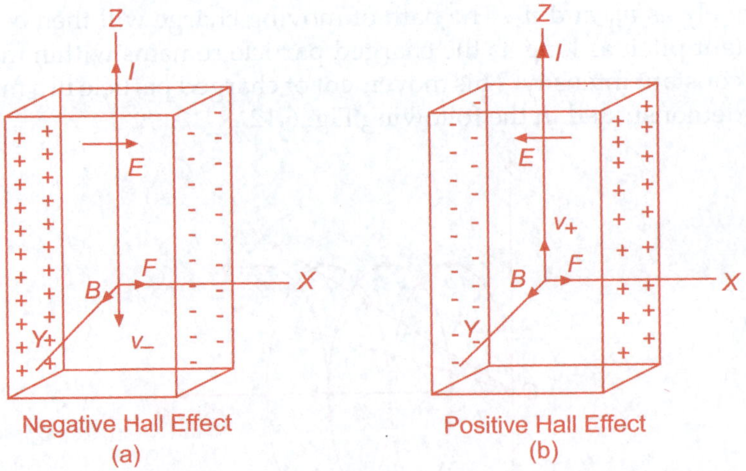


**Fig. 6.13 (a) and (b)** Negative and positive hall effects

In (a) current $I$ is due to the movement of conducting electrons in the direction opposite to $I(v_-)$ and are subjected to Lorentz force $F = -e(v_- \times B)$. Though $(v_- \times B)$ is directed to $-X$ direction but when multiplied with the negative sign due to negative charge of electrons, the $F$ is directed in the $+X$ direction resulting in to accumulation of negative charge on the right side and the Hall field $E$. In (b), the positive Hall effect the current is due to the movement of positively charged "holes" and due to Lorentz force holes have increased concentration on the right side which results in to field $E$ shown which is opposite to that of (a).

Now, this Hall Effect and Hall current differentiate a $n$-type of semi-conductor from $p$-type considering the direction of Hall voltage and also the carrier mobility considering the Hall current.

## 6.5 FARADAY'S LAW

We have seen that a moving charge induces a magnetic field called magnetic field of induction. As all natural phenomena have its mirror counterpart that is if current which is essentially a change in electric field produces a magnetic field, a changing magnetic field through a closed loop produces a current in the loop. The strength of the current is proportional to the rate of change of the magnetic flux through the closed loop and this effect i.e. the resulting current in the loop will always oppose the cause of it i.e. the changing magnetic flux.

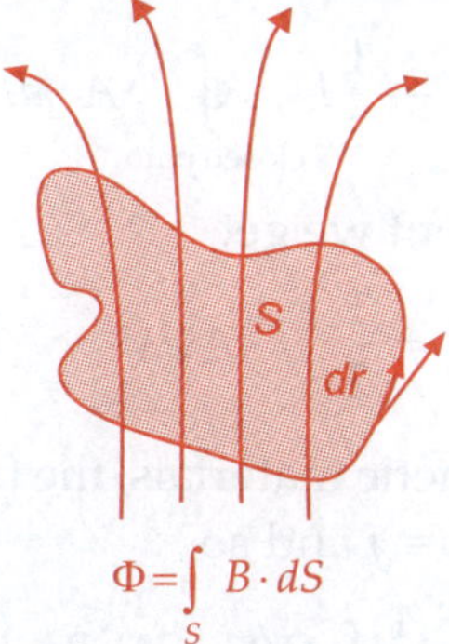


**Fig. 6.14** Magnetic flux $\Phi$ through a closed loop of wire and of area $S$ and faraday's law

Based on this observation the Faraday's Law states that "The resulting e.m.f. around a closed path is minus the time rate of change of the magnetic flux over an arbitrary surface bounded by the path".

Now, if $\Phi$ represents the total magnetic flux through the arbitrary area $S$ of the closed loop and $B$ being the magnetic field then:

$$\Phi = \int_S B \cdot dS \qquad \ldots(6.27)$$

Now, as $B = \nabla \times A$ where $A$ is the magnetic vector potential then using Stoke's law equn. (6.27) can be written as:

$$\Phi = \int_S (\nabla \times A) \cdot dS = \oint_{\text{closed path}} A \cdot dl$$

Now, the electro motive force (e.m.f.) is given by

$$\text{e.m.f} = \oint_{\text{closed path}} E \cdot dr \qquad \ldots(6.28)$$

Equating the equns. (6.27) and (6.28), we get the mathematical statement of Faraday's Law:

$$\oint_{\text{closed path}} E \cdot dr = -\frac{d}{dt} \int_S B \cdot dS \qquad \ldots(6.29)$$

## 6.5.1 Energy in Magnetic Field

The energy stored in a circuit having "self induction" $L$ is equal to the work done against the back e. m. f. in order to maintain the current $I$ is given as: $dU = -(\text{e.m.f.}) \, I dt = L I d\,I$ which turns out as follows for the current changing from zero to I as

$$U = \frac{1}{2} L I^2, \quad \text{Now as} \quad F = LI \text{ and so,}$$

$$LI = \oint_{\text{closed path}} A \cdot dl$$

Therefore, $\quad U = \dfrac{1}{2} I \displaystyle\oint_{\text{closed path}} A \cdot dl = \dfrac{1}{2}\oint I \cdot A dl$ using $Idl = Idl$

For any system of current we get

$$U = \frac{1}{2} \int_V j \cdot A \, dV.$$

In the presence of magnetic materials, the free current density $j$ is given by Ampere's Law as: $\nabla \times H = j$ and so,

$$U_H = \frac{1}{2} \int_V (\nabla \times H) A dV = \frac{1}{2} \int_V H(\nabla \times A) dV - \int_V \nabla \cdot (A \times H) dV.$$

The second term in the right hand side vanishes and so,

$$U_H = \frac{1}{2} \int_V (H \nabla \times A) dV \text{ and as } B = \nabla \times A, \text{ the energy becomes:}$$

$$U_H = \frac{1}{2} \int_V H \cdot B.$$

Therefore, the energy density associated with magnetic field is given by

$$U_H = \frac{1}{2} H \cdot B \qquad \ldots(6.30)$$

## REVIEW QUESTIONS

1. Considering a hydrogen atom consisting of a proton at the centre of the atom and an electron circulating round it in circular orbit and the following relations: radius of the orbit $a$ $\dfrac{\hbar}{me^2}$, the speed of the electron $v = \dfrac{e^2}{\hbar}$ where $e$ the charge, $m$ the mass and $\hbar = \dfrac{h}{2\pi}$. Using the above expressions find the current the circulating charge is equivalent to and the strength of the magnetic field at proton arising out of the motion of the electron.

2. The vector potential $A$ is related to the magnetic field $B$ as $B$ is related to current density $J$ so that curl $A = B$ and curl $B = (4p/c)\,J$. What statement about A corresponds to the statement that the line integral of $B$ around any close path equals $4p/c$ times the current enclosed by the path? Using the relations as mentioned above, find the vector potential associated with the field of an infinitely long solenoid.

3. In electric circuits wires carrying current in opposite directions are often twisted together. What is the advantage for doing this?

4. Since parallel current elements attract each other why not current flowing through a long straight conductor concentrates at the central axis of the cylinder instead of getting distributed evenly through the cross section?

5. If two electrons in a cathode ray tube move parallel to each other with same speed as $v$ and the distance between them is $r$, what is the force that acts on each of them, due to the presence of the other?

## REFERENCES


1. M. Alonso and E. J. Finn - Fundamental University Physics (Vol. II), Addison-Wesley Pub. Co. Reading, Mass, 1968.

2. R. A. Serway and J. W. Jewett – Principle of Physics, Thomson, Singapore.

3. S. B. Palmer and M. S. Rogalski – Advanced University Physics, Gordon Breach Pub. 1996.

4. E. M. Purcell – Berkeley Physics Course, McGraw-Hill Co. New York.

5. J. H. Fewkes and J. Yarwood – Electricity, Magnetism and Atomic Physics (Vol. I), University Tutorial Press Ltd., London.

# Electrodynamics

## 7.1 INTRODUCTION TO MAXWELL'S EQUATIONS

The theory of Electrodynamics is unique and novel in the sense that this elegant theory and its superstructures are based on some physical laws and some relations observed and verified experimentally. These physical laws and the relations which form the foundation of electrodynamics are known as – Maxwell's equations. These equations are derived from the domains of electrostatics and magnetostatics i.e. Coulomb's law, Gauss's Law and magnetic induction field and establishes the elegant fact that they are inter-related phenomena.

An important kind of interaction among fundamental particles composing matter is the one called electromagnetic interaction. It is related with the characteristic property of these fundamental particles as electric charge. In order to describe these interactions the electromagnetic fields noted by two vectors representing electric field $E$ and magnetic field $B$ are involved and this involvement is expressed in the force applied on a moving charge $q$ introduced before as Lorentz force and given as

$$F = q(E + v \times B)$$

The electric $E$ and magnetic $B$ fields are determined by the position of the charge and its motion. The separation of the electric and magnetic components depends on the relative motion of the charge and the observer.

## 7.2  MAXWELL'S EQUATIONS

Let us recall the Gauss's law in electrostatics in differential form from Chapter 5 in the section 5.2.1 as

$$\text{div } E = \frac{\rho}{\varepsilon_0} \qquad \text{or,} \qquad \nabla \cdot E = \frac{\rho}{\varepsilon_0} \qquad \qquad \dots (7.1)$$

This equation is noted as "First Maxwell's Equation".

As the concept of monopole is a remote concept, therefore unlike electrostatic field we must adopt the view that every magnetic field line that enters a closed surface must also exit the surface and mathematically this fact has already been introduced in Chapter 6, Ampere's law (equn. 6.15) as

$$\text{div } \boldsymbol{B} = \nabla \cdot \boldsymbol{B} = 0. \qquad \qquad ...(7.2)$$

This equation is noted as "Second Maxwell's Equation".

Now, if we recall the Faraday's Law given in Chapter 6 (equn. 6.29) which states $\displaystyle\oint_{closed\,path} \boldsymbol{E} \cdot dr = -\frac{d}{dct} \int_S \boldsymbol{B} \cdot dS.$ Now dividing both sides by surface area $S$ and using the limit $S \to 0$ we get:

$$\text{curl } \boldsymbol{E} = \lim_{S \to 0} \left\{ \frac{1}{S} \oint \boldsymbol{E} \cdot dr \right\}$$

$$\boldsymbol{B} = \lim_{S \to 0} \left\{ \frac{1}{S} \int_S \boldsymbol{B} \cdot ds \right\} \text{ and therefore we can write:}$$

$$\text{curl } \boldsymbol{E} = \nabla \times \boldsymbol{E} = -\frac{\partial}{\partial ct} \boldsymbol{B} \qquad \qquad ...(7.3)$$

This is noted as "Third Maxwell's Equation".

Now, from Ampere's Law, in Chapter 6 the magnetic induction field $\boldsymbol{B}$ is related to the current density $j$ (equn. 6.16 ) as:

$$\oint_L \boldsymbol{B} \cdot d\boldsymbol{L} = \mu_0 \int_S j \cdot dS \text{ and also,}$$

$$\nabla \times \boldsymbol{B} = \mu_0 j. \qquad \qquad ...(7.4)$$

Now, this Ampere's Law does not contain any time dependent flux of the electric field as it has been derived under static condition and therefore, this Ampere's Law needs revising when it is applied to time dependent fields.

Now, for a closed surface we know that

$$I = \int_S j \cdot dS \text{ and as } I = -\frac{dq}{dt}, \text{ therefore, } -\frac{dq}{dt} = \oint_S j \cdot dS$$

From Gauss's Law , the total charge within a closed surface is expressed in terms of electric field as:

$$q = \varepsilon_0 \oint_S \boldsymbol{E} \cdot dS \text{ and so, } \frac{dq}{dt} = \varepsilon_0 \frac{d}{dt} \oint_S \boldsymbol{E} \cdot dS \text{ and substituting this in above}$$

equation we get

$$\oint_S j \cdot dS + \varepsilon_0 \frac{d}{dt}\oint_S E \cdot dS = 0.$$ This is the principle of conservation of charge and incorporates the Gauss's Law. For static field however, $\oint_S j \cdot dS = 0$ and for time dependent field the Ampere's Law expressed in Chapter 6 and recalled here in equn. (7.4) is to be written as: $\oint_L B \cdot dL$

$$= \mu_0 \oint_S j \cdot dS + \mu_0 \varepsilon_0 \frac{d}{dt}\oint_S E \cdot dS$$ this can be written in the vector form as

$$\text{curl } B = \nabla \times B = \mu_0\left[ j + \mu_0 \frac{d}{dt} E \right] \qquad \qquad ...(7.5)$$

This is "Fourth Maxwell's Equation".

Now, these Maxwell's equations for Electromagnetic field can be made more explicit in the following Table 7.1.

Table 7.1 Maxwell's equations for electromagnetic field

| | Law | Integral form | Differential form |
|---|---|---|---|
| 1. | Gauss's Law for Electric Field | $\oint_S E \cdot dS = \dfrac{q}{\varepsilon_0}$ | $\text{div } E = \dfrac{\rho}{\varepsilon_0}$ |
| 2. | Gauss's Law for Magnetic Field | $\oint B \cdot dS = 0$ | $\text{div } B = 0$ |
| 3. | Faraday's Law | $\oint_L E \cdot dl = -\dfrac{d}{dct}\oint_S B \cdot dS$ | $\text{curl } E = -\dfrac{\partial}{\partial ct} B$ |
| 4. | Ampere-Maxwell Law | $\oint_L B \cdot dl = \mu_0 I + \varepsilon_0\mu_0 \dfrac{d}{dt}\int E \cdot dS$ | $\text{curl } B = \mu_0 j + \mu_{00}\, \varepsilon \dfrac{\partial}{\partial t} E$ |

## 7.3  GENERAL PROPERTIES AND APPLICATIONS OF MAXWELL'S EQUATIONS

The synthesis of electromagnetic interactions as expressed by Maxwell's equations is one of the greatest achievements in Physics. They are the equations which mathematically enable the understanding of the electromagnetic interactions.

Now to further establish the interdependence of electric and magnetic fields, let us recall the following equations of Faraday's law, discussed in Chapter 6 and also given above as

$$\Phi = \int_S \boldsymbol{B} \cdot d\boldsymbol{S} \quad \text{and} \quad \oint_{\text{closed path}} \boldsymbol{E} \cdot d\boldsymbol{r} = -\frac{d}{dct} \int_S \boldsymbol{B} \cdot d\boldsymbol{S}$$

Where as stated $\Phi$ is the total magnetic flux and $\boldsymbol{B}$ magnetic induction over the closed surface $S$ Since Faraday's law deals empirically with the electric field induced changing magnetic flux, taking into mind that the time derivative must include two terms one intrinsic time variance term and a convective term due to relative motion of the observer.

Therefore,
$$\frac{d}{dt} = \frac{\partial}{\partial t} + \boldsymbol{v} \cdot \nabla$$

In the laboratory frame, the Faraday's equations, we can write

$$-\frac{1}{c}\frac{\partial \Phi}{\partial t} = -\frac{1}{c}\left(\frac{\partial}{\partial t} + \boldsymbol{v} \cdot \nabla\right) \int_S \boldsymbol{B} \cdot \boldsymbol{n} dS = \oint_{\text{closed path}} \boldsymbol{E} \cdot d\boldsymbol{r} \qquad \ldots(7.6)$$

In a frame of reference moving with respect to the laboratory frame, the Faraday's law can be correctly written as

$$\oint_r \boldsymbol{E}' \cdot d\boldsymbol{r}' = -\frac{1}{c}\int_S \left(\frac{\partial}{\partial t} + \boldsymbol{v} \cdot \nabla\right) \boldsymbol{B} \cdot \boldsymbol{n} dS .$$

This velocity of the frame of reference with respect to observer is not very high so that relativistic transformation is unnecessary here and we can assume the spatial derivative of $\boldsymbol{B}$ is zero.

We know from vector relation: $(\boldsymbol{v} \cdot \nabla) \boldsymbol{B} = \nabla \times (\boldsymbol{B} \times \nabla) + \boldsymbol{v}(\nabla \cdot \boldsymbol{B}) - \boldsymbol{B}(\nabla \cdot \boldsymbol{v}) + (\boldsymbol{B} \cdot \nabla) \cdot \boldsymbol{v}$ and writing $\nabla \cdot \boldsymbol{B} = 0$, we get: $(\boldsymbol{v} \cdot \nabla) \boldsymbol{B} = \nabla \times (\boldsymbol{B} \times \boldsymbol{v})$ and therefore,

$$\oint_r \boldsymbol{E}' \cdot d\boldsymbol{r}' = -\frac{1}{c}\oint_S \frac{\partial}{\partial t} \boldsymbol{B} \cdot \boldsymbol{n} \, dS + \frac{1}{c}\oint_r \boldsymbol{v} \times \boldsymbol{B} \cdot d\boldsymbol{r} \qquad \ldots(7.7)$$

Combining equns. (7.6) and (7.7) we can conclude:
$$\boldsymbol{E}' = \boldsymbol{E} + (\boldsymbol{v} \times \boldsymbol{B})/c$$

This interprets the velocity dependent term as the transformed change in $\boldsymbol{E}$.

The Maxwell's equations state that the time derivatives of the fields are sources of their counterpart that is whenever one of either $\boldsymbol{E}$ or $\boldsymbol{B}$ changes with time the counterpart field results, the theoretical basis of electromagnetic induction. One field transforms into other due to their relative motion. It has been established from experimental evidence that when a conductor moves in a magnetic field, an electromotive force is induced in the conductor and induced electric current flows through it. The direction of this current is determined by Fleming's right hand rule.

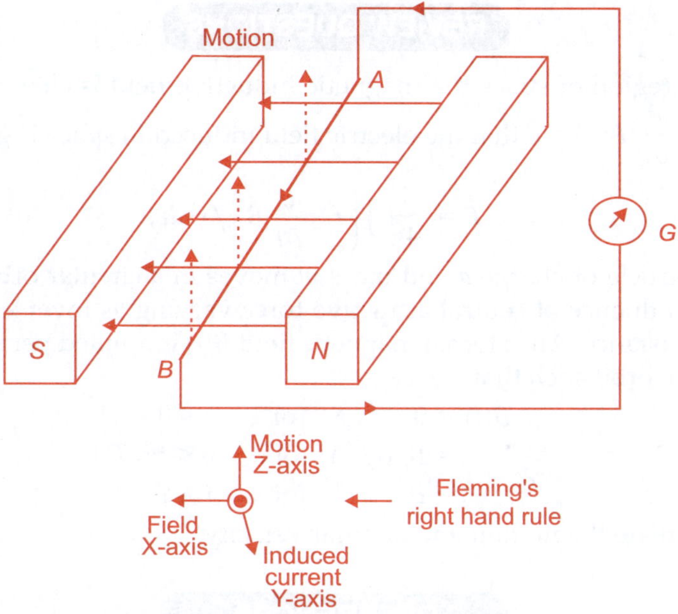



**Fig. 7.1** Electromagnetic induction of current in a varying magnetic field

The following figure demonstrates such an experimental evidence. Due to motion of the conductor the magnetic flux linked with it changes and as a result an electric field is induced. Conversely if electric field is changed within a conductor, a magnetic field is induced within another conductor placed near it. Therefore it shows an interrelation between these two fields established by Maxwell's equations.

We also should know that the Maxwell's equations as they are derived have some limitations. They though explain satisfactorily the electromagnetic interactions between large aggregate of charges such as beams of ionized atoms or molecules, but they fail to explain the interactions between fundamental particles and their interactions are to be treated in different way known as quantum electrodynamics.

## REVIEW QUESTIONS

1.  In a region of space the magnetic induction field is changing at the rate $\dfrac{\partial}{\partial t} \boldsymbol{B}$, show that the electric field induced in space is given by

$$E = \frac{1}{4\pi} \int \left( \boldsymbol{r} \times \frac{\partial}{\partial t} \boldsymbol{B} \right) / r^3 \, dv.$$

2.  A particle of charge $q$ and mass $m$ moves in a circular orbit $r$ under the influence of central attractive force varying as inverse square of the distance. An external magnetic field $\boldsymbol{B}(t)$ is applied perpendicular to the orbit such that

$$
\begin{aligned}
\boldsymbol{B}(t) &= 0 & \text{for} \quad & t < 0 \\
&= \boldsymbol{B}_0 \, (t/T) & \text{for} \quad & 0 < t < T \\
&= \boldsymbol{B}_0 & \text{for} \quad & t > T
\end{aligned}
$$

Calculate the motion and angular velocity.

## REFERENCES


1.  W. K. H. Panofsky and M. Phillips – Classical Electricity and Magnetism, Addison Wesley Pub. Co. Reading, Mass, 1962.

2.  F. Melia – Electrodynamics, The University of Chicago Press, Chicago, 2001.

# Elastic Waves

## 8.1 INTRODUCTION

When a bell is struck, it starts ringing and then it can be heard at a distance. At this instance we tell that, on being struck the surface of the bell is set to vibrate which is a forward and backward movement of the surface and as a result of the surface compresses and rarefies the air with which it is in contact. This compression and rarefaction of the air is called sound and it moves in all possible direction. Moreover, this alternate compression and rarefaction i.e. movement of the constituent of the "medium" is generally periodic at least in an ideal condition and is known as wave. The same is true when a pebble is thrown in to the still water surface of a pond where we observe the crest and trough move on the surface of water. These air and water and other solid material mediums having mass and elastic property act here as elastic medium and the wave created in such mediums are termed as elastic wave. Depending on the elastic parameter of such mediums solid, liquid or gas(air), the mediums set in to convenient mode of vibration i.e. either the vibration of the sections of the mediums are transverse or longitudinal with respect to the direction of the propagation of the wave concerned. When a solid rode is struck on its cross section in the direction of its length a longitudinal wave propagates along its length, when a string is plucked, the transverse wave is created along the string, when the still water surface of the pond is disturbed the surface of water is set to vibrate in the transverse direction and the air or gases are disturbed so that a longitudinal wave propagates. All these natures of the waves depend on the elastic parameter which becomes responsible to counter the distortions created in the medium.

So, several types of waves are possible and the difference of one from other depends on the nature of motion of the particles. For a simple sinusoidal wave, when the displacement of all the particles at a time is considered it gives space-displacement curve and when the displacement of any particular particle with time is considered it shows time displacement curve. For sinusoidal wave both of these two curves are similar and are sine wave.

### 8.1.1 Wave Equation and Mathematical Description of its Propagation

Let us consider a function $\xi$ given by $\xi = f(x)$ where $f$ determines the manner in which $\xi$ depends on $x$. Now, if the nature of dependence with $x$ remains same and if $\xi$ is given as: $\xi = f(x - a)$ then to maintain the left side same $x$ is actually to be $x + a$. In the following (Fig. 8.1) the curve (1) represents $\xi = f(x)$ and the curve (2) then will represent $\xi = f(x - a)$ after a translation of $x$ by an amount a and this can be said that the function $\xi$ has travelled a distance a along $x$ so that it can represent now the curve (2) Similarly, if $\xi = f(x + a)$ then $x$ is changed to $x - a$, to represent the curve (3) and the function then has to travel in the opposite (left) side by a distance $a$.

Now, this travel of the curve $\xi = f(x)$ towards right or left can be better explained if we consider $a = vt$, a the travel length or distance, $v$ the velocity of the travel and $t$ is the time. Then, the function $\xi$ can be written as

$$\xi(x, t) = f(x \pm vt) \qquad \qquad \ldots(8.1)$$

and now we can conclude that the above expression equn. (8.1) is adequate to explain the situation of travel or propagation without distortion of the function in both positive and negative direction. This is what is called "Wave Motion". The quantity $\xi(x, t)$ then may represent widely different physical quantities like deformation in solid, pressure in a gas and electric and magnetic field etc.

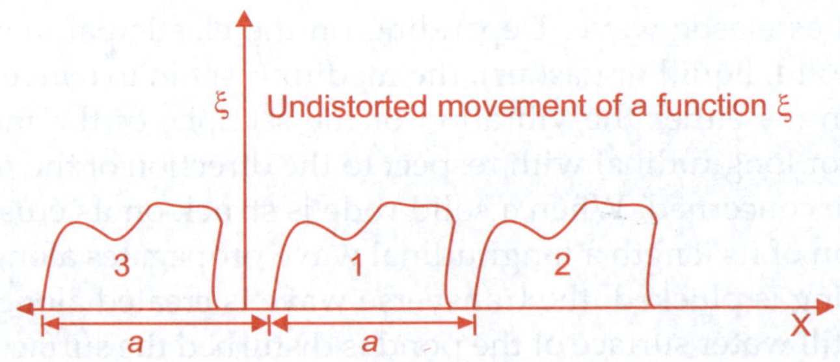


**Fig. 8.1**  Undistorted propagation of a wave

Now, if $f$ is defined as a sinusoidal relation then the equn. (8.1) can be written as:

$$\xi(x, t) = \xi_0 \sin k(x - vt) \qquad \qquad \ldots(8.2)$$

The quantity $k$ in the above equation has a special meaning. It does not alter the function $\xi(x, t)$ when $x$ is changed to $x + 2p/k$ as

$$\xi\left(x+\frac{2\pi}{k},t\right)=\xi_0\sin k\left(x+\frac{2\pi}{k}-vt\right)=\xi_0\sin\left[k(x-vt)+2\pi\right]$$

$$=\xi_0\sin k(x-vt)\quad=\xi(x,t)\qquad\qquad\ldots(8.3)$$

Then $\lambda=\dfrac{2\pi}{k}$ where $l$ is the "space period" i.e. the distance after which the curve repeats itself and is known as "Wave-Length". The following Fig. 8.4 shows a sine wave.

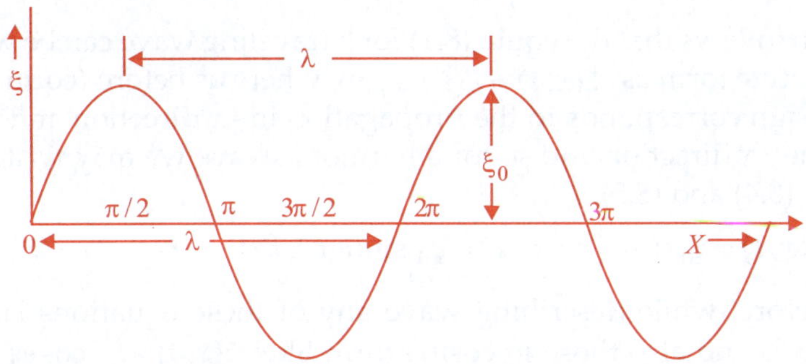


**Fig. 8.2** Harmonic wave showing wavelength and amplitude

When $\lambda$ is introduced as wavelength, $k = 2p/\lambda$ represents the number of wavelengths in the distance $2p$ and is called "Wave number". Now recalling equn. (8.3) we get

$$\xi(x,t)=\xi_0\sin k(x-vt)=\xi_0\sin\frac{2\pi}{\lambda}(x-vt)=\xi_0\sin(kx-\omega t)\qquad\ldots(8.4)$$

where $\omega=kv=\dfrac{2\pi v}{\lambda}$ and so $\omega$ gives the angular frequency of the wave.

Now, as velocity of the propagation of wave $v=\lambda v$, where $n$ is the frequency. Introducing the period of oscillation $T$ as $T=2\pi/\omega$, then from equn. (8.4)

$$\xi=\xi_0\sin 2\pi\left(\frac{x}{\lambda}-\frac{t}{T}\right)\qquad\ldots(8.5\text{ a})$$

$$\xi=\xi_0\sin 2\pi\left(\frac{x}{\lambda}+\frac{t}{T}\right)\qquad\ldots(8.5\text{ b})$$

While equn. (8.5 a) represents a harmonic sinusoidal wave moving in the $+x$ direction, equn. (8.5 b) represents that moving in $-x$ direction.

Now, recalling again equn. (8.4) as $\xi=\xi_0\sin k(x-vt)$ and taking partial derivatives with respect to space and time we get

$$\frac{\partial \xi}{\partial x} = k\xi_0 \cos k(x - vt), \qquad \frac{\partial^2 \xi}{\partial x^2} = -k^2 \xi_0 \sin k(x - vt);$$

$$\frac{\partial \xi}{\partial t} = -kv\xi_0 \cos k(x - vt), \qquad \frac{\partial^2 \xi}{\partial t^2} = -k^2 v^2 \xi_0 \sin k(x - vt) \quad \ldots(8.6)$$

Therefore, from the above equation, we can write the differential equation of the motion of the harmonic wave as:

$$\frac{\partial^2 \xi}{\partial t^2} = v^2 \frac{\partial^2 \xi}{\partial x^2} \qquad \ldots(8.7)$$

It also follows that the equn. (8.1) for a traveling wave can be written in an alternative form as: $\xi(x, t) = F(t \pm x / v)$, where as before (equn. 8.1), the positive sign corresponds to the propagation in $-x$ direction and negative sign in the $+x$ direction and so for a harmonic wave we may write instead of equns. (8.4) and (8.5),

$$\xi(x, t) = \xi_0 \sin \omega(t \pm x / v) = \xi_0 \sin(\omega t \pm kx) \qquad \ldots (8.8)$$

Therefore, while describing wave any of these equations i.e. equns. (8.4) or (8.8) and also those in cosine form like: $\xi(x, t) = \xi_0 \cos \omega(t \pm x / v) = \xi_0 \cos(\omega t \pm kx)$ can be used.

## 8.2 ELASTIC WAVE IN DIFFERENT MEDIA

The elastic waves in different media solid, liquid or gases are dependent on the material elastic constants, density or mass etc., and so these waves are called elastic waves.

We will discuss the propagation of such waves in solid, surface wave on liquid and the waves through a gas.

### 8.2.1 Elastic Wave in Solid Rod

When we hit one end of a rod say by a hammer the disturbance elastic in nature propagates along the length and reaches the other end.

Let us consider a uniform rod of cross section A subjected to a stress along the axis resulted by the force $F$. The force $F$ is not necessarily same in all sections and may vary along the length of the rod. However each section will be subjected to two equal and opposite forces as $F$ and $F'$. The force $F$ is the pull by the left part of the rod on the right part and $F'$ is the force on the left part by the right part.

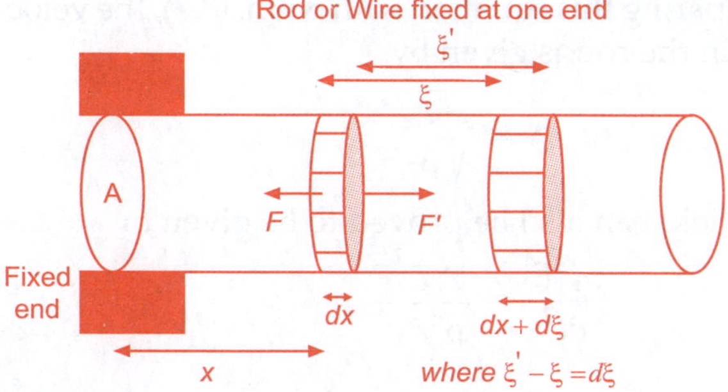



**Fig. 8.3** Propagation of elastic wave through a rod fixed (equilibrium) at one end

The normal stress $\sigma$ at a section of the rod is defined as force per unit area and is given by: $\sigma = \dfrac{F}{A}$ and the strain $\varepsilon$ which is the deformation $\chi$ along the axis per unit length is given by $\varepsilon = \dfrac{\partial \xi}{\partial x}$.

Within the elastic limit for this longitudinal deformation we know: $\sigma = Y\varepsilon$ and using the above relations we get

$$F = YA\frac{\partial \xi}{\partial x}$$

Therefore, $\dfrac{\partial F}{\partial x} = YA\dfrac{\partial^2 \xi}{\partial x^2}$

From this expression $\displaystyle\int_0^{\xi} d\xi = \frac{F}{YA}\int_0^{x} dx$ and $\xi = \dfrac{F}{YA}x$.

Now, using the dynamical relation, the force $F$ in an element of length $dx$ is equal to mass of the element multiplied with acceleration and so where $\acute{A}$ is defined as density

$$\frac{\partial F}{\partial x}dx = (\rho A\, dx)\frac{\partial^2 \xi}{\partial t^2}$$

$$\frac{\partial F}{\partial x} = \rho A\frac{\partial^2 \xi}{\partial t^2}$$

or, ... (8.9)

Comparing this equation with equn. (8.8) and equating the right hand side we get

$$\frac{\partial^2 \xi}{\partial t^2} = \frac{Y}{\rho}\frac{\partial^2 \xi}{\partial x^2} \qquad \text{... (8.10)}$$

Now, comparing this equation with equn. (8.7), the velocity of the elastic wave motion in the rod is given by

$$v = \sqrt{\frac{Y}{\rho}} \qquad \qquad \text{... (8.11)}$$

The force field can also be proved to be given by

$$\frac{\partial^2 F}{\partial t^2} = \frac{Y}{\rho}\frac{\partial^2 F}{\partial x^2}.$$

### 8.2.2 Elastic Transverse Wave in a String

When a string under tension is plucked, bowed or struck, it starts vibrating and an elastic wave transverse in nature and of different shapes depending on how the wire is set under vibration. Irrespective of the mode of setting the wire in vibration each section of the wire is set to vibrate perpendicular to the direction of wave propagation. In the following Fig. 8.4 which shows the forces acting on a small section of the wire $AB$.

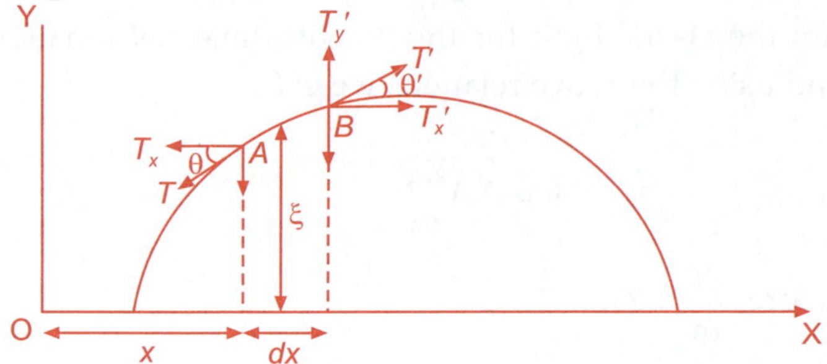


**Fig. 8.4** A section of the deformed wire of general shape AB and the forces on it

The tension $T$ will remain uniform along the length of the wire and is directed tangential to every section of it. The vertical components of each of this at the two ends of the section $AB$ are respectively $T\sin\theta$ and $T\sin\theta'$. Now, as the angles $\theta$ and $\theta'$ are different because of the shape of the deformed wire, the net upward force acting on this segment is:

$F_Y = T\left(\sin\theta' - \sin\theta\right)$ and as the curvature of the wire segment is not large the expression can be approximated as

$$F_Y = T\left(\sin\theta' - \sin\theta\right) = Td(\tan\theta) = T\frac{\partial}{\partial x}(\tan\theta)\,dx$$

The partial derivative is used as $\tan\theta$ depends both on time and position $x$. As $\tan\theta$ is the slope of the curvature of the string, therefore $\tan\theta = \dfrac{\partial \xi}{\partial x}$ and the above expression can be equated to

$$F_Y = T\frac{\partial}{\partial x}\left(\frac{\partial \xi}{\partial x}\right)dx = T\left(\frac{\partial^2 \xi}{\partial x^2}\right)dx.$$

This force must then be equated to mass of the segment multiplied with the acceleration and if $m$ be the mass per unit length of the wire, then

$$(m\,dx)\frac{\partial^2 \xi}{\partial t^2} = T\frac{\partial^2 \xi}{\partial x^2}dx \quad \text{or} \quad \frac{\partial^2 \xi}{\partial t^2} = \frac{T}{m}\frac{\partial^2 \xi}{\partial x^2} \qquad \text{...(8.12)}$$

This is the wave equation for the transverse elastic wave propagating in the stretched wire. Comparing this equation with equn.(8.7), the velocity of this elastic transverse wave in the wire is given by

$$v = \sqrt{\frac{T}{m}}\,. \qquad \text{...(8.13)}$$

### 8.2.3  Surface Wave in a Liquid

Let us consider a liquid section $ABCD$ in a liquid at rest in a large reservoir. The surface wave of large wave length is created on the surface (Fig. 8.5). Considering the liquid as incompressible, the volumes of the open section and shaded section are equal and so

$$Lhdx = L(h+h_1)\,(dx+d\xi) = L(h\,dx + h\,d\xi + h_1\,dx + h_1\,d\xi)$$

where, $h$ is the depth of the liquid in the undisturbed region at a distance $x$, $L$ is the breadth of the liquid section and $h_1$ is the difference of height of the shaded section.

Now, as both $h_1$ is very small compared with $h$ and $d\,\xi$ with $dx$ we can equate after canceling the last term $h_1\,d\,\xi$ which is the product of two small quantities we get: $h_1\,dx + h\,d\xi = 0$ and then $h_1 = -h\dfrac{\partial \xi}{\partial x}$. This relates the vertical surface displacement to the horizontal displacement of the incompressible liquid.

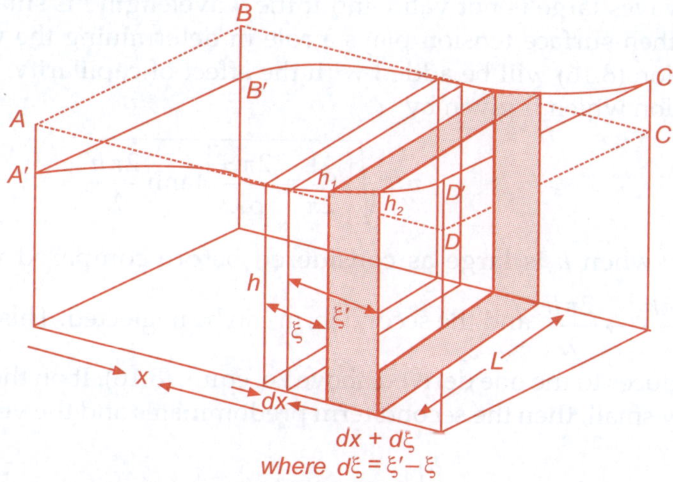


**Fig. 8.5** A section of liquid in a large liquid reservoir. The surface wave of large wave length created on the surface is shown by bold out line

Now, if $p_{av}$ is the average pressure on the shaded surface from left and $p'_{av.}$ that from right and a is the area of cross section of the section then force to the right from left is given as: $-(p'_{av} - p_{av}) = -a\,dp_{av}$ and if $\rho$ is the density of the liquid, then equating this force from dynamical relation

$$(\rho\,a\,dx)\frac{\partial^2 \xi}{\partial t^2} = -a\,dp_{av.} \quad \text{or} \quad \rho\frac{\partial^2 \xi}{\partial t^2} = -\frac{\partial p_{av.}}{\partial x} \qquad \text{...(8.14)}$$

and as $p_{av.} = \rho\,gh$ so, $dp_{av.} = \rho g\,(h_2 - h_1) = \rho g\frac{\partial h_1}{\partial x}\,dx.$

The partial derivative is taken as $h$ the depth of the disturbed surface is dependent both on position $x$ and time $t$. Therefore, $\dfrac{\partial p_{av.}}{\partial x} = \rho g\dfrac{\partial h_1}{\partial x}$

Then comparing with equn. (8.14) we get:

$\dfrac{\partial^2 \xi}{\partial t^2} = -g\dfrac{\partial h_1}{\partial x}$ Now, differentiating the expression $h_1 = -h\dfrac{\partial \xi}{\partial x}$ we get

$\dfrac{\partial h_1}{\partial x} = -h\dfrac{\partial^2 \xi}{\partial x^2}$ and eliminating $\dfrac{\partial h_1}{\partial x}$ from above two expressions

$$\frac{\partial^2 \xi}{\partial t^2} = gh\frac{\partial^2 \xi}{\partial t^2}. \qquad \text{... (8.15)}$$

Comparing this equn. (8.15) with equn. (8.7), we get the velocity of surface wave on liquid is given by

$$v = \sqrt{gh}. \qquad \text{...(8.16)}$$

Now, if the assumption that has been made here i.e. the wavelength of the waves large is not valid and if the wavelength $l$ is small compared with $h$ then surface tension plays a role in determining the velocity. The expression (8.16) will be added with the effect of capillarity. The velocity expression which is given by

$$v = \sqrt{\left[\frac{g\lambda}{2\pi} + \frac{2\pi S}{\rho\lambda}\right]\tanh\frac{2\pi h}{\lambda}}. \qquad \text{...(8.17)}$$

But, when $h$ is large as considered before compared with $\lambda$ then,

$\tan h\dfrac{2\pi h}{\lambda} \rightarrow \dfrac{2\pi h}{\lambda}$ and the second term can be neglected. This equn. (8.17),

then reduces to the one derived above i.e. equn. (8.16). If on the other hand, $\lambda$ is very small, then the second term predominates and the velocity is then given by

$v = \sqrt{\dfrac{2\pi S}{\rho\lambda}}$ and the wave then is called capillary wave, otherwise for large $l$ the equn. (8.16) is termed as Gravity wave.

## 8.2.4 Pressure Wave in Gases

We will now consider the elastic wave in a gas which results in pressure variation. Unlike solid or liquid the gases are compressible and so with variation of pressure the density of the gas changes. Consider a gas enclosed in a cylindrical pipe as given in the following Fig. 8.6.

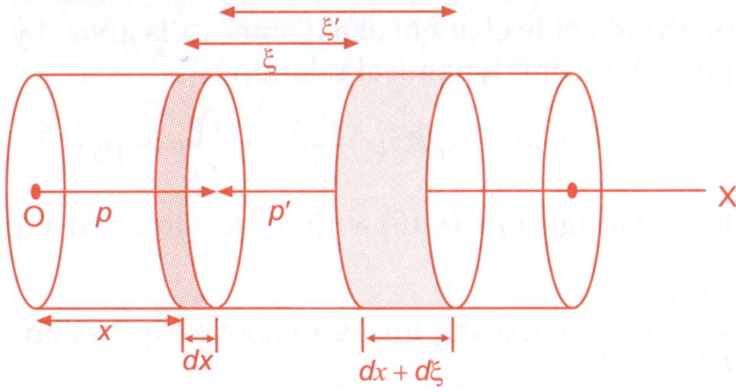


**Fig. 8.6** Compression wave in a gas column

Let $p$ and $p'$ be two pressures on two sides of the left side element and as $p$ is greater than $p'$, the element's left side moves right through a distance of $\xi$ and right side of the element through $\xi'$, so that thickness of the element changes after deformation to $dx + d\xi$.

If the density in the *un* deformed state be $\rho_0$ and that in the deformed state be $\rho$, then as the mass is conserved: $\rho A(dx + d\xi) = \rho_0 A dx$, where $A$ is the area of cross section of the elements

or, $\qquad \rho\left(1 + \dfrac{\partial \xi}{\partial x}\right) = \rho_0$ and solving $\rho = \dfrac{\rho_0}{1 + \dfrac{\partial \xi}{\partial x}}$

Now, as $\dfrac{\partial \xi}{\partial x}$ is small we can approximate $\left(1 + \dfrac{\partial \xi}{\partial x}\right)^{-1} = 1 - \dfrac{\partial \xi}{\partial x}$ and can write

$$\rho = \rho_0\left(1 - \dfrac{\partial \xi}{\partial x}\right) \qquad\qquad \text{...(8.18)}$$

Now as pressure $p = f(\rho)$, we can apply Taylor's expansion and retain only first two terms considering small change of density as

$$p = p_0 + (\rho - \rho_0)\left(\frac{dp}{d\rho}\right)_0$$

The bulk modulus is defined as: $\kappa = \rho_0 \left(\dfrac{dp}{d\rho}\right)_0$ then we can write the above expression as

$$p = p_0 + \kappa \left(\frac{\rho - \rho_0}{\rho_0}\right), \text{ using the equn. (8.18)}$$

$$p = p_0 - \kappa \frac{\partial \xi}{\partial x} \qquad \qquad \dots(8.19)$$

As force on the left side element at a distance $x$ is given by $- A\,dp$ where $dp = p' - p$ is given by from dynamical relation

$$-A\,dp = (\rho_0\,A\,dx)\frac{\partial^2 \xi}{\partial t^2} \quad or \quad \frac{\partial p}{\partial x} = -\rho_0 \frac{\partial^2 \xi}{\partial t^2}$$

Now, differentiating equn. (8.19) with respect to $x$ and as $\rho_0$ is constant, we get:

$$\frac{\partial p}{\partial x} = -\kappa \frac{\partial^2 \xi}{\partial x^2}$$ and comparing this with above expression we get

$$\frac{\partial^2 \xi}{\partial t^2} = \frac{\kappa}{\rho_0}\frac{\partial^2 \xi}{\partial x^2}. \qquad \qquad \dots(8.20)$$

Comparing with equn. (8.7) we get velocity with which disturbance due to pressure difference propagates:

$$v = \sqrt{\frac{\kappa}{\rho_0}} \qquad \qquad \dots(8.21)$$

For sound wave which makes adiabatic disturbance as its velocity is high, it can be shown that it is given as: $v = \sqrt{\gamma p / \rho}$.

### 8.2.5 Sound Wave and Doppler's Effect

The sound wave which is also an elastic wave propagates through gases by creating longitudinal wave i.e. alternate compressions and rarefaction. This is possible because a gas is a fluid having bulk modulus. The introductory discussion of sound wave generated by a vibrating object is however not discussed here. We will discuss however the change of frequency of the emitted and transmitted wave when there is a relative motion between the source and the receiver. This change of frequency of the wave is not only applicable for elastic wave like sound but also for electromagnetic wave like light. Below is presented a brief discussion on this property.

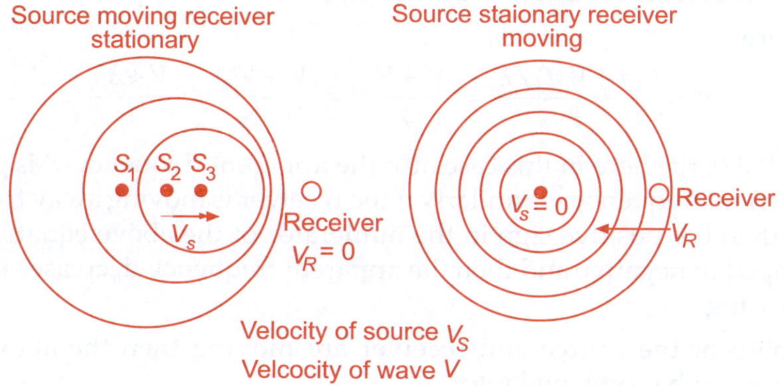


**Fig. 8.7** The doppler's effect in two cases source moving and receiver moving

## Case I  Source moving but Receiver is stationary

Let first the source is moving with velocity $V_S$ but receiver is stationary i.e. $V_R = 0$ and let $S_1$, $S_2$, $S_3$ …. are the subsequent positions of the source. The spherical waves are also continuously shifting towards the receiver. The distance between the spherical surfaces are the wavelength of the wave. If the velocity of the wave is $V$ and when the source and receiver both are stationary, the frequency n is given by

$$\nu = \frac{V}{\lambda}.$$

But, when the source is moving, the changed frequency due to change of wavelength is given by

$$\nu' = \frac{V}{\lambda'} = \frac{V}{VT - V_S T} = \frac{V}{V/\nu - V_S/\nu} = \nu \frac{V}{V - V_S} \qquad …(8.22)$$

where $T$ is the time period and is given by $T = \dfrac{1}{\nu}$.

As $V_S < V$, the apparent frequency $\nu' > \nu$, the original frequency. The negative sign in the denominator will be changed to positive, if the source with velocity $V_S$ is moving away from the stationary receiver and in that case the $\nu' < \nu$.

## Case II Source stationary but receiver is moving

Now, in this case if source is stationary but receiver is moving towards the source with velocity $V_R$, then the receiver will be encountering the waves more frequently than when it is stationary. In time $t$, the number of waves intercepted by the receiver with velocity $V_R$ will then be $(Vt + V_R t)/l$ and the rate of such interception is given by the apparent frequency. Therefore,

$$v' = \frac{(Vt + V_R t)/\lambda}{t} = \frac{V + V_R}{\lambda} = \frac{V + V_R}{V/v} = v\frac{V + V_R}{V} \qquad \dots(8.23)$$

As $V_R{}^1\,0$, the here in this case also the apparent frequency $v'$ is greater than original frequency $v$. Similarly if the receiver is moving away from the source then the positive sign in the numerator of the above equation will be changed to negative and then the apparent frequency decreases instead of increasing.

If both of the source and receiver are moving then the above two equations can be combined into

$$v' = v\,\frac{V \pm V_R}{V \mp V_S}.$$

The upper signs both in numerator and denominator represent when the source and receiver are both moving towards each other and lower signs when they move away from each other.

There is an important application of this effect in Astrophysics, if a distant star is moving towards earth the light received will suffer frequency increase and the light reaching earth will have its colour a blue shift and if it is receding, it will have a red shift.

## 8.3  WAVE ENERGY AND PROPAGATION

So far we have discussed the elastic waves in different mediums and have come to the conclusion that the wave equns. (8.8) in a special case of solution can be written as $\xi(x,\,t) = \xi_0 \cos\{\omega t \pm (kx + \phi)\}$ where, $\omega$, $k$ are already introduced.

Now, expanding the cosine of the above equation gives a linear combination of two waves in quadrature:

$$\xi(x,t) = \xi_1 \cos k(x \pm \upsilon t) + \xi_2 \sin k(x \pm \upsilon t) \quad \text{where } \lambda = \upsilon T = 2\pi\frac{\upsilon}{\omega} = \frac{2\pi}{k}$$

$$\xi_1 = \xi_0 \cos\varphi \;\text{ and }\; \xi_2 = -\xi_0 \sin\varphi \;\text{ or }\; \tan\varphi = -\frac{\xi_2}{\xi_1}.$$

The above equation of $\xi(x,\,t)$ can be generalized in the form

$$\sum_{j=1}^{n} \xi_j \cos(kx \pm \omega t + \varphi_j) = \xi_0 \cos(kx \pm \omega t + \varphi), \text{ the quantity } \varphi \text{ is introduced}$$

as "Phase Factor" and is given as

$$\tan \varphi = \frac{\sum\limits_{j=1}^{n} \xi_j \sin \varphi_j}{\sum\limits_{j=1}^{n} \xi_j \cos \varphi_j}. \qquad \ldots(8.24)$$

The formation of wave theory is essentially simplified by means of the complex linear combination of two waves in quadrature having same amplitude and frequency

$$\xi(x, t) = \xi_0 \cos(kx \pm \omega t + \varphi) + i\,\xi_0 \sin(kx \pm \omega t + \varphi) = \xi_0\, e^{i(kx \pm \omega t + \varphi)}.$$

From this mathematical relation it can be concluded that:

"A wave may be represented in an equation in which case the real part of the complex function is to be taken during the operation of mathematical operations like addition, multiplication etc. by real or with respect to a real variable".

We have already discussed different types of waves propagating in different mediums and all these types of waves constitute certain kinds of motion of atoms or molecules of the medium. The atoms or molecules do not actually move but they vibrate around one mean position either perpendicularly or longitudinally with respect to the "movement" of the wave. Then the question still remains that what really propagate? As the matter of the medium does not propagate, it is their state of motion or vibration or more precisely their dynamical condition is the thing that propagates. As by dynamical condition of any matter we understand their momentum or energy, then by wave propagation essentially we mean the transfer of momentum or energy from one place to other of the medium. Now, let us consider the propagation of elastic wave longitudinal in character in a solid rod (Section 8.2.1) where the right side of the rod pulls the left side with a force F and the left side pulls the right side with a force–F

and a particular section is displaced with a velocity, $\dfrac{\partial \xi}{\partial t}$. The power or the work per unit time that the left side transmits to the right side of the section is:

$$\frac{\partial W}{\partial t} = (-F)\frac{\partial \xi}{\partial t} \qquad \ldots(8.25)$$

Now, if we consider a sinusoidal elastic wave of the form $x = x_0 \sin(\omega t - kx)$.

Then, $\dfrac{\partial \xi}{\partial t} = \omega \xi_0 \cos(\omega t - kx)$ and we know from section 8.2.1 that

$F = YA \dfrac{\partial \xi}{\partial x}$ and putting the value of the derivative:

$$F = -Y A \xi_0 k \cos(\omega t - k x)$$

Putting these values in equn. (8.25) we get:

$$\frac{\partial W}{\partial t} = -F \omega \xi_0 \cos(\omega t - k x), \text{ but as}$$

$$F = -Y A \xi_0 k \cos(\omega t - k x)$$

and $\qquad \omega = kv$ and $v = \sqrt{\dfrac{Y}{\rho}}$

So, $\dfrac{\partial W}{\partial t} = YA\omega k \xi_0^2 \cos(\omega t - kx) = \left(\rho v^2\right) A \left(\omega^2 / v\right) \xi_0^2 \cos^2(\omega t - kx)$

$$= v A [\rho w^2 \xi_0^2 \cos^2(\omega t - k x)]. \qquad \qquad \dots(8.26)$$

One important physical consequence of this equation is that as $\cos^2(\omega t - kx)$ is involved the transfer of power is always positive and as it is time dependent this transfer will be fluctuating and also satisfy the wave equation. Therefore, this equn. (8.24) may be taken as an "Energy Equation".

Taking time average on both sides and as $[\cos^2(\omega t - kx)]_{average} = 1/2$ so,

$$\left[\frac{\partial W}{\partial t}\right]_{average} = v A \left(\frac{1}{2}\rho \omega^2 \xi_0^2\right). \qquad \qquad \dots(8.27)$$

Now, as the total energy per unit volume or energy density in the rod due to oscillation resulting from wave motion is given in terms of amplitude $\xi_0$ and mass density $\rho$ as

$$E = \frac{1}{2}\rho \omega^2 \xi_0^2, \text{ and substituting this in equn. (8.27) we get}$$

$$\left[\frac{\partial W}{\partial t}\right]_{average} = vAE, \text{ Now as } v \text{ is the velocity of propagation, we have}$$

that $vE$ is the energy flow through the unit area per unit time which is

$$vE = \frac{1}{A}\left[\frac{\partial W}{\partial t}\right] = I, \text{ where } I \text{ is the intensity of the wave. Therefore, we may}$$

conclude that in wave motion energy and momentum are transferred from one place to other along the wave.

## 8.4 PHASE VELOCITY AND GROUP VELOCITY

Now, if we recall the equn. (8.7) as $\dfrac{\partial^2 \xi}{\partial t^2} = v^2 \dfrac{\partial^2 \xi}{\partial x^2}$ or, $\dfrac{\partial^2 \xi}{\partial x^2} - \left[\dfrac{k}{\omega}\right]^2 \dfrac{\partial^2 \xi}{\partial t^2}.$

where $v = \dfrac{\omega}{k}$, the $\omega$ is the number of radians of the wave that pass a given location per unit time and $1/k$ is the spatial length of the wave per radian. It follows that $v = \dfrac{\omega}{k}$ is the speed with which the shape of the wave is moving i.e. the speed at which any fixed phase of the cycle is displaced. Consequently this is called the "Phase Velocity" of the wave, henceforth will be denoted by $v_p$. In terms of cycle velocity and wave length we then say that $v_p = \lambda v$. Here we may repeat that with this phase velocity $v_p$ the wave pattern moves yet not causing any movement of the material particle of the medium in the lateral direction.

Since a general wave or a wave like phenomenon needs not only represent the flow of any physical effects, there exists obviously no upper limit on the possible phase velocity of the wave. However, for a genuine physical wave i.e. a chain of sequentially dependent events, the phase velocity does not necessarily correspond to the speed at which energy or information is propagating. This is partly a semantic issue, because in order to actually convey information, a signal can not be a simple periodic wave and so we must consider non periodic signals leading to the concept of phase a little bit ambiguous. A signal implies something that begins at a certain time and ends at a certain later time and a wave with such a shape is called "pulse". Therefore, if we measure the velocity with which a signal is transmitted we are essentially implying the velocity with which this pulse travels. A single section as pointed out by arrow is a pulse and together such pulses constitute a signal.

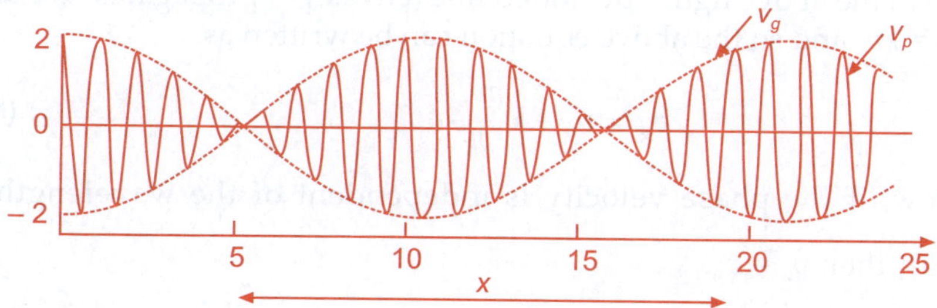


**Fig. 8.8** Group velocity and phase velocity in a wave pulse

In practice and common usage, though we tend to define the pulse of a signal with respect to the intervals between consecutive local maxima or minima to illustrate, consider a signal consisting of two superimposed sine waves with slightly different frequencies and wave lengths i.e. a signal with amplitude function $\xi(x, t)$ having two superimposed sinusoidal waves of same amplitude but with slightly different frequencies as $\omega$ and $\omega'$ so that is very small.

$$\xi(x, t) = \xi_0 \sin (kx - \omega t) + \xi_0 \sin (k'x - \omega' t)$$

$$= \xi_0 \left[ \sin (kx - \omega t) + \sin (k'x - \omega' t) \right]$$

$$= 2\xi_0 \cos \frac{1}{2}\left[(k' - k)x - (\omega' - \omega)t\right] \sin \frac{1}{2}\left[(k' + k)x - (\omega' + \omega)t\right]$$

Now, since $\omega$ and $\omega'$ are small and also $k$ and $k'$ are almost equal, we may approximate:

$\frac{1}{2}(\omega + \omega')$ as $\omega$ and $\frac{1}{2}(k + k')$ as $k$ and so,

$$\xi(x, t) = 2\xi_0 \cos \frac{1}{2}\left[(k' - k)x - (\omega' - \omega)t\right]\sin(kx - \omega t). \qquad ...(8.28)$$

The above equation represents a wave motion whose amplitude is modulated. The modulation is given by the factor

$$2\xi_0 \cos \frac{1}{2}\left[(k' - k)x - (\omega' - \omega)t\right].$$

The modulated amplitude corresponds to a wave motion propagating with velocity $v_g$, introduced as "Group Velocity". This is also shown in the above figure. Now, this Group velocity is given by

$$v_g = \frac{\omega' - \omega}{k' - k} = \frac{d\omega}{dk}. \qquad ...(8.29)$$

This group velocity is the velocity with which the amplitude wave shown in the above figure by dotted line (envelope) propagates. We know that $\omega = kv_p$ and so the above equation can be written as

$$v_g = v_p + k\frac{dv_p}{dk}. \qquad ...(8.30)$$

Now, if the phase velocity is independent of the wavelength i.e. $\frac{dv_p}{dk} = 0$, then $v_g = v_p$.

Therefore, in non dispersive media there is no difference between phase velocity and group velocity but in dispersive medium the group velocity may be smaller or larger than the phase velocity. The maximum of the pulse in the above figure propagate with the group velocity $v_g$. Therefore, in dispersive medium the signal velocity is the group velocity. To be more explicit the propagation of information or energy in a wave always occurs as a change in the wave. The most obvious example is changing the wave from being absent to being present, which propagates at the speed of the leading edge of a wave train. Generally, some modulation of the frequency or amplitude of a wave is required in order to convey information and it is this modulation that represents the signal content. Hence the actual speed of the content in the situation described above is $\dfrac{d\omega}{dk}$. This is the phase velocity of the amplitude wave, but since each amplitude wave contains a group of internal waves, this speed is actually called the group velocity. The physical waves of a given type in a given medium generally exhibit a characteristic group velocity as well as a characteristic phase velocity. This is because within a given medium there is a fixed relationship between the wave number and frequency of waves. In a transparent optical medium the refractive index $n$ is defined as the ratio $c/v_p$ where $c$ is the speed of light in vacuum and $v_p$ is the phase velocity of light in that medium. Since $v_p = \omega/k$, we have $\omega = kc/n \cdot$ Taking the derivative of $\omega$ as

$$\frac{d\omega}{dk} = \frac{c}{n} - \frac{ck}{n^2}\frac{dn}{dk}$$

Hence any modulation of an electromagnetic wave in this medium will propagate at the group velocity

$$v_g = v_p\left[1 - \frac{k}{n}\frac{dn}{dk}\right].$$

Now, in a medium whose refractive index is constant, independent of frequency i.e. as vacuum, we have $\dfrac{dn}{dk} = 0$, the group velocity equals the phase velocity but in commonly known dispersive media like air, water, glass etc. where refractive index is dependent on frequency (wave number), the group velocity of light is less than the phase velocity.

An example of a physical application of phase and group velocity lies in the propagation of electromagnetic waves through a hollow magnetic conductor often called a waveguide. A waveguide imposes a cutoff frequency $\omega_0$ on any propagating electromagnetic waves based on the geometry of the tube and will not sustain waves of any lower frequency. As a result the dominant wave pattern of a propagating wave with a frequency of $w$ will have a wave number $k$ given by:

$$k = \frac{1}{c}\sqrt{\omega^2 - \omega_0^2}$$

As we have seen the phase velocity is $\dfrac{\omega}{k}$, this implies that the phase velocity in a waveguide with cutoff frequency $\omega_0$ is

$$v_p = \frac{c}{\sqrt{1 - \left[\dfrac{\omega_0}{\omega}\right]^2}}.$$

Hence, not only is the phase velocity generally greater than $c$, it approaches infinity as $\omega$ approaches the cutoff frequency $\omega_0$. However, the speed at which information and energy actually propagates down a waveguide is the group velocity which as we have seen is given by $\dfrac{d\omega}{dk}$. Taking derivative of the earlier expression of $k$ we get

$$\frac{dk}{d\omega} = \frac{\omega}{c\sqrt{\omega^2 - \omega_0^2}}.$$

So, the group velocity in a waveguide with cutoff frequency $\omega_0$ is

$$v_g = \frac{d\omega}{dk} = c\sqrt{1 - \left[\frac{\omega_0}{\omega}\right]^2},$$

This is always less than or equal to the speed of light $c$.

As another interesting example, consider the case of surface wave in a liquid having wavelength long and the depth is very great compared with wavelength. The phase velocity of such surface wave is given as (from equn. 8.17):

$$v_p = \sqrt{\frac{g\lambda}{2\pi}} = \sqrt{\frac{g}{k}} \quad \text{as} \quad k = \frac{2\pi}{\lambda}$$

Now, $$\frac{dv_p}{dk} = -\frac{1}{2k}\sqrt{\frac{g}{k}} = -\frac{v_p}{2k}$$

Recalling equn. (8.28). $v_g = v_p + k\dfrac{dv_p}{dk}$ and putting the value of $\dfrac{dv_p}{dk}$ as derived for the surface wave, we get: $v_g = \dfrac{1}{2}v_p$.

Therefore, the group velocity is just half of the phase velocity. This observation implies that if a long wave disturbance is created in water the initial disturbance is distorted in such a way that the components of longer wave length moves out from the disturbance with a velocity twice as large as that of group velocity which is the velocity of the peak of the disturbance.

## REVIEW QUESTIONS

1.  Let a wave pulse moving towards right along $x$-axis be represented by wave function

$$y(x, t) = \frac{3.0}{(x - 5.0t)^2 + 1}.$$

    Show that the wave pulse is a progressive wave function and is a solution to the linear wave equation.

2.  Establish that elastic waves created in different media are transverse or longitudinal depending upon their elastic moduli.

3.  If a vibrating wire is firmly supported at its two ends and set to vibrate, explain the creation of standing waves and terms 'harmonics'

4.  In surface wave in liquid explain the role of two factors, gravity and surface tension.

5.  Introducing some daily experiences explain the 'Doppler Effect'. Mention its use in finding stellar motions.

6.  Explain Group velocity and Phase velocity. Can they be different in a medium?

## REFERENCES


1.  S. B. Palmer and M. S. Rogalski – Advanced University Physics, Gordon and Breach Publishers, U. K., 1996.

2.  R. P. Feynman, R. B. Leighton and M. Sands – The Feynman Lectures on Physics, Addison Wesley Pub. Co., Reading, Mass, Narosa Pub. House New Delhi, 2003.

# Physical Optics - I

## 9.1 LIGHT WAVE AND SPECTRA

After introducing the wave motion in elastic medium, here in this chapter we will discuss the wave nature of light and its various aspects. In order to explain various physical phenomena caused by light, two points of view have emerged and each of these points of view have their place in the history of physics. Almost simultaneously in the second half of seventeenth century the corpuscular theory was developed by Newton and the wave theory by Huygens. Some basic phenomena like rectilinear propagation, reflection and refraction can be explained by both the theories but in explaining the interference and diffraction of light, the corpuscular theory failed and these could only be explained by Huygens' wave theory. Two light beams meeting together and creating darkness can only be explained when light is considered as a wave. The success of Maxwell's electrodynamics in the nineteenth century, which interprets light as electromagnetic wave established finally the wave nature of light. Discovery of photo electric effect by Heinrich Hertz in 1887, however brought back the particle concept of light once introduced by Newton. This development of the theory of light as electromagnetic radiation led ultimately to the view that light is to be considered either as particles or waves, depending on the specific problem considered. The "particles" of light are called "Quanta of light" or "Photons", the wave packets and this is the co-existence of waves and particles together called wave-particle dualism an aspect we will consider in some latter chapters. Here in this chapter we will concentrate only on the wave aspects of light and the related experiments which establish this aspect of light.

The wave lengths of visible light extend between about $4 \times 10^{-5}$ cm. (4000 $A°$ or 400 nm) for the extreme violet and $7.2 \times 10^{-5}$ cm (7200 $A°$ or 720 nm) for the deep red. This range of wavelength of light is designated as visible region as human eye fails to "see" the light beyond this region as human ear fails to recognize and sound frequency beyond some region on both sides. The following Fig. 9.1 gives the spectrum of the electromagnetic radiation in terms of wavelength though the boundaries between different regions as they are named are not demarcated exactly.

It will be seen that visible light covers an almost insignificant fraction of the entire spectrum. There is no basic difference between radiations of the electromagnetic waves but differing only in wavelength. The term "Light" is conventionally extended only to the adjacent portion of the spectrum namely Ultra violet and Infra red. Though there is no basic difference between different types of radiation except that they are generated and detected in different ways. While X-rays are invisible and are detected on their property of fluorescence and ionization, the visible region can be detected by human eye. Nichols and Tear in the year 1917 produced infrared waves having wavelengths up to $42 \times 10^4$ nm and radio waves down to $22 \times 10^4$ nm. The two regions are said to overlap though the waves themselves remain of the same nature. The same is valid for the boundaries of the all other regions of the spectrum.

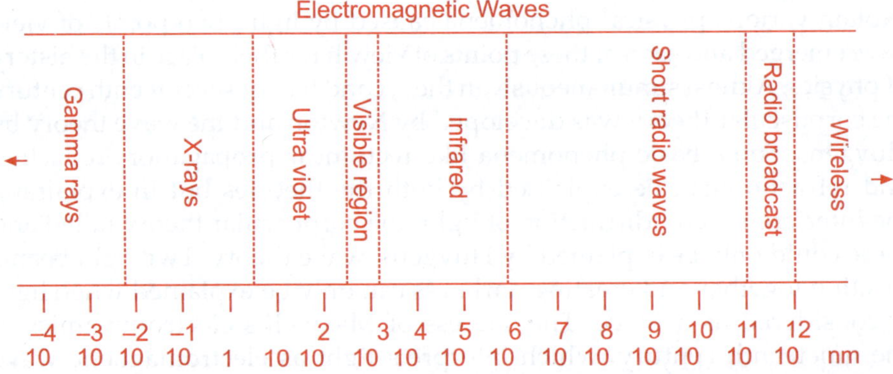


**Fig. 9.1** The spectrum of electromagnetic waves. The boundaries of each region can only be tentatively defined, as no well defined boundaries exist. The spectra also extends indefinitely both in longer and shorter wavelength directions

Like sound and other elastic waves a change in wavelength occurs when the source has a translational motion. The waves sent out in the direction of motion are shortened and in the opposite direction are lengthened. No change is however produced in the velocity of sound. A stationary observer receives a frequency which is larger or smaller than that of the source. If on the other hand the source is at rest and the observer is under motion, a change of frequency is also observed but for different reason. Here there is no change of wavelength but the frequency is altered by the change in relative velocity of the waves with respect to the observer. This observed change in frequency of the wave is known as Doppler Effect and is most commonly experienced in sound as changes in the acoustic pitch. Doppler's principle has become a powerful method of studying the radial velocities of stars. The spectra of other galaxies like spiral nebulae show displacement toward the red, which for the most distant ones amount to several hundreds of angstrom units. Such values would indicate recessional velocities of tens of thousands of kilometers per second, and have been so interpreted. It is rather interesting that there is enough reddening to change the colour of the object, as postulated by Doppler, but in this case it occurs for objects far too faint to be seen by the naked eye.

## 9.2 INTERFERENCE AND COHERENCE

In Huygens's principle a source of light at a large distance is thought of emitting light of plane wave. A section of a circle or sphere of very large radius may be safely considered as a plane. The wave emitted by a point source or a narrow opening is considered to emit spherical wave. The points on a single wave surface exist in the same phase and the plane generated by joining them is said as Wave front. In the following figure we consider the plane wave from a source at a long distance reaching a narrow slit S. The slit S in turn emits spherical wave fronts and the points on the wave front $S_1$, $S_2$ and $S_3$ and all other act as secondary sources and the surface generated by joining them is another spherical surface and is taken as a new position of spherical wave front.

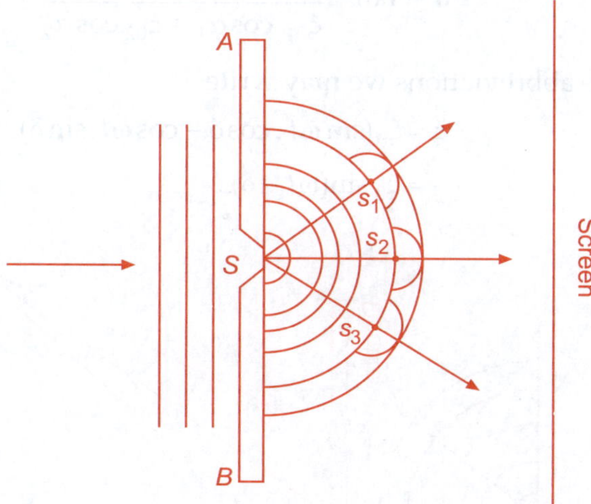


**Fig. 9.2** *A B* is the slit with opening at *S* acting as a point source The plane wave from a distant source and incident on the slit emerges as spherical waves. Each point on the spherical wavefront acts as a source

## 9.2.1 Young's Experiment and Interference

There are two types of interference phenomenon and these are due to Division of wave front and Division of amplitude. This classification is based on the causes of the phenomenon concerned. Let us first start with the earlier type i.e. division of wave front.

We shall now derive the expression of intensity of light reaching at a point $P$ on the screen. A double slit is placed before the spherical wave front coming from a source and the two sinusoidal waves emerging out from the slits $S_1$ and $S_2$ are given by the expressions as

$$\xi_1 = \xi_{01} \sin(\omega t - kS_1P) \text{ and } \xi_2 = \xi_{02} \sin(\omega t - kS_2P)$$

where $\xi_{01}$ and $\xi_{02}$ are the amplitudes of the waves and $S_1P$ and $S_2P$ are the paths traveled by the waves from the respective slits to the point $P$ on the screen. If the resultant wave is given by $\xi$ where $\xi = \xi_1 + \xi_2$ then

$$\xi = \xi_{01} \sin(\omega t - k\alpha_1) + \xi_{02} \sin(\omega t - k\alpha_2) \text{ where } \alpha_1 = kS_1P \text{ and } \alpha_2 = kS_2P$$
$$= \xi_{01} \sin\omega t \cos\alpha_1 - \xi_{01} \cos\omega t \sin\alpha_1 + \xi_{02} \sin\omega t \cos\alpha_2 - \xi_{02} \cos\omega t \sin\alpha_2$$
$$= (\xi_{01} \cos\alpha_1 + \xi_{02} \cos\alpha_2)\sin\omega t - (\xi_{01} \sin\alpha_1 + \xi_{02} \sin\alpha_2)\cos\omega t$$

Now, writing $\left(\xi_{01} \cos\alpha_1 + \xi_{02} \cos\alpha_2\right) = \xi_0 \cos\delta$ and

$$\left(\xi_{01} \cos\alpha_1 + \xi_{02} \cos\alpha_2\right) = \xi_0 \sin\delta \text{ where,}$$

$$\xi_0^2 = \xi_{01}^2 + \xi_{02}^2 + 2\xi_{01} \xi_{02} \cos(\alpha_1 - \alpha_2) \text{ and}$$

$$\delta = \tan^{-1} \frac{\xi_{01} \sin\alpha_1 + \xi_{02} \sin\alpha_2}{\xi_{01} \cos\alpha_1 + \xi_{02} \cos\alpha_2}.$$

Using these abbreviations we may write

$$\xi = \xi_0(\sin\omega t \cdot \cos\delta - \cos\omega t \cdot \sin\delta)$$
$$= \xi_0 \sin(\omega t - \delta). \qquad \qquad \dots(9.1)$$

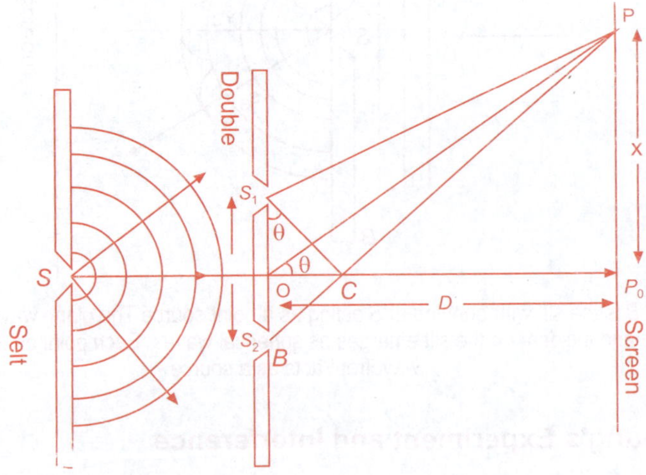


**Fig. 9.3** Young's 'double slit' experiment

The optical path difference between the waves coming out from $S_1$ and $S_2$ and reaching the point $P$ on the screen $S_2P - S_1P = \Delta = d\sin\theta$ and when $\theta$ is very small we can approximate $\sin\theta = \tan\theta = X/D$ also as $D$ is thousand time larger than $d$ or $x$.

Then, the phase difference $\delta = 2\pi/\lambda(S_2P - S_1P) = (2/\lambda)\,Xd/D$ and as intensity I is proportional to the square of the amplitude so,

$$I \sim \xi_0^2 = 2\xi_{00}^2(1+\cos\delta) = 4\xi_{00}^2\cos^2\frac{\delta}{2}, \text{ considering } \xi_{01} = \xi_{02} = \xi_{00} \quad \ldots(9.2)$$

Now, the intensity I have maximum values equal to $4\xi_{00}^2$ whenever $d$ is an integral multiple of $2\pi$ and according to $\delta = 2\pi/\lambda\,(S_2P - S_1P) = (2\pi/\lambda) \times d/D$, this will occur when the path difference is an integral multiple of $l$ and so for:

**Bright fringes:**

$$\frac{Xd}{D} = 0, \lambda, 2\lambda, 3\lambda\ldots = m\lambda \quad \text{or,} \quad X = m\lambda\frac{D}{d}. \qquad \ldots(9.3a)$$

The minimum value of the intensity is zero and this occurs when $\delta = \pi$, $3\pi, 5\pi,\ldots$ and for these points on the screen

**Dark fringes:**

$$\frac{Xd}{D} = \frac{\lambda}{2}, \frac{3}{2}\lambda, \frac{5}{2}\lambda, \ldots = (m+\frac{1}{2})\lambda \quad \text{or,} \quad X = (m+\frac{1}{2})\lambda\frac{D}{d} \qquad \ldots(9.3b)$$

The whole number $m$ which characterizes a particular bright fringe is called the order of interference and so fringes with $m = 0, 1, 2, \ldots.$ are called the zero, first, second etc. orders. Now, from equn. (9.2 a ) or (9.2 b) the distances between two successive fringes on the screen can be obtained by changing $m$ by unity and subtracting one from other. This distance on the screen $\Delta X = \dfrac{\lambda D}{d}$ and therefore its magnitude is directly proportional to the slit-screen distance and inversely proportional to the slit separation d and

also directly proportional to the wavelength $\lambda$. Knowing the fringe separation it is therefore possible to measure the wavelength of the monochromatic light as other parameters remain as instrument constants. These maxima and minima of intensity exist throughout the space behind the slits and a lens is not required to produce them but as the fringes is so fine that a magnifier lens must be used to see them visually. This distribution of fringe intensities is shown in the following Fig. 9.4. The variation of intensity shows that it varies between $4\xi_{00}^2$ and zero. Now each beam acting separately would contribute $\xi_{00}^2$ and so without interference we would have a uniform intensity 2, as indicated by the dotted line Fig. 9.4. The average value of the square of the cosine $(\cos^2\delta/2)$ is $1/2$ and this gives from the above equation for intensity I as $I \sim 2 \cdot$ Therefore, it shows that the distribution of this intensity does not violet conservation of energy.

There are two other examples of this type of interference i.e. division of wave front and the more important experiments are (1) Fresnel's biprism and (2) Lloyd's mirror. The following diagrams explain these two methods of forming interference.

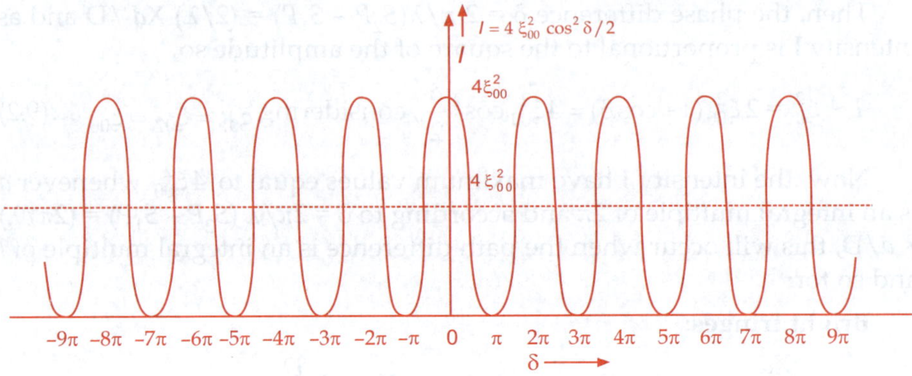


**Fig. 9.4** The distribution of intensity on the screen due to interference.

## 9.2.2  Division of Wave Front: Fresnel Biprism and Lloyd Mirror

In the Fresnel's biprism experiment as above the wave front from the source S is divided in to two parts due to refraction from up and down parts of the biprism. This results the division of real source S into two virtual sources $S_1$ and $S_2$. The shaded portion BC on the screen is the interference region. In Lloyd's mirror the original source S and the virtual source $S_1$ send waves which interfere in the region AB of the screen.

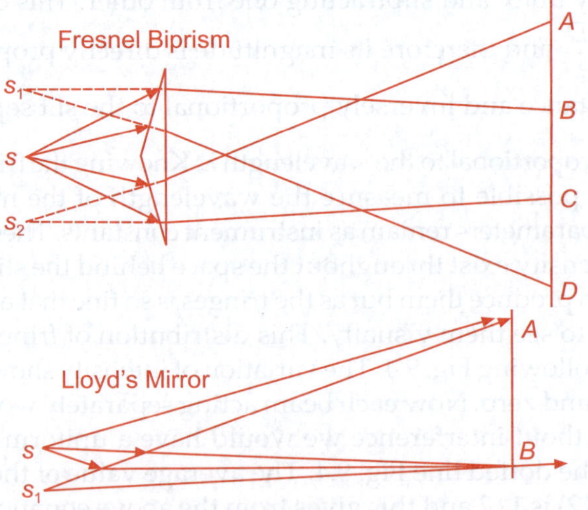


**Fig. 9.5** Fresnel's biprism and Lloyd's Mirror experimental set up

## 9.2.3 Division of Amplitude

**(i) Newton's Ring:** The multiple reflections from a thin film is an example of the formation of interference pattern by division of amplitude. Let us consider the light from a source $S$ be incident on a thin film at A. It then refracting in the film suffers multiple reflection on surfaces and 1, 2, 3, 4,…. are the beams emerging out of the front surface. These rays are made to converge at the point $P$.

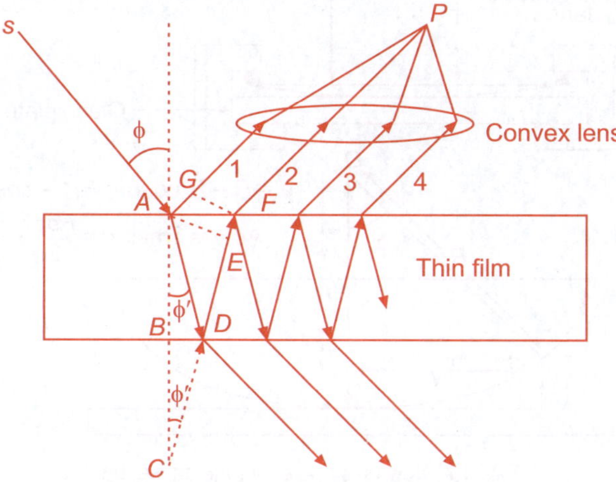


**Fig. 9.6** Multiple reflections from a thin film

When we consider only the pair of beams (1) and (2), $AE$ and $GF$ are the two subsequent wave fronts of the waves suffering internal and external reflections, and then the optical path difference between (1) and (2) is:

$\Delta = n(ADF) - AG$, where $n$ is the refractive index of the medium of the film.

Now, from geometry, $\Delta = nCF - AG = n(CE + EF) - AG$,

Now, as optical path $n\,EF = AG$ then

$\Delta = nCE = n(2d \cos \phi')$ where $d$ is the thickness of the film. Therefore, when (1) and (2) meet at $P$ the path difference 2nd cos φ' satisfy the following conditions we get interference maxima and minima taking into account the fact that the ray suffers a phase change by $p$ but the ray (2) does not as the latter is an internal reflection and so the conditions of maxima and minima are reversed.

**Minima:** $2nd \cos \phi' = m\lambda$ and

**Maxima:** $2nd \cos \phi' = (m + \frac{1}{2})\lambda$. ...(9.4)

In Newton's ring experiment described in the following Fig. 9.7 a part of the incident ray is reflected from the inner convex surface of the lens and a part after refracting out suffers reflection from the plane surface of the glass slide.

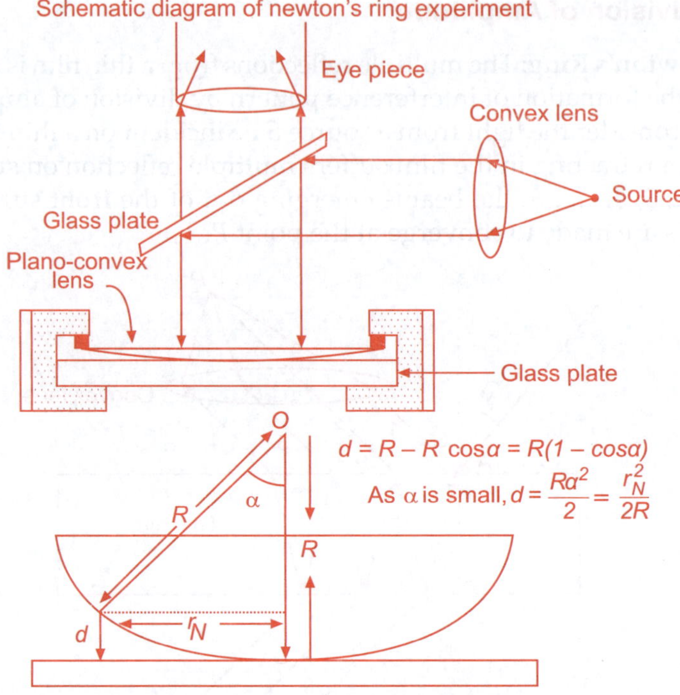



**Fig. 9.7** Newton's ring experimental set up

Now, for Newton's ring set up the light is incident normally on the plane surface of the plano-convex lens and so, $\phi = 0$ and also $\phi' = 0$, the interference condition from equn. (9.3) transform into

$2nd = m\lambda$ for Minima and $2nd = (m + \frac{1}{2})\lambda$ for Maxima.

This $d$, the depth of the wedge shaped gap is given as: $d = \dfrac{r_N^2}{2R}$, which

remains same along the circumference of the circle of radius $r_N$ of Nth number of fringe.

**(ii) Fabrey-Perot Interferometer:** Febrey-Perot interferometer is one of the most important application of interference by division of amplitude. The working diagram is given below (Fig. 9.8)

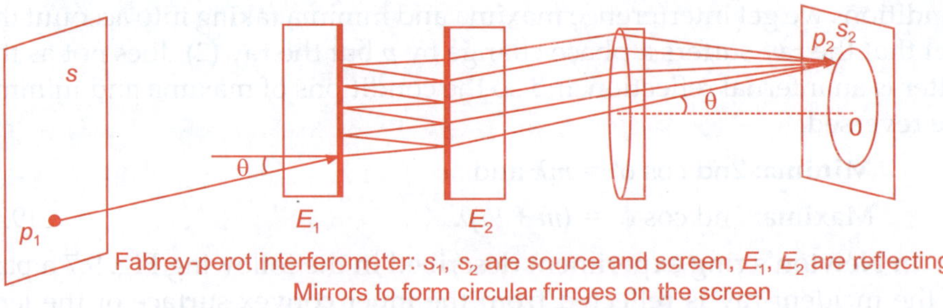



Fabrey-perot interferometer: $s_1$, $s_2$ are source and screen, $E_1$, $E_2$ are reflecting Mirrors to form circular fringes on the screen

**Fig. 9.8** F-P Interferometer: $P_1$ is the point on the extended source and $P_2$ is the corresponding point on the screen. A circular fringe is resulted after multiple reflection and refraction

The condition of maxima is $2nd \cos \theta = m\lambda$ which is different from equn. (9.3) as here the refracted beams overlap, the other parameters are same.

## 9.3 FRAUNHOFER DIFFRACTION

### 9.3.1 Diffraction of Light Waves: An Introduction

So far we have seen in interference of light from Huygens's principle that every point on the approaching wave front being spherical (for near source) or plane (for distant source) generate secondary wavelets which are being spherical travel in all directions. Now, if two of them are selected and rests are obstructed then the two selected waves after traveling different distances assume a phase difference and create interference pattern (maxima and minima) if they meet together. These are achieved in double slit and biprism experiments which are interference by division of wave front and also by dividing the amplitude of light from the same wave front by multiple reflections which are achieved in thin parallel films or Newton's ring experiments.

Now, if a part of the approaching wave front is obstructed, the rest which is exposed generates secondary wavelets and two of which meet at a particular point with different phases even though the selected point lies within the geometric shadow part of the obstacle. This is the reason for observing no sharp out line of the shadow of the obstacle, which can never be explained without the wave characteristics of light. This is another important aspect of wave theory and is known as diffraction of light, which is the sole reason for getting light in the geometric shadow part of any obstacle and which can never be explained on the basis of assumption that light is the energy which simply travels in geometric straight line. The phenomenon of "Diffraction" is classified into two types, the one which is applicable for plane approaching waves from a source at infinity and is known as Fraunhofer Diffraction and the other is applicable for spherical wave front from a near source and this is known as Fresnel Diffraction. The mathematical interpretations of these two types are different and are known after the names of the Physicists who developed their mathematical interpretations. However, strictly speaking the Fresnel diffraction is the actual case of diffraction as the light waves coming out from a source is spherical in general but for a source at a considerable distance the wave fronts may be approximated as plane and so Fraunhofer diffraction can be considered without much error involved.

### 9.3.2 Fraunhofer Diffraction by a Single Opening

Let a plane wave front from a distant source is approaching a single slit opening as shown in the following Fig. 9.8. The different secondary wavelets after passing through the opening of width say $b$ are made to meet together at points on the screen.

In the Fig. 9.8 a plane wave is approaching a single slit of width b from a distant object. Let $dS$ be an element of a section of wave front at a distance $S$ from the center and $P$ be any point on the screen.

Now, as the energy from a source is uniformly distributed over the surface of the spherical wave and as the area of the surface varies as the square of the radius of the sphere i.e. the distance from the source, the energy density varies inversely as the square of the distance (radius). The amplitude of the wave then should vary inversely as the distance (radius). Now, if $\xi_0$ is the amplitude per unit area reaching the slit, then amplitude at point $P$ from an element $dS$ of the wave front will be $\dfrac{\xi_0 dS}{x}$ at a distance of $x$ and the equation of wave will be given by

$$d\xi = \frac{\xi_0 dS}{x} \sin(\omega t - kx) \qquad \qquad ...(9.5)$$

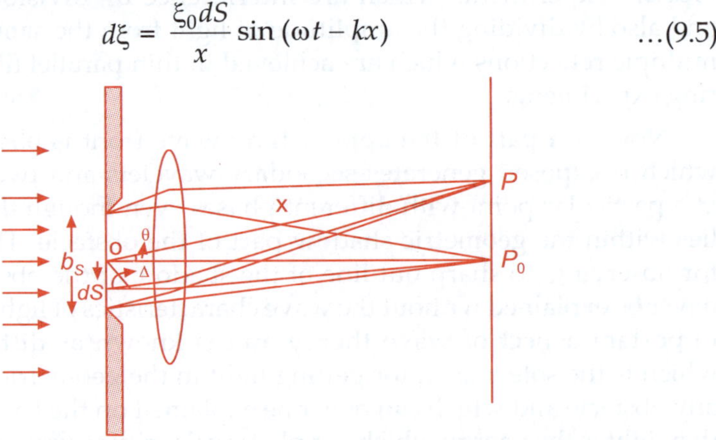

Fig. 9.9 Fraunhofer diffraction for a single opening (single slit)

The displacement $d\xi$ will vary both in magnitude and phase as $\xi$ changes with different positions of the point of observation $P$ on the screen. Now, taking the element of area $dS$ on the wave front at a distance s down the centre 0 an additional path difference $\Delta$ results and the equn. (9.4) is modified as

$$d\xi_{+S} = \frac{\xi_0 dS}{x} \sin(\omega t - k[x+\Delta]) = \frac{\xi_0 dS}{x} \sin(\omega t - kx - kS\sin\theta) \qquad ...(9.6\ a)$$

And similarly for an element dS up the center 0

$$d\xi_{-S} = \frac{\xi_0 dS}{x} \sin(\omega t - k[x-\Delta]) = \frac{\xi_0 dS}{x} \sin(\omega t - kx + kS\sin\theta). \qquad ...(9.6\ b)$$

Now, taking the combined effect from two such wave front elements by adding equns. (9.5 a) and (9.5 b) we get:

$$d\xi = d\xi_{+S} + d\xi_{-S}$$

$$= \frac{\xi_0 dS}{x}\left[\sin(\omega t - kx + kS\sin\theta) + \sin(\omega t - kx - kS\sin\theta)\right]$$

which is of the form: $\sin\alpha + \sin\beta = 2\cos\frac{1}{2}(\alpha - \beta)\sin\frac{1}{2}(\alpha + \beta)$

and so $d\xi = \dfrac{\xi_0 dS}{x}\left[2\cos(kS\sin\theta)\sin(\omega t - kx)\right].$

Now, integrating for the entire wave front from $S = 0$ to $\beta/2$ we get

$$\xi = \frac{2\xi_0}{x}\sin(\omega t - kx)\int_0^{b/2}\cos(kS\sin\theta)\,dS$$

$$= \frac{2\xi_0}{x}\left[\frac{\sin kS\sin\theta}{k\sin\theta}\right]_0^{b/2}\sin(\omega t - kx)$$

$$= \frac{\xi_0 b}{x}\left[\frac{\sin(\frac{1}{2}kb\sin\theta)}{(\frac{1}{2}kb\sin\theta)}\right]\sin(\omega t - kx).$$

The resultant vibration will therefore be a simple harmonic one, the amplitude of which varies with $\theta$ *i.e.* with different positions of point P on the screen. Now, abbreviating $\dfrac{1}{2}kb\sin\theta$ or $\dfrac{\pi b\sin\theta}{\lambda}$ as $\beta$ the above equation is transformed in to

$$\xi = \frac{\xi_0 b}{x}\left[\frac{\sin\beta}{\beta}\right]\sin(\omega t - kx) \qquad\qquad \ldots(9.7)$$

Now, writing the resultant amplitude as $A$ and $\dfrac{\xi_0 b}{x}$ as $A_0$ we get

$$A = A_0\frac{\sin\beta}{\beta} \qquad\qquad \ldots(9.8\ a)$$

Therefore, as intensity $I \sim A^2$ so, it may be written as $I = A_0^2\dfrac{\sin^2\beta}{\beta^2}$ $\ldots(9.8\ b)$

In the following Fig. 9.10 the variation of both amplitude and also intensity with $\theta$ are shown

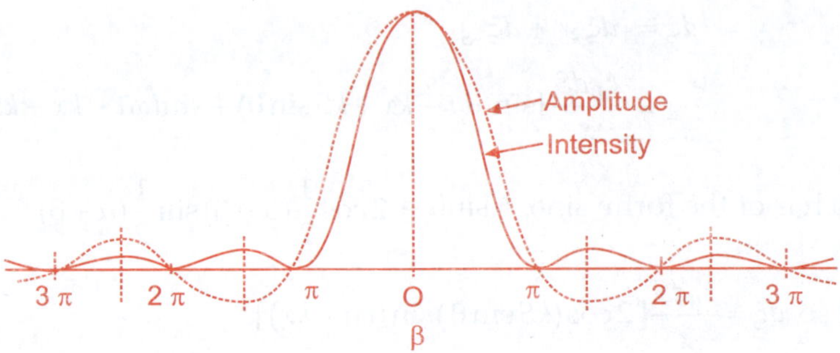


**Fig. 9.9** Variation of intensity and the amplitude with θ which is dependent on slit with and angle θ (which depends on positions P on the screen)

### 9.3.3 Fraunhofer Diffraction by a Double Slit (Opening)

Without going in to experimental details, if a double slit is placed before a source at a distance, then interference fringes giving its intensity expression by Equn. 9.2 and 9.3 and variation by Fig. 9.4 along with diffraction from each slit given by equn. 9.8 a and 9.8 b. These two phenomena i.e. the interference from two coherent waves and diffraction from different places of each wave fronts combine together to give the net results from a double slit. It will be then in effect the superposition of two graphs showing intensity variation.

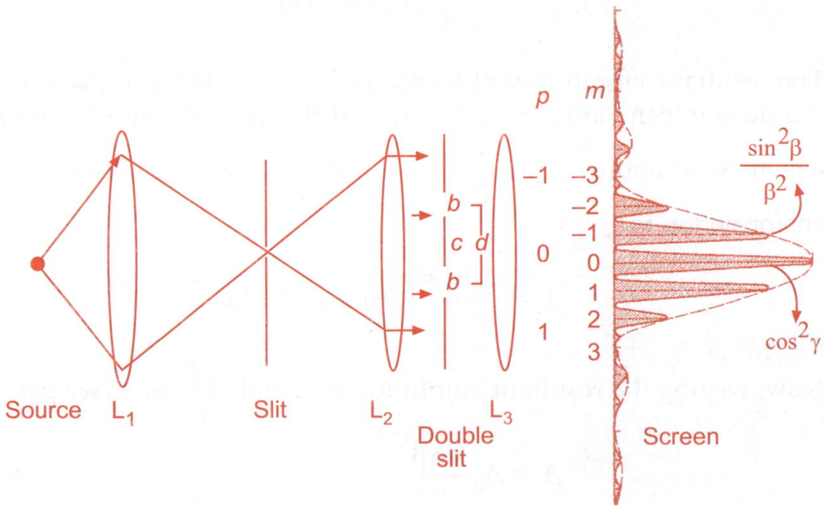


**Fig. 9.10** The out line of the experimental set up for diffraction phenomenon by a double Slit. $L_2$ is used to result parallel rays as position of slit is at its focus and $L_3$ is to converge the diffracted wave fronts on the screen to view the diffraction pattern, $b$ is the width of each opening and $c$ is the opaque part which separates the two slits

Now, the integration limits used while deriving the equn. 9.7 for single slit i.e. from 0 to $b/2$ is to be modified now in to $s = d/2 - b/2$ to $d/2 + b/2$. Recalling and inserting the modified integration limits:

$$\xi = \frac{2\xi_0}{x} \sin(\omega t - kx) \int_{d/2-b/2}^{d/2+b/2} \cos(kS\sin\theta)\,dS \text{ and this gives}$$

$$\xi = \frac{2\xi_0}{xk\sin\theta}\left[\sin\left\{\frac{1}{2}k(d+b)\sin\theta\right\} - \sin\left\{\frac{1}{2}k(d-b)\sin\theta\right\}\right]\sin(\omega t - kx) \quad \cdots(9.9)$$

Now, the quantity with the bracket is of the form $\sin(a+b) - \sin(a-b)$, which when expanded we get

$$\xi = \frac{2\xi_0 b}{x}\frac{\sin\beta}{\beta}\cos\gamma\,\sin(\omega t - kx) \qquad \cdots(9.10)$$

Where as before $\beta = \frac{1}{2}kb\sin\theta$ where $k = \frac{2\pi}{\lambda}$ and

$$\gamma = \frac{1}{2}k(b+c)\sin\theta = \frac{\pi}{\lambda}d\sin\theta \cdot$$

The intensity is equal to the square of the amplitude and so using as before $A_0 = \frac{\xi_0 b}{x}$.

We get $\qquad\qquad I = 4A_0^2\,\frac{\sin^2\beta}{\beta^2}\cos^2\gamma \qquad\qquad \cdots(9.11)$

It can be emphasized here that $\dfrac{\sin^2\beta}{\beta^2}$ term which represents the

$d$the diffraction part and is the envelope (dotted) of the curve of the total diffraction pattern Fig. 9.11 and $\cos^2\gamma$ represents the interference effect from two slits and are shown by the shaded pattern within the envelope.

Therefore, the diffraction part (the envelope) shows minima at values $\beta = \pi, 2\pi, 3\pi, \ldots \left\{\beta = \frac{1}{2}kb\sin\theta = \frac{\pi}{\lambda}b\sin\theta\right\}$ and the second part due to interference will give minima at values

$$\gamma = \pi/2, \frac{3}{2}\pi, \frac{5}{2}\pi, \ldots \left\{\gamma = \frac{1}{2}k(b+c)\sin\theta = \frac{\pi}{\lambda}d\sin\theta\right\}.$$

The final diffraction minima will be shown when either of the above two conditions are satisfied. It can be shown that $\beta$ and $\gamma$ are not independent and in terms of the dimensions of the slits,

$$\frac{\gamma}{\beta} = \frac{d}{b}.$$

Now, a study of the diffraction profile shows that certain orders are missing or at least the intensities are reduced to minimum. These so called missing orders result where the condition for a maximum of the interference and for minima for diffraction are both fulfilled for the same value of $\theta$, that is for

$$d \sin \theta = m\lambda$$

$$b \sin \theta = p\lambda$$

$$\frac{d}{b} = \frac{m}{p} . \qquad \qquad \qquad ...(9.12)$$

Therefore, when $d/b = 3$, orders 3, 6, 9... will be missing Fig. 9.11.

### 9.3.4 Fraunhofer Diffraction by an Ideal Grating

Now, in place of using the sine or the cosine to represent a simple harmonic wave, one may write the wave equation in the exponential form $y = ae^{i(wt - kx)} = ae^{iwt}e^{-i\delta}$ where $\delta = kx$ and is constant at a particular point in space and exp $(iwt)$ is the time varying factor. We may use this complex representation and at the end of the problem take either the real (cosine) or the imaginary (sine) part of the resulting expression. This procedure may be used for the diffraction analysis from a system of say $N$ number of openings. Nevertheless, we can also calculate the diffraction from $N$ number of slits by following the same procedure of integration but this would involve cumbersome process. If we follow the imaginary exponential and designate the amplitudes from each slit by '$a$' and the change of phase from slit to the other by $d$, the resultant complex amplitude is then will be given by the sum of the series:

$$A e^{i\theta} = a(1 + e^{i\delta} + e^{i2\delta} + e^{i3\delta} + ..........)$$

$$= a\frac{1 - e^{iN\delta}}{1 - e^{i\delta}} .$$

To find intensity we are to multiply the above expression with its complex conjugate

$$A^2 = a^2 \frac{(1 - e^{iN\delta})(1 - e^{-iN\delta})}{(1 - e^{i\delta})(1 - e^{-i\delta})}$$

$$= a^2 \frac{1 - \cos N\delta}{1 - \cos \delta}$$

$$= a^2 \frac{\sin^2 (N\delta/2)}{\sin^2 (\delta/2)} = a^2 \frac{\sin^2 N\gamma}{\sin^2 \gamma} . \qquad ...(9.13)$$

Now, as for the double slit $\gamma = \sigma/2 = (\pi d \sin \theta)/\lambda$. This expression includes the result of diffraction by a single slit given in equn. (9.8 b) and so the intensity expression from an ideal diffraction grating is given by

$$I \sim A^2 = A_0^2 \frac{\sin^2 \beta}{\beta^2} \frac{\sin^2 N\gamma}{\sin^2 \gamma} \qquad \dots(9.14)$$

In the above equation if we put $N = 2$ (double slit) the equation is transformed in to that for the double slit diffraction (9.11).

### *Principal Maxima*

The new factor in the equn. (9.14) $\dfrac{\sin^2 N\gamma}{\sin^2 \gamma}$ represents the interference term

for $N$ slits. It will represent maximum values equal to $N^2$ for $\gamma = 0, \pi, 2\pi, 3\pi,$ .... as the limit of the function:

$$\left[ \lim_{\gamma \to m\pi} \left( \frac{\sin N\gamma}{\sin \gamma} \right) \right]^2 = [N]^2 . \text{ Now as seen before, } \gamma = \frac{1}{2}k(b+c)$$

$= \dfrac{\pi}{\lambda} d \sin\theta$ for the double slit.

Principal maxima are obtained at $d \sin\theta = 0, \lambda, 2\lambda, 3\lambda, \dots = m\lambda$ ...(9.15)

However, the maxima are more intense and they are in the ratio of the square of the number of slits.

The relative intensities of the different orders $m$ are in all cases governed by single slit diffraction envelope $\dfrac{\sin^2 \beta}{\beta^2}$. Therefore, the relation between

$\beta$ and $\gamma$ in terms of slit width and slit separation remains unchanged and also the conditions of missing orders.

### *Minima and Secondary Maxima*

Now, to find the minima of the function $\dfrac{\sin^2 N\gamma}{\sin^2 \gamma}$, we can note that the

numerator becomes zero more often than the denominator and this occurs at values $N\gamma = 0, \pi, 2\pi, \dots$ or $p\pi$. In the special cases for $p = 0, N, 2N, \dots \gamma$ will be $0, \pi, 2\pi, \dots$; so for these values both numerator and denominator vanish simultaneously and we get as a result Principal Maxima as above and for other values of $p$ the denominator does not vanish with numerator and we get zero intensity. Hence the condition for a minimum is: $\gamma = p\pi/N$, excluding those values of $p$ for which $p = mN$, $m$ being the order.

Between other minima the intensity rises again, but the secondary maxima thus produced are of much smaller intensity than the principal maxima. Fig. 9.12 shows the interference curve for grating having $N = 20$ number of slits.

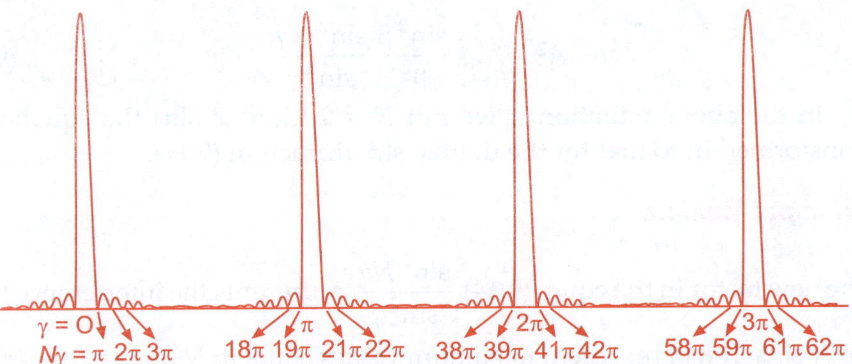

**Fig. 9.12 a** Intensity profiles showing the principal maxima and secondary maxima (adopted from fundamentals of optics, jenkins and white)

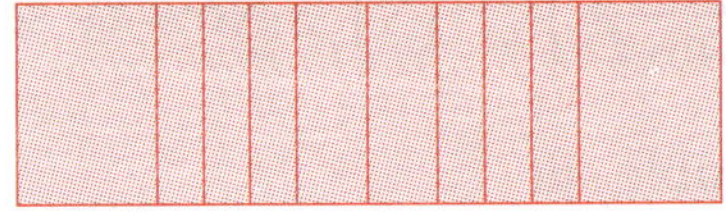

**Fig. 9.12 b** Replica of diffraction pattern for same grating with $N = 20$ Slits. The principal maxima are narrow and intense whereas the higher order secondary maxima are almost not visible

Now, the equn. (9.15) i.e. $d \sin \theta = m\lambda$ is normally known as grating equation is to be modified for light incident on the grating at an angle $i$ and the resulting equation will then be more general. This can be seen from the following figure.

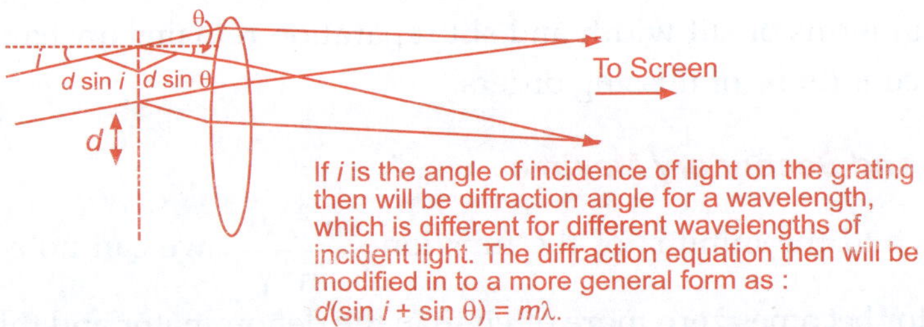


If $i$ is the angle of incidence of light on the grating then will be diffraction angle for a wavelength, which is different for different wavelengths of incident light. The diffraction equation then will be modified in to a more general form as :
$d(\sin i + \sin \theta) = m\lambda$.

The grating equation $d(\sin i + \sin \theta) = m\lambda$ is individually valid for each value of the wavelengths of the incident radiation and so if we have in the incident light two or more wavelengths we will get maxima of the corresponding wavelength at different value of $\theta$ and so we will have maxima for the different wave lengths (different colours) in sets of entirely separate line spectrum. The separation of any two colours such as say for the wavelengths $\lambda_1$ and $\lambda_2$ increases with $m$ the order number. To express this separation between different colours the term frequently used is known as "Angular dispersion", which is the rate of change of angle with change of wavelength. Now, taking the finite increment of the grating equation $d(\sin i = \sin \theta) = m\lambda$, we get for constant angle of incidence $i(\sim 0)$.

$$\Delta(d \sin\theta) = \Delta(m\lambda)$$
$$d\cos\theta\,\Delta\theta = m\,\Delta\lambda$$

or, $$\frac{\Delta\theta}{\Delta\lambda} = \frac{m}{d\cos\theta}.$$ ...(9.16)

Therefore, larger will be the separation between two colours at higher order values of $m$ for a fixed value of grating space, $d$ and at higher angles i.e. lesser values of $\cos\theta$.

The real advantage of grating over prism lies not in its large dispersion, however, but in the high resolving power it affords.

The chromatic resolving power of a grating is defined as:

Chromatic Resolving Power = Angular Dispersion × Width of the emergent beam

$$\frac{\lambda}{\Delta\lambda} = \frac{\Delta\theta}{\Delta\lambda} \times N\,d\cos\theta = mN$$

Now, substituting the value of $m$ from equation $d(\sin i = \sin\theta) = m\lambda$, we get

$$\frac{\lambda}{\Delta\lambda} = \frac{d(\sin i + \sin\theta)}{\lambda}N = \frac{W(\sin i + \sin\theta)}{\lambda}.$$ ...(9.17)

Here, $W = N\,d$ is the total width of the grating. At given values of $i$ and $\theta$, the resolving power is therefore independent of the number of lines (rulings) in the distance $W$. Theoretically, the ideal maximum resolving power obtainable with any grating occurs when $i = \theta = 90°$ and it equals $2\,W/\lambda$. However, one can only attain about $2/3^{rd}$ of this ideal maximum.

### *Measurement of Wavelength by Grating: An important use*

The most important application of diffraction grating is the determination of unknown wave length of the incident light. Using the general grating equation $s(\sin i + \sin\theta)$ and setting the grating surface normal corresponding to the incident light. This is done first by observing the reflected light from the surface of grating and then rotating the prism table through 45° further. Under this alignment incident angle $i$ will be zero and the grating equation will be simplified into $d\sin\theta = m\lambda$. As $d$, the grating space is supplied, then measuring the diffracting angle $\theta$ and the order of diffraction line the $\lambda$ can be calculated. In the following Fig. 9.13 these two arrangement of the grating on the prism spectrometer are shown in $A$ and $B$.

However, the spectrometric method (by Prism Spectrometer) is not absolutely accurate as the lenses used in the telescopes in the spectrometer are not absolutely free from chromatic aberration. To avoid this difficulty Rowland invented the concave grating, in which the focusing is done by a concave mirror and thus not requiring any lens system either in the collimator or in the telescope.

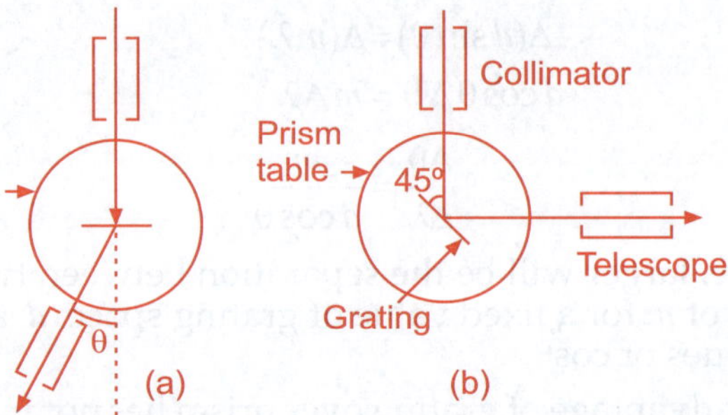


**Fig. 9.13** The Spectrometer arrangement for the determination of wave length of light

If the grating instead of being ruled on plane surface, is ruled on a concave spherical mirror of metal, it will then diffract and also focus light at the same time. This eliminates the requirement of any lens and also chromatic aberration. Besides this Concave grating made on metal surface can also be used in the ultra violet region of the spectrum where the conventional grating etched on glass is not applicable due to absorption in the glass.

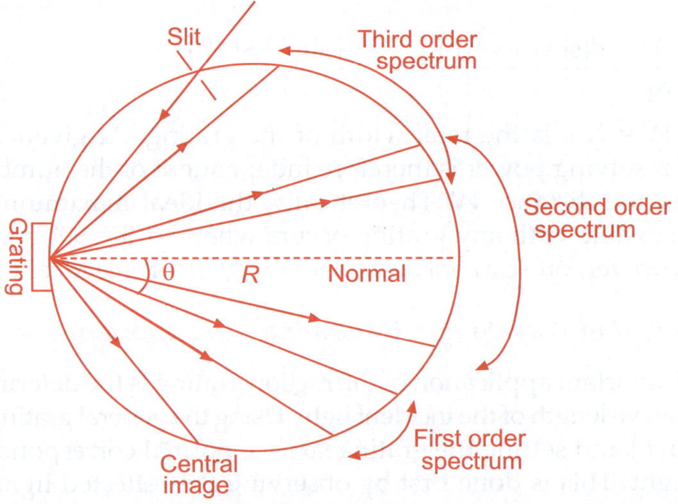


**Fig. 9.14** Concave grating, with slit and the orders of diffraction are focused on the circle of radius $R$. The circle is known as rowland circle

## 9.4 FRESNEL DIFFRACTION

It has been stated earlier that the diffraction phenomenon observed when either the source of light or viewing screen or both are at a finite distance is known as Fresnel Diffraction. It has also been stated that in this type of diffraction, we do not require any lens system either to render the light parallel from a point source at finite distance or to focus the diffracted light on the screen and thus make it much simpler experimental set up than that of Fraunhofer diffraction. Since Fresnel diffraction is the easiest to observe, it was historically the first type to be investigated, although its explanation requires much more mathematical theory than that for Fraunhofer diffraction.

It should be understood by this time that it is diffraction phenomenon due to which the light from a source reaches in to the geometrical shadow part of the obstacle placed before it. However, it is observed that a dark shadow is formed though of slightly diffuse boundary. According to Fresnel, the limits of geometrical shadow the secondary wavelets arrive with phase relations such that they interfere destructively and so an obstacle produces practically dark shadow. There is variation of amplitude with direction, which is known as "obliquity factor" and due to this the amplitude varies as $1 + \cos \theta$.

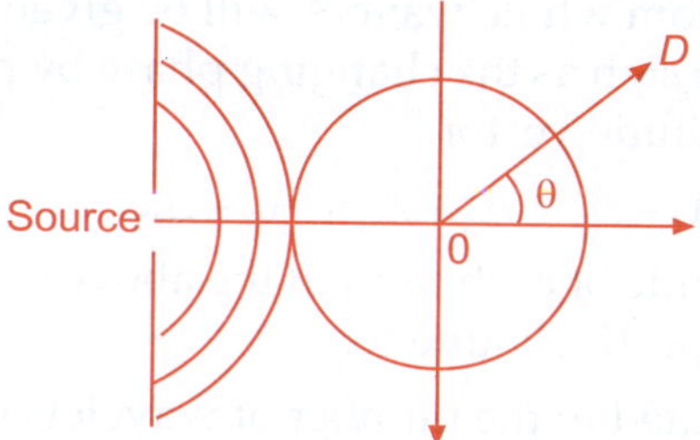


Following Fresnel approach to the diffraction problem let us now find the effect of this reduction of diffracted intensity due to the effect of obliquity from a slightly divergent spherical wave. In the following Fig. 9.15 *ABCD* represents a section of spherical wave front of monochromatic light traveling from a source toward the right. Every point on this spherical section may be thought of as the origin of secondary wavelets and we wish to find the resultant effect on a point *P*. The spherical section is divided in to circular sections called zones around the point *O* which is the foot of perpendicular drawn from the point *P* on the section. Each such sections or zones are a distances as $S_1$, $S_2$ and $S_3$, .... $S_m$ etc., and also each such zones is half wave length further from *P*. As the distance from *O* to *P* is taken as *b*, the zone circles will then be at distances:

$$b + \lambda/2, \, b + 2\lambda/2, \, b + 3\lambda/2, ........, b + m\lambda/2.$$

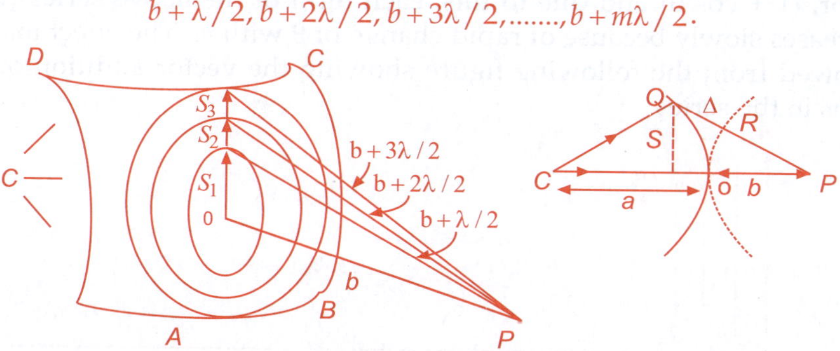


**Fig. 9.15** A shows the half period zones on the section of spherical wave front and *B* shows the path difference from the pole of the spherical wave *i.e.* $CQP - COP = \Delta$

The distances $S$ is very small compared to $a$ and $b$ and so $S$ may be considered as vertical distance of $Q$ above the axis and $\Delta$ may be equated to the sum of the sagittas of the two arcs $OQ$ and $OR$. From the by sagitta formula we have

$$\Delta = \frac{S^2}{2a} + \frac{S^2}{2b} = S^2 \frac{a+b}{2ab}.$$

As each zone is at a distance $\lambda/2$ further from the point $P$, the successive zones will produce resultant at $P$ which differ by $\pi$. This difference by a half period in the vibrations from each zone is the origin of the name "Half period Zone". If now $A_m$ gives the resultant amplitude from $m^{\text{th}}$ zone, then the total amplitude from whole wave $A$ will be given by a series with each term having alternate sign as the changing phase by p means the reversing the direction of amplitude vector.

$$A = A_1 - A_2 + A_3 - A_4 + A_5 - \ldots\ldots\ldots + (-1)^{m-1} A_m. \quad \ldots(9.18)$$

Now, the magnitude of each term in the above series depends on three factors which are respectively due to:

1. Each term is related to the number of wavelets generated from it and as area increases slowly, so is the number of wavelets and the value of each term.

2. Since amplitude decreases with the increase of distance from the source at $P$, the magnitude of each term decreases with increase of the order of terms $m$.

3. Because of the increase of obliquity the magnitude also decreases.

Therefore, the amplitude due to $m^{\text{th}}$ term, $A_m$ is given as: $A_m = \text{const.}$ $\dfrac{S_m}{d_m}(1 + \cos\theta)$, where $d_m$ is the average distance to $P$ and $\theta$ is the angle at which light leaves the zone.

Practically, the amplitude from any zone depends on the obliquity factor, $(1 + \cos\theta)$ and due to this, each term of the above series (9.18) decreases slowly because of rapid change of $\theta$ with $m$. The effect may be followed from the following figure showing the vector addition of the terms in the series.

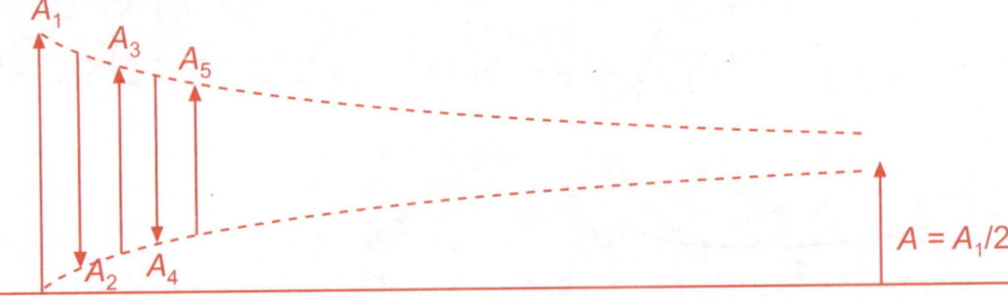


The vectors $A$'s represent the terms of the series (9.18). Like these shown for only five terms if all other terms are drawn, the resultant amplitude $A$

is shown by height of the final arrow head above the base line. Here the tail of each vector is put at the same height as the head of the previous one.

### 9.4.1  Small Circular Opening and Obstacle

If now as an example we consider a very small opening having radius of the opening equal to say $S_1$ the amplitude will be $A_1$ which is twice the amplitude of the unscreened wave and thus the intensity will be four times of the intensity if the screen is removed and entire wave front is allowed to pass. If now the radius of the hole is increased so as to include first two zones $S_1$ and $S_2$ the intensity will almost vanish. This may also be done by changing the distance observing screen from the opening and as a result we will observe alternate maxima and minima along the axis of the aperture. This is definitely an interesting conclusion from Fresnel diffraction and more specifically, smaller the aperture better is the intensity of diffracted image considering the distance between the aperture and the observation screen or photographic plate. This concept though explained much later, was implemented in pin-hole camera centuries ago and many beautiful and historically important photographs were taken without use of any lens system.

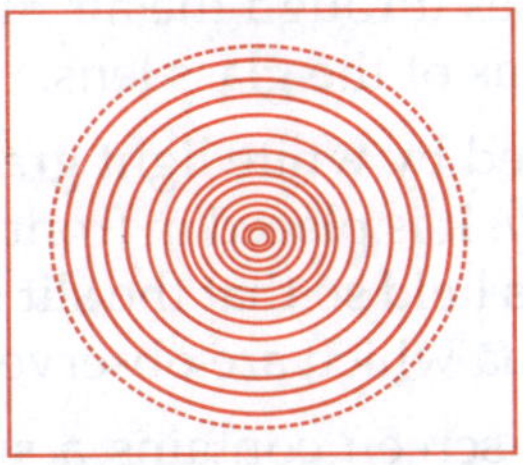

**Fig. 9.16** The replica of the diffraction pattern from a small circular opening

Now, if an obstacle is placed to block off the light from every other half period zones, the result will be to remove either every positive or negative terms of the series (9.18). In both cases the amplitude at point $P$ (Fig. 9.14) will be increased to many times in its value. No further discussions on Fresnel diffraction are included in this section as these are beyond the scope of this book. Readers are requested to go through the references given.

## REVIEW QUESTIONS

1. If a system of vibration differing in amplitude and period is superimposed. Show that the resulting displacement is periodic and that its period is the least common multiple of the periods of the different vibrations.

2. Explain the similarity and differences between interference produced by division of wave front and by division of amplitude.

3. What are the differences between 'Interference' and 'Diffraction'? Establish that both of them can only be explained by wave theory of light.

4. What are the differences between Fraunhofer and Fresnel diffraction?. What is more general?

5. If the number of openings on a stop are more in number but small in size, as in diffraction grating, the diffraction pattern is drastically changed from a single opening. — Explain.

6. Newton's rings are formed between a plane surface of glass and a lens. The diameter of the fifth black ring is 9 mm. When sodium light is used and light passes through the air film at an angle of 30° to the normal. Find the radius of the glass lens.

7. A soap film illuminated by white light gradually becomes thinner as the liquid drains away. It is placed in front of the slit of a direct vision spectroscope, which is held so that the slit is horizontal. Describe and explain the phenomena which are observed.

8. A very large opaque screen contains a small rectangular opening. Parallel monochromatic light incident normally on the screen, passes through the aperture. Investigate diffraction phenomena produced.

## REFERENCES


1. R.A. Houston- A treatise on Light, Longmans, Green and Co. New York, 1947.

2. F.A. Jenkins and H.E. White – Fundamentals of Optics, 4th Edition McGraw Hill international, 1982.

3. K.D. Moller-Optics, University Science Book, Mill valley, 1988.

4. S. B. Palmer and M. S. Rogalski- Advanced University Physics, Gordon and Breach Pub, 1996.

# Physical Optics - II

## 10.1 POLARIZATION OF LIGHT AND STATE OF POLARIZATION

The phenomena of interference and diffraction of light just discussed are the verification of the fact that light is a wave and now the question comes what type of wave? Is it an elastic waves like sound wave or the wave that propagate in solids, liquid and gaseous medium? The phenomenon, known as polarization gives the answer to this question. The fact that light can pass through vacuum and does not require any material medium like sound or other elastic waves. This proves that light is an electromagnetic wave having electric and also magnetic vectors perpendicular to the direction of propagation and so these vectors lie on the plane of the wave front of light rendering the light as a transverse electromagnetic wave. We will in this chapter discuss the important properties of transverse wave like polarization. The vibrations of the say electrical vector lies on the plane of the wave front and the resultant vibration in most general case is elliptical, of which the linear and circular vibrations are two extreme cases. When the phase difference between two vectors unequal in magnitude is $\pi/2$, the resultant is elliptical vibration and it will be circular if the magnitudes are equal and when the phase difference is integral multiple of $\pi$, the resultant is linear. It may here be emphasized that sound wave being a longitudinal wave must necessarily be symmetrical about the direction of propagation, light wave under certain conditions shows dissymmetry and under that condition light is known as polarized.

Before we proceed for discussions on polarization, it is better to introduce and clarify the meaning of "Optic Axis". Calcite and Quartz are examples of anisotropic crystals in which the physical properties vary with direction. If there is a single direction, perpendicular to which the physical property, say heat conductivity remains same in any direction and at other angles to the direction, the property changes, then that direction is called "Optic Axis". Further more, this direction is also the axis of symmetry of the molecules and also of the crystal form.

Polarization may take place from the effects called (a) Reflection (b) dichroism (c) Double refraction and (d) Scattering.

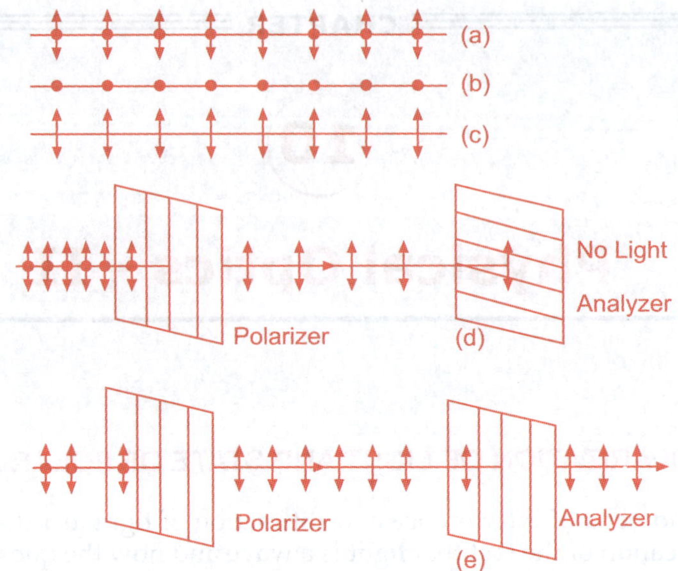



**Fig. 10.1** Pictorial representations of ordinary light (a), plane polarized with vibrations perpendicular on the plane of the diagram (b) and parallel to the plane of the fig.(c). (d) and (e) are the effects of plane polarized light through polarizer and analyzer

The above figure shows in (d) and (e) that when ordinary light gets polarized in the plane of the paper after passing through a "polarizer", having say vertical gates, interact with another similar gate but with opening perpendicular to the earlier one will fail to pass through and will pass when the gates are parallel. This is a simplified pictorial representation of the phenomenon and it's outcome.

### Polarization by reflection and refraction: Brewster's Law

It has been observed that when unpolarized light is reflected from a glass plate at an angle of incidence 57° and the reflected light is again reflected by a parallel glass plate the final reflected light will decrease in intensity with the rotation and may be cut off if the second glass plate is rotated by 90°.

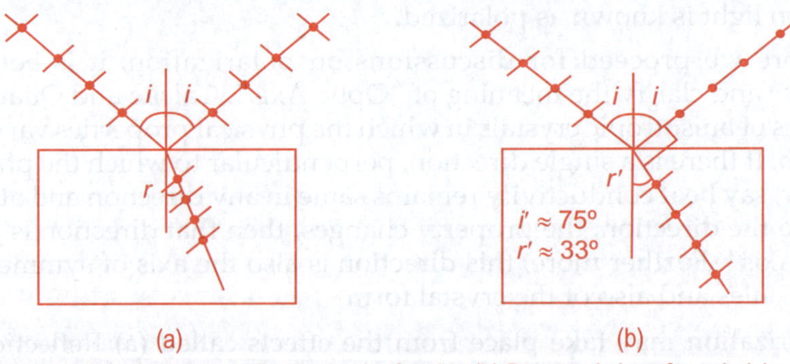



**Fig. 10.2** (a) Polarization by reflection and refraction (b) Brewster's Law for polarizing angle

It was observed by Brewster that when unpolarized light ordinary light is incident on glass surface the reflected light is partially polarized and when the incidence angle is about 57° the reflected light is fully polarized and at that angle of incidence the reflected light makes 90° with the refracted light. This relates the polarizing angle with the index of refraction which is known as Brewster's Law.

$$\frac{\sin i}{\sin r} = n \, (\text{refractive Index})$$

$$\frac{\sin i'}{\sin r'} = \frac{\sin i'}{\sin(180° - 90° - i')} = \frac{\sin i'}{\sin(90° - i')} = \frac{\sin i'}{\cos i'} = \tan i' = n$$

Therefore,    $i' = \tan^{-1} n$.

When light is incident on the glass surface, the plane of vibration which lies on the plane of the figure and refracted in the glass can not have its component at right angle and so the reflected beam at Brewster's angle of incidence will be polarized perpendicular to the figure. The refracted beam will contain both planes of vibration but upon subsequent refractions through refracting piles it may be also be fully polarized.

Again when a narrow pencil of unpolarized light is incident on some mineral crystals like "Tourmaline", the transmitted light is polarized. The phenomenon is demonstrated in Fig. 10.1, where both "Polarizer" and "Analyzer" are tourmaline crystals.

## Double Refraction

The crystals of Calcite ($CaCO_3$) and Quartz ($SiO_2$), the former belonging to crystal classes as rhombohedra and have the molecules having axis of symmetry are placed in such a regular fashion that the crystal as a whole have a net axis of symmetry named as Crystal axis or Optic axis. Calcite can be cleaved into thin slices and can be used for polarization by refraction. The electric vector of light wave can be resolved in the direction parallel to the optic axis and perpendicular to the optic axis of the crystal. These two components are named as Extra ordinary ($E$), which is parallel and Ordinary ($O$) which is perpendicular to the optic axis and they travel with different velocities within the crystal. As the velocities are different so are the refractive indices and as a result, we get two refractive rays (Fig. 10.3 b). This is known as "Double Refraction".

$$n^{\circ} - \frac{\sin \varphi}{\sin \varphi_1} \quad \text{and} \quad n^{E} = \frac{\sin \varphi}{\sin \varphi_2}$$

$$As\, \varphi_1 \langle \varphi_2 \ so, \sin \varphi_1 \langle \sin \varphi_2 \quad \text{and} \quad n^{\circ} \rangle n^{E}$$

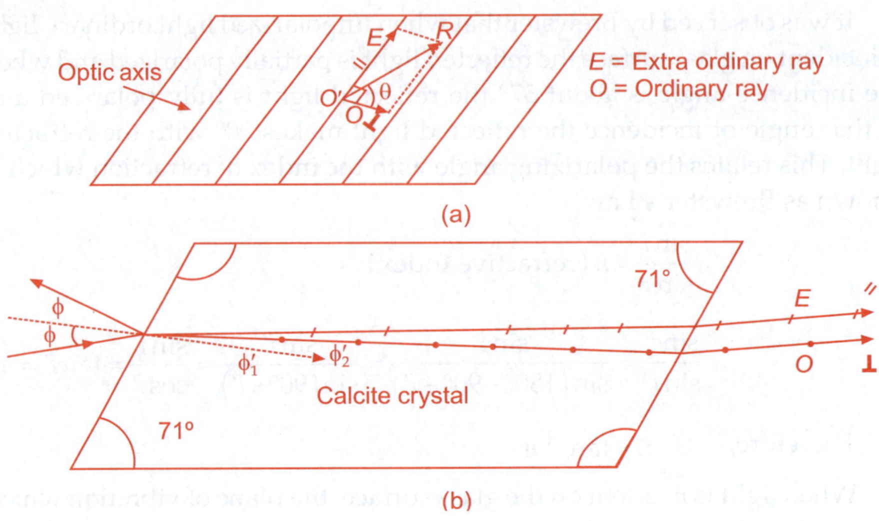


**Fig. 10.3 A.** Represents a slice of Calcite crystal, the parallel lines show the optic axis of the plane of crystal. The electric vector $O'R$ incident on such crystal at an angle, may be resolved in the direction perpendicular to the optic axis, called $O$ ray (Ordinary Ray) and parallel to the optic axis called as $E$ ray (Extra ordinary). **B.** Two separated rays of light separated due to refraction at different angles

From the stand point of optics, doubly refracting crystals are classified as either "uniaxial" or "biaxial". In uniaxial crystals the velocities of $O$ and E waves become equal along the only and unique direction introduced as optic axis. But in biaxial crystals there are two directions in which the velocity of plane wave is independent of the orientation of the incident vibration. However, the uniaxial crystals may be assumed to be a special case of biaxial crystal where the angles between the axes are zero.

Uniaxial crystals are divided into two classes negative and positive. In negative uniaxial crystals the extraordinary index of refraction ($n^E$) is less than ordinary index ($n°$) $\cdot$ Quartz being a positive uniaxial crystal ordinary refractive index $n°$ is less than $n^E$. Let us now start with negative uniaxial crystal.

## Nicol Prism

If such Calcite crystal slice is polished so as to reduce one edge angle from 71° to 68° and is cut diagonally into two pieces, polished and then joined together by Canada balsam (an adhesive cement having refractive index almost same as glass), device can be used as polarizer and also as analyzer using the double refraction phenomenon.

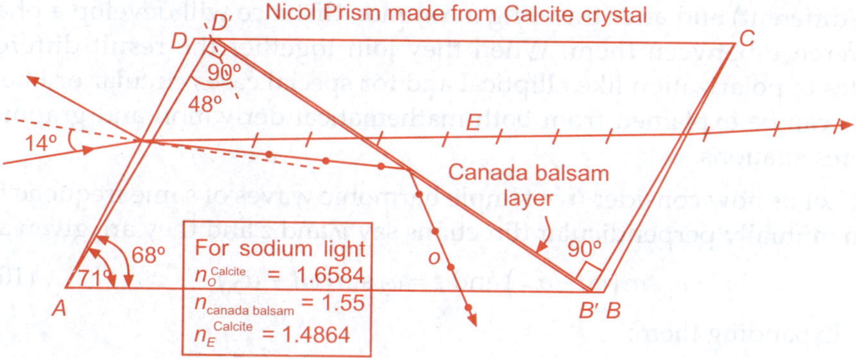



**Fig. 10.4** Nicol Prism, showing the removal of ordinary ray (O) from extra ordinary ray (E)

If ordinary un polarized ray is incident on the side at an angular range of 14° (near normal), the ray is divided into two parts, ordinary $O$ and extra ordinary $E$. The refractive indices for $O$ and $E$ rays in Calcite medium are different and are respectively as: $n_O^{Calcite} = 1.6584$ and $n_E^{Calcite} = 1.4864$. The refractive index for Canada balsam which is used to join two parts is $n_{Canada\ balsam} = 1.55$. Therefore, when $O$ and $E$ rays are incident on the Canada balsam thin layer, for $E$ ray it is from a rarer medium to denser medium and so the E ray will refract into the second part of calcite but for $O$ ray it is from a denser medium to rarer and for the geometry of the prism, the incident angle for $O$ ray will be greater than critical angle and so $O$ ray will suffer total internal reflection and thus get separated from $E$ ray. The emerging ray will then be totally plane polarized on the plane of optic axis of the crystal.

Now, if this $E$ ray coming out from the Nicol prism, called polarizer is again incident on another such prism then $E$ ray will also pass through the second one if and only if the optic axes of these two prisms are parallel. If now the second one, called Analyzer is rotated about the direction of the $E$ ray, the optic axis of the analyzer will also rotate and when the angle between the optic axes of these two prisms becomes any value other than zero, the intensity of the final emergent ray decreases and finally it will be zero when this angle becomes 90°. The situation is same as Fig. 10.1 a.

## State of Polarization

The electric vector $E$ of the light which is electromagnetic wave may be assumed to vary sinusoidally and when it is incident on a polarizer it is resolved into two components one parallel and other perpendicular to the optic axis. Now these two components assumed to be sine waves travel with different velocities (as refractive indices in any medium for them are different) and after traveling a common distance will develop a phase difference between them. When they join together the result different states of polarization like elliptical and for special cases circular or linear. This can be explained from both mathematical derivation and graphical representations.

Let us now consider two simple harmonic waves of same frequency in two mutually perpendicular directions say $y$ and $z$ and they are given as

$$y = a_1 \sin(\omega t - \alpha_1) \text{ and } z = a_2 \sin(\omega t - \alpha_2) \qquad \text{...(10.1)}$$

Expanding them:

$$\frac{y}{a_1} = \sin\omega t \cos\alpha_1 - \cos\omega t \sin\alpha_1 \quad \text{and} \quad \frac{z}{a_2} = \sin\omega t \cos\alpha_2 - \cos\omega t \sin\alpha_2.$$

Multiplying the first by $\sin\alpha_2$ and the second by $\sin\alpha_1$, and subtracting the first from second, we get

$$-\frac{y}{a_1}\sin\alpha_2 + \frac{z}{a_2}\sin\alpha_1 = \sin\omega t(\cos\alpha_2 \sin\alpha_1 - \cos\alpha_1 \sin\alpha_2) \quad \text{...(10.2 a)}$$

Similarly, multiplying first by $\cos\alpha_2$ and second by $\cos\alpha_1$ and subtracting second from first, we get

$$\frac{y}{a_1}\cos\alpha_2 - \frac{z}{a_2}\cos\alpha_1 = \cos\omega t(\cos\alpha_2 \sin\alpha_1 - \cos\alpha_1 \sin\alpha_2) \quad \text{...(10.2 b)}$$

Now, squaring both equn. (10.2 a) and (10.2 b) and adding we eliminate $t$.

$$\sin^2(\alpha_1 - \alpha_2) = \frac{y^2}{a_1} + \frac{z^2}{a_2} - \frac{2yz}{a_1 a_2}\cos(\alpha_1 - \alpha_2) \qquad \text{...(10.2 c)}$$

Now, if the phase difference $(\alpha_1 - \alpha_2) = \delta$, then above equation is simplified as:

$$\sin^2\delta = \frac{y^2}{a_1} + \frac{z^2}{a_2} - \frac{2yz}{a_1 a_2}\cos\delta. \qquad \text{...(10.2 d)}$$

This is the equation for the resultant path. The following figure gives the graphical representations of the resultant vibrations.

If $\delta$, the phase difference between two sine waves vibrating on two planes, mutually perpendicular to each other is $m\,p$, where $m$ is an integer 0, 1, 2, 3... etc. then resultant vibration is linear and polarized light is linearly polarized. If $d$ is $m\,p$ and m is odd numbers like 1, 3, 5,... etc. then the resultant vibrations are circular if the amplitudes $a_1$ and $a_2$ are equal in all other cases as shown in the Fig. 10.5 the resultant vibration are elliptical. The resultant light wave is then elliptically polarized. This elliptical polarization is then a general case and the circular and linear polarization states are only special.

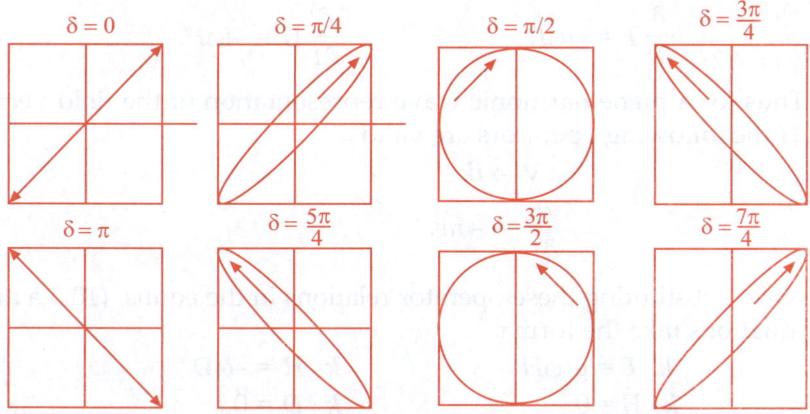



**Fig. 10.5** Combination of two waves at right angles having different phase. Differences showing different states of polarization

## 10.2  OPTICAL CLASSIFICATION OF ANISOTROPIC MEDIA

Recalling the Maxwell's equations as discussed in Chapter 7 i.e. equns. (7.1), (7.2), (7.3) and 7.5.

$$\nabla \cdot E = \frac{\rho}{\varepsilon_0}$$

$$\nabla \cdot B = 0,$$

$$\text{curl } E = \nabla \times E = -\frac{\partial}{\partial t} B, \quad \text{and}$$

$$\text{curl } B = \nabla \times B = \mu_0 \left[ j + \varepsilon_0 \frac{\partial}{\partial t} E \right].$$

Now, if we consider only the non magnetic media i.e. $B = 0$ and $\mu_r = 1$ (from $B = \mu_r H$), which contains no volume charge $r = 0$ and no conduction current $j = 0$, then the above Maxwell's field equations (recalled) reduce to :

$$\nabla \cdot H = 0 \text{ and } \nabla \cdot D = 0 \text{ (As } B = \mu_0 H \text{ and } D = \varepsilon_0 E) \qquad \text{...(10.3 a)}$$

$$\nabla \times E = -\mu_0 \frac{\partial}{\partial t} H \text{ and } \nabla \times H = \frac{\partial}{\partial t} D \qquad \text{...(10.3 b)}$$

Now, assuming that this media sustains monochromatic plane wave, propagating in any direction $k = n\,k$ (wave vector) then the field vectors $E$, $D$ and $H$ may then be given the harmonic representation as :

$$E = E_0\, e^{i(k \cdot r - wt)}, \; H = H_0\, e^{i(k \cdot r - wt)} \text{ and } D = D_0\, e^{i(k \cdot r - wt)}$$

Therefore, the field variation with both position and time is harmonic and as the wave vector $k = nk$ the following properties may be derived

$$\nabla \cdot E = i\,k \cdot E \qquad\qquad \nabla \cdot H = ik \cdot H$$
$$\nabla \times E = i\,k \times E \qquad\qquad \nabla \times H = ik \times H \text{ and}$$

$$\frac{\partial}{\partial t} E = -i\omega E \qquad\qquad \frac{\partial}{\partial t} H = -i\omega H.$$

Thus for a plane harmonic wave representation of the field vectors $E$ and $H$ the following operators are valid

$$\nabla \rightarrow i k$$

$$\frac{\partial}{\partial t} \rightarrow - iw.$$

Now, substituting these operator relations in the equns. (10.3 a and b), the equations take the form :

$$k \cdot E = \mu_0 \omega H \qquad\qquad k \cdot H = -\omega D$$
$$k \cdot H = 0 \qquad\qquad k \cdot D = 0. \qquad\qquad ...(10.4)$$

Now if we eliminate $H$ from equn. (10.4), the plane wave equation can be formulated as

$$k \times (k \times E) = \mu_0 \omega (k \times H) = -\mu_0 \omega^2 D. \qquad\qquad ...(10.5)$$

Now, the relation between $D$ and $E$ as from Maxwell's equations as $D = \varepsilon_0 \varepsilon_r E = \varepsilon E$ cannot be always used or expressed with scalar dielectric constant only as for anisotropic medium. We then must assume the general form of a linear relation between the field components with respect to a Cartesian axes fixed in the medium

$$D_i = \varepsilon_0 \sum_{j=1}^{3} \varepsilon_{ij} E_j, \text{ where } i, j \text{ stand for } x, y \text{ and } z.$$ There are nine coefficients

$\varepsilon_{ij}$, which form the elements of dielectric tensor $\hat{\varepsilon}_r$ so that the relation with $D$ can be written as

$$\begin{pmatrix} D_x \\ D_y \\ D_z \end{pmatrix} = \varepsilon_0 \begin{pmatrix} \varepsilon_{xx} & \varepsilon_{xy} & \varepsilon_{xz} \\ \varepsilon_{yx} & \varepsilon_{yy} & \varepsilon_{yz} \\ \varepsilon_{zx} & \varepsilon_{zy} & \varepsilon_{zz} \end{pmatrix} \begin{pmatrix} E_x \\ E_y \\ E_z \end{pmatrix}.$$

Now, if all the elements of the matrix $\hat{\varepsilon}_r$ are real the matrix is symmetric about the diagonal and describes a non active medium which exhibits no optical activity, where as in any optically active medium some elements $\varepsilon_{ij}$ must be complex. For real symmetric matrices there exists always a set of Cartesian axes, called principal axes, so that dielectric matrix takes the diagonal form

$$\hat{\varepsilon}_r = \begin{pmatrix} \varepsilon_x & 0 & 0 \\ 0 & \varepsilon_y & 0 \\ 0 & 0 & \varepsilon_z \end{pmatrix} = \begin{pmatrix} n_x^2 & 0 & 0 \\ 0 & n_y^2 & 0 \\ 0 & 0 & n_z^2 \end{pmatrix}.$$

Where, $\varepsilon_x$, $\varepsilon_y$ and $\varepsilon_z$ are principal dielectric constants and $n_x$, $n_y$ and $n_z$ are principal refractive indices. Substituting these conditions in equn. (10.5), we get the general plane wave equation in anisotropic media:

$k \times k \times E + (w^2/c^2)\, \hat{\varepsilon}_r E$ and using the vector relation:

$$k\,(k \cdot E) - k^2 E + (\omega^2/c^2)\, \hat{\varepsilon}_r E = 0 \qquad \text{...(10.6)}$$

If now we assume that $k$ lies on the $x - z$ plane, then $k_y = 0$ and $k^2 = k^2_x + k^2_z$, then the scalar components are

$$\left(\frac{\omega^2}{c^2} n^2_x - k^2_z\right) E_x + k_x k_z E_z = 0$$

$$\left(\frac{\omega^2}{c^2} n^2_y - k^2_x - k^2_z\right) E_y = 0$$

$$k_z k_x E_x + \left(\frac{\omega^2}{c^2} n^2_z - k^2_x\right) E_z = 0. \qquad \text{...(10.7)}$$

The second reduces to the equation of a circle i.e.

$$k^2_x + k^2_z = \frac{\omega^2}{c^2} n^2_y \qquad \text{...(10.8)}$$

and from first and third we get equation of an ellipse.

$$\frac{k^2_x}{\left(\omega n_z / c\right)^2} + \frac{k^2_z}{\left(\omega n_x / c\right)^2} = 1 \qquad \text{...(10.9)}$$

In general case we must assume that the three principal refractive indices are different and for convenience, we can order them as $n_x > n_y > n_z$. The end points of the possible wave vectors in the $x$-$z$ plane are distributed on the circle equn. (10.8) and ellipse equn. (10.9).

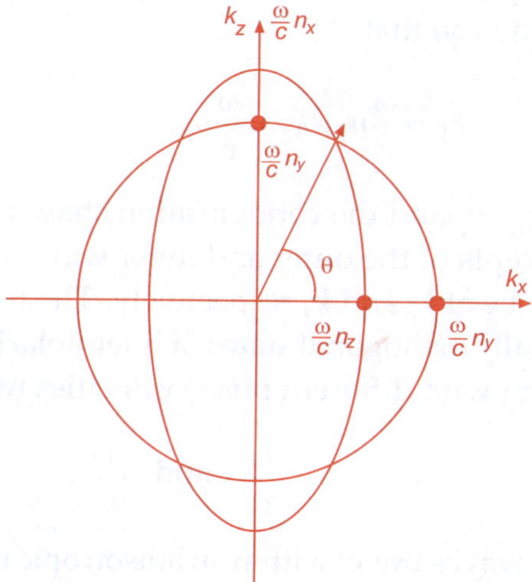


**Fig. 10.6** The solid circle represents the wave vector surface (equn. 10.8) and the broken ellipse represents the equn. (10.9) with $x$-$z$ plane

## 10.3 BIREFRINGENCE

The velocity of light in any refracting medium depends on the different colour components of white light. This phenomenon is known as dispersion but this is not only dependence of velocity of light, it also depends on the planes of polarization resulting into different refractive indices in different planes of polarization. This latter phenomenon is known as Birefringence. Let us discuss this in a bit more details.

The optical properties of anisotropic medium as derived in the earlier section can be simplified if we consider wave propagation along one of the principal axes of dielectric tensor, say the $y$-axis such that $k_x = k_z = 0$ and $k_y = k$, the equn. (10.7) derived from Maxwell's equations then take the form

$$\left(\frac{\omega^2}{c^2}n_x^2 - k^2\right)E_x = 0$$

$$\frac{\omega^2}{c^2}n_y^2\,E_y = 0$$

$$\left(\frac{\omega^2}{c^2}n_z^2 - k^2\right)E_z = 0\,. \qquad\qquad …(10.10)$$

The second equation states that there is no electric vector in the $y$-axis as the light cannot have any longitudinal component as we have considered that wave travels in the $y$-axis direction. So, $E_y = 0$ but $E_x \neq 0$ and also $E_z \neq 0$. Therefore from equations first and third of (10.10), as we are considering anisotropic medium these two equations must have different values of $k$ say $k_1$ and $k_2$ so that

$$k_1 = \frac{\omega}{c}n_x,\; k_2 = \frac{\omega}{c}n_z\,.$$

If we consider $n_x \rangle n_z$ and the configuration shown in the Fig. 10.6, we can obtain the intercepts of the outer and inner wave vector surfaces with the principal axis to be at $k_1$ and $k_2$ respectively. The two wave vectors $(k)$ correspond to mutually orthogonal states of liner polarization, in the $x$ and $z$ directions, traveling with different phase velocities respectively as

$$\frac{\omega}{k_1} \quad \text{and} \quad \frac{\omega}{k_2}\,.$$

When these two waves travel within an anisotropic medium they follow different paths and due to this:

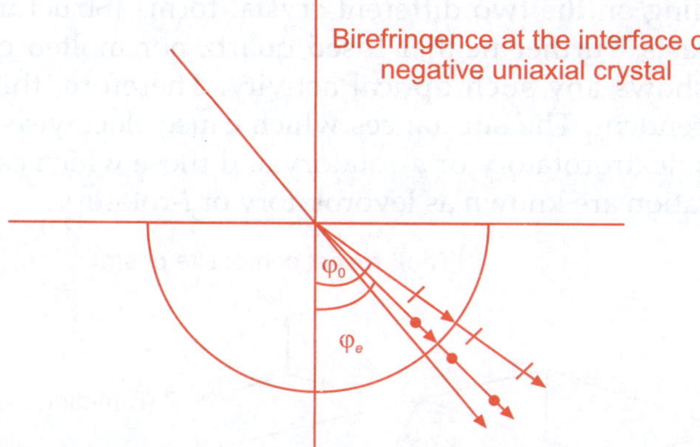



For negative uniaxial crystal (for positive it is just opposite)

$$n_0 \sin\theta = \sin\varphi_0, \; n_e \sin\theta = \sin\varphi_e \text{ and so, } \frac{1}{n_0} < \sin\theta < \frac{1}{n_e}.$$

There are certain media like quartz, which are found to rotate the plane of linearly polarized electromagnetic waves propagating through them and are said to be "optically active". If for an electromagnetic wave propagating in $z$ direction has the $E$ and $H$ components in $x$ and $y$ directions and is incident on a quart crystal block of thickness $\Delta Z$, the $E$ and $H$ get rotated either clockwise or anti clockwise through certain angle say $\varphi$ (depending on $\Delta Z$) after it emerges out. This is shown in the following figure.

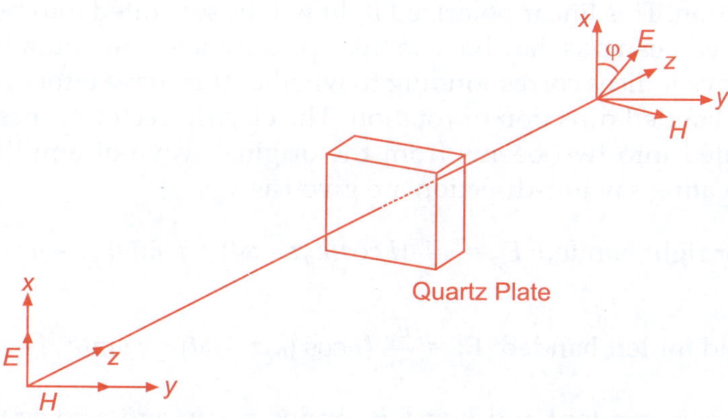



As early as in 1811 Dominique. F. J. Arago, a French physicist first observed that a plane polarized light undergoes a continuous rotation of its plane of polarization as it passes through a quartz plane in direction of its optic axis. This is the phenomenon now known as "Optical Activity". It has been observed that when we look towards the source, the plane of polarization either rotates right (clockwise) or left direction (Anticlock wise) depending on the two different crystal forms (Structure) in which quartz can exists. Further neither fused quartz nor molten quartz being amorphous shows any such optical activity. Therefore, this activity is structure dependent. The substances which cause clockwise rotation are referred to as dextrorotatory or *d*-rotatory and those which cause counter clockwise rotation are known as levorotatory or *l*-rotatory.

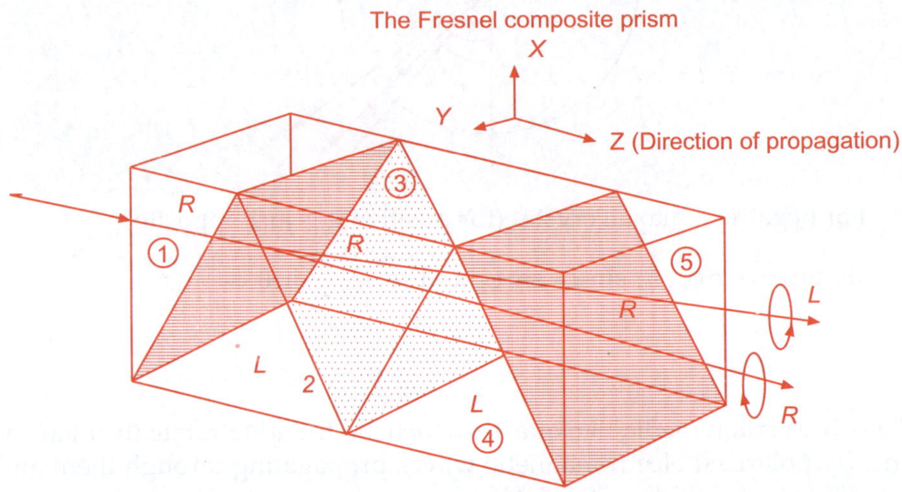


Fresnel composite prism is an important study of circularly polarization in clockwise and anti clockwise directions. A linear polarized light is incident from left of the composite prism which is made up of aligning say five prisms in alternate order having left handed and right handed direction of rotation. This linear polarized light will be separated into two circularly polarized beams as they have in this optically active medium two different refractive indices corresponding to whether they have either right handed or left handed direction of rotation. The electric vector of these two fields separated into two beams from the original wave of amplitude $E_0$ and propagating say in $z$-direction are given as

For right handed: $E_R = \dfrac{E_0}{2} \{i \cos(k_R z - \omega t) + j \sin (k_R z - \omega t)\}$

and for left handed: $E_L = \dfrac{E_0}{2} \{i \cos (k_L z - \omega t) + j \sin(k_L z - \omega t)\}$

as $w$ is constant and $k_R = k_0 n_R$ and $k_L = k_0 n_L$ and also as the resultant disturbance E is given by

$$E = E_R + E_L.$$

The $E$ can then be found after some simple trigonometric operations and that is

$$E = E_0 \cos[(k_R + k_L) z/2 - \omega t] [i \cos (k_R - k_L)z/2 + j \sin(k_R - k_L) z/2]$$

It can also be followed that at the point of entrance i.e. $z = 0$, the above expression reduces to: $E = E_0 i \cos \omega t$, which the equation of linearly polarized light. Keeping this in mind we can understand that the field at point $z$ makes an angle of $\beta$ with respect to the original orientation as: $\beta = (n_L - n_R)z/2$ and for a medium of thickness $d$ the plane of vibration rotates through an angle of $\beta = \dfrac{\pi d}{\lambda_0}(n_L - n_R)$ for $d$-rotatory $n_L > n_R$ and for $l$-rotatory $n_R > n$.

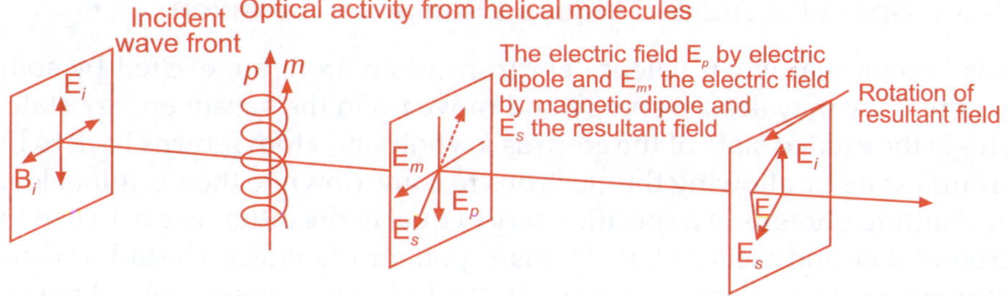


**Fig. 10.7** Propagation of linear polarized light through optically active medium and its rotation

In quartz the silicon and oxygen atoms are known to be arranged in either right or left handed spirals about the optic axis and because of this the incoming wave will interact oppositely with the specimen depending on whether it "sees" the molecules as right or left handed helices. Let us consider the molecule has its optic axis parallel to the time dependent electric vector of the incident wave ($E_i$). Now this field will drive charges up and down along the length of the molecule and produces a time variant electric dipole moment $p(t)$ parallel to the axis. Now, due to the spiraling motion of the electrons a current may be considered to exist in spiral motion about the axis. This current will result magnetic dipole moment $m(t)$. If the helical axis is parallel to the time dependent magnetic field vector $B_i$, then the varying magnetic field would have produced induced electric current circulating the molecule. This would again produce oscillating axial electric and magnetic dipole moments. In either case the $p(t)$ and $m(t)$ will be parallel or anti parallel depending on whether the molecule has its atoms spiral right handed or left handed. Now the molecule considered as oscillating dipole will absorb energy from the incident field and radiate. The electric field radiated by the electric dipole $E_p$ and the electric field radiated by the magnetic dipole $E_m$ will be perpendicular to each other. The resultant of these two fields, $E_S$ will however be not parallel to the incident field $E_i$. The same is also true for the magnetic field. The plane of vibration of the resultant transmitted wave will then be $E_S + E_i = E$ and will be rotated in the direction determined by the sense of helix. The amount of rotation will however, vary with the orientation of each molecule.

## 10.5 LIGHT AMPLIFICATION BY STIMULATED EMISSION OF RADIATION (LASER)

Laser is the abbreviation of the title of this section. The explanation of the origin and emission needs quantum mechanical approach. Though this is an microscopic phenomena, it is included in this part of the book as the quantum mechanical treatment is avoided and the phenomenon is described rather qualitatively.

### 10.5.1 Spontaneous and Induced Radiative Transition

It is known that when valence electrons of an atom are excited by some energy, they may absorb energy and move up in the vacant energy states. This is the excited state of the atom as a whole and atom returns back to the ground state by allowing the electrons to come down to their original level by emitting photons of a specific energy equal to the difference between the ground state and excited state. This is a spontaneous effect. The induced and spontaneoustransitions are shown in the following energy level diagram, (Fig. 10.8).

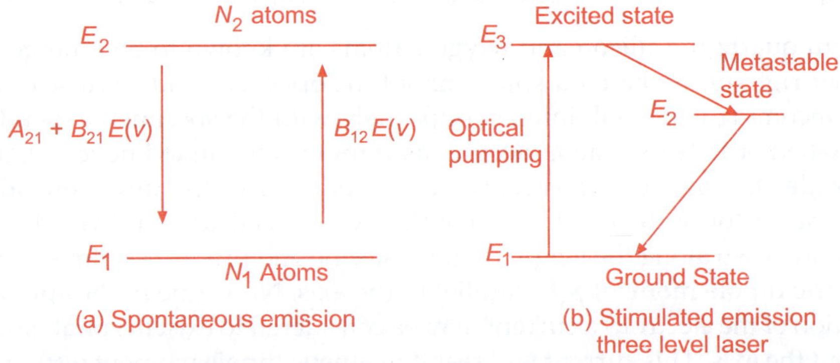


**Fig. 10.8** Spontaneous emission and stimulated emission (three level laser)

First let us consider the spontaneous emission described in Fig. 10.8 a. $N_1$ and $N_2$ are the number of atoms in energy levels $E_1$ and $E_2$. The photon energy corresponding to the transition between these two levels is $hv = E_2 - E_1$. Atoms from higher level $E_2$ may spontaneously jump to $E_1$ and emit the photon. Now, let $A_{21}$ represent the probability of this transition. If the radiation of energy $E = hv$ is present, then it may be responsible for submitting this energy to the $E_1$ level atoms to get excited to the level $E_2$. Now, if $B_{12}$ is the transition probability per unit time per unit intensity of radiation present, then $B_{12} E(v)$ will give the induced absorption transition probability per unit time. But the radiation because of its interaction with excited atoms at $E_2$, also produces emission transition from $E_2$ to $E_1$ with an induced emission transition probability per unit time as $B_{21} E(v)$. Therefore, total emission per atom per unit time from $E_2$ to $E_1$ is $A_{21} + B_{21} E(v)$. Now, if there are $N_2$ number of atoms in $E_2$, number of such transition per unit time is given by $[A_{21} + B_{21}\,E(v)]N_2$. At the same time the number atoms that jump from $E_1$ to $E_2$ is given by $B_{12}\,E(v)\,N_1$. Therefore the net change in the number of atoms at $E_2$ level per unit time is equal to net gain by absorption minus the rate of loss due to emission.

$$\frac{dN_2}{dt} = B_{12}\,E(v)\,N_1 - [A_{21} + B_{21}\,E(v)]N_2 .$$

Now, when equilibrium is established between atoms and radiation, we have $\dfrac{dN_2}{dt} = 0$. This leads to: $B_{12}\,E(v)\,N_1 = [A_{21} + B_{21}E(v)]\,N_2$. This implies that the number of absorption and emission per unit time between these two levels are same. Now, if the atoms are in thermal equilibrium and follow Maxwell-Boltzmann statistics (discussed in Chapter 12), then

$$\frac{N_1}{N_2} = e^{(E_2 - E_1)/kT} = e^{hv/kT} .$$

So that, $\qquad B_{12}\,E(v)\,e^{hv/kT} = A_{21} + B_{21}\,E(v)$

or that, $\qquad\qquad E(v) = \dfrac{A_{21}/B_{12}}{e^{hv/kT} - B_{21}/B_{12}}$ $\qquad\qquad$ ...(10.11)

Now, comparing this with Planck's radiation equation discussed in Chapter 12, (equn. 12.54) which will be derived as : $E(n) = \dfrac{8\pi h v^3}{c^3}\dfrac{1}{e^{hv/kT} - 1}$

and comparison with the above equn. (10.11), we find that

$$\frac{A_{21}}{B_{12}} = \frac{8\,\pi v^3}{c^3} \quad \text{and} \quad \frac{B_{21}}{B_{12}} = 1.$$

This expression is known after Einstein equation.

Using the above equations we get

$$\frac{\text{Spontaneous emission probability}}{\text{Induced emission probability}} = \frac{A_{21}}{B_{21}E(v)} = e^{hv/kT} - 1. \qquad ...(10.12)$$

Now, two cases may arise either $hv \gg kT$ or $hv \ll kT$. In the first case spontaneous emission is much more probable than induced (stimulated) radiation but in the second case where $hv \ll kT$, the induced or stimulated emission become prominent. As this induced radiation is the result of the action of incoming radiation on atoms and so this forced atomic oscillation bears a constant phase difference relative to the incoming radiation. This means that all atoms radiate in phase and as a result the stimulated radiation is coherent. This is the one of the most important property of stimulated emission or Laser. All the wave phenomena of light like interference, diffraction and polarization etc., can be more efficiently demonstrated with Laser. On the other hand, the spontaneous radiation occurs at random with no correlation with time and as the phases of atomic radiation are distributed randomly, the spontaneous emission is incoherent. But not only that the stimulated radiation is coherent it is very intense and strictly monochromatic. When several sources radiate in phase, the resultant amplitude is the sum total of the individual amplitude. If now all the sources are alike, we have;

Resultant amplitude of coherent radiation = $Nx$ source amplitude

where $N$ is the number of sources.

As intensity is the square of amplitude, we get :

Resultant intensity of coherent radiation = $N^2x$ source intensity. If the number of sources is large, we get stimulated radiation of very high intensity. Therefore, major importance of Laser is its perfect coherency and very high intensity. So, it can be effectively used in optical communication, photoluminescence and in many industrial applications.

## REVIEW QUESTIONS

1. It is found that when sodium light is incident on the surface of a certain glass plate at an angle of 58°18, that the reflected light is plane polarized. What is the refractive index of the glass?

2. Explain the reason for polarization and the different types of it like circularly polarized, elliptically polarized etc.

3. What is the reason for double refraction, Explain the function of Nicol's prism.

4. The two refracted rays of one single incident ray travel with different velocities in the medium. Narrate the consequence of this phenomenon and also explain the Birefringence.

5. Explain the differences between spontaneous emission and stimulated emission.

6. Explain the phenomena Birefringes and mention the use of this in liquid crystal devices.

## REFERENCES

1. F. A. Jenkins and H. E. White – Fundamentals of Optics, 4th Edition McGraw Hill international, 1982.

2. D. Halliday, R. Resnick and J. Walker-Fundamentals of Physics, Asian Books Pvt. Ltd, New Delhi, 1993.

# Electromagnetic Waves

## 11.1 INTRODUCTION

We have seen that a particle's momentum given by $p = mv$ depends on its inertia, its velocity and also its kinetic energy and angular momentum. As such a particle's dynamical behaviour can change as $v$ changes along its path. For particles a complete description of the dynamics requires a solution for their trajectory, which provides $v(t)$ and $r(t)$ subject to the initial conditions. In contrast to this particle behaviour, when waves are moving in vacuum, their speed is invariant since it is measured by all observers as equal to velocity of light. The field amplitude may be different. Therefore for fields, the momentum and angular momentum, densities and also the energy flux change because the field intensities $E$ and $B$ vary in time and space. To complete the description of field dynamics, we need to know the time dependent and space dependent behaviour of the field amplitudes which should include the information on wave vector or propagation direction. This is what we will discuss in this chapter. We shall see that the fields may either be "attached" to the sources or become completely "detached" in the radiation zone. These two categories result in to the difference in their dynamical properties.

## 11.2 ELECTROMAGNETIC WAVE

In the region of space where there are no free sources, Maxwell's equations as introduced in Chapter 7 reduce to simple form as given by:

$$\nabla \cdot D = 0$$
$$\nabla \cdot B = 0$$

$$\nabla \times E = -\frac{1}{c}\frac{\partial}{\partial t}B$$

$$\nabla \times H = \frac{1}{c}\frac{\partial}{\partial t}D. \qquad \qquad \dots(11.1)$$

We will assume here that the medium is non conducting, because otherwise from Ohm's Law, $J = \sigma E$ results in a current density $J$ and this in turn would act as a source of additional fields.

Considering linear isotropic media,

$$D = \varepsilon E,$$

$$H = \mu' B, \text{ where } \varepsilon \text{ and } \mu' = 1/\mu \text{ are constants so that}$$

their derivatives are zero.

Substituting these in the above equn. (11.1), they can be written as:

$$\nabla \cdot E = 0 \qquad\qquad\qquad\qquad\qquad (a)$$

$$\nabla \cdot B = 0 \qquad\qquad\qquad\qquad\qquad (b)$$

$$\nabla \times E = -\frac{1}{c}\frac{\partial}{\partial t} \cdot B \qquad\qquad\qquad\qquad (c)$$

and

$$\nabla \times B = \frac{\mu\varepsilon}{c}\frac{\partial}{\partial t} E. \qquad\qquad\qquad ...(11.2)(d)$$

These simultaneous equations are first order differential equations. Now, if we take the curl of equn. (11.2 d) and add to it the time derivative of (11.2 c), we get

$$\nabla \times (\nabla \times B) = -\frac{\mu\varepsilon}{c^2}\frac{\partial^2}{\partial t^2} B$$

But since $\nabla \times (\nabla \times B) = \nabla(\nabla \cdot B) - \nabla^2 B$ and as $\nabla \cdot B = 0$, we get

$$\nabla^2 B - \mu\varepsilon \frac{\partial^2}{\partial (ct)^2} B = 0.$$

This hyperbolic partial differential equation is commonly known as wave equation and one of its solutions is

$$B = B_0 \exp\{i(k \cdot x - \omega t)\}. \qquad\qquad\qquad ...(11.3)$$

Here $\omega$, the angular frequency is related to the magnitude of the wave vector $k$ by the relation

$$|k| = \frac{\omega}{v} \text{ where } v = c/\sqrt{\mu\varepsilon} \text{ is a constant of the}$$

medium and will be less than $c$ but will be equal to $c$ in vacuum where $\mu = \varepsilon = 1$.

A similar reduction may be made for electric field $E$, which produces a second wave equation

$$\nabla^2 E - \mu\varepsilon \frac{\partial^2}{\partial (ct)^2} E = 0$$

whose solution is $\quad E = E_0 \exp\{i(k \cdot x - \omega t)\} \qquad\qquad ...(11.4)$

Now, the presence of sources in the system i.e. where $\rho \neq 0$ and $J \neq 0$ modifies the character of these wave equations considerably. In that case, a complete solution to the corresponding wave equations with sources on the right hand side will include waves of the fields associated with the charges themselves and these are not planar. However, the discussion on this aspect is beyond the scope of the present book.

Now, recalling the above two wave equns. (11.3) and (11.4), where $E_0$ and $B_0$ are constant vectors,

$$k \cdot k = \left(\frac{\omega}{v}\right)^2 \quad \text{where } v = \frac{c}{\sqrt{\mu\varepsilon}} \qquad \qquad \ldots(11.5)$$

let us now examine this two solutions (11.3) and (11.4) and see what other properties are suggested by Maxwell's equations. The electric field must satisfy Gauss's law $\nabla \cdot D = 0$ and this with $\varepsilon = $ constant, leads to the condition:

$$\nabla \cdot E = (i\,k_1\,E_{01} + i\,k_2\,E_{02} + i\,k_3\,E_{03})\,\exp\{i(k \cdot x - \omega t)\} = 0.$$

That is $k \cdot E_0 = 0$ and also $k \cdot B_0 = 0$ as the latter $B$ must also satisfy the same constraint.

This shows that the electromagnetic wave in non conducting medium must be transverse to the propagation vector $k$. This result can be shown in the following Fig. 11.1.

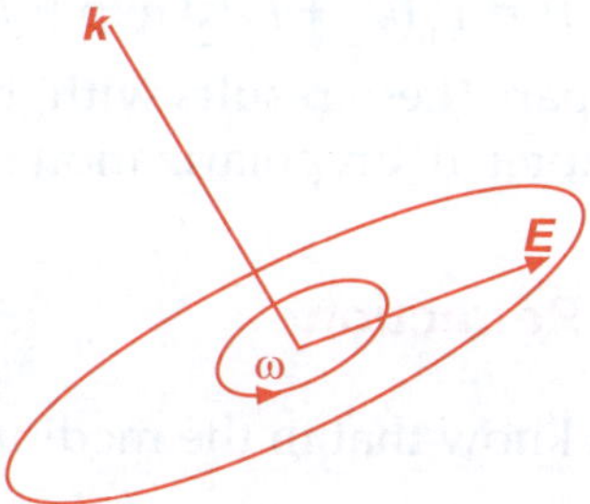


**Fig. 11.1** The electric vector *E* rotates around *k*, the direction of propagation with angular frequency, say w in circular polarized electromagnetic wave

## 11.2.1 Polarization

Since $E$, $B$ and $k$ form an orthogonal set, the more general way to express the electric field vector is

$$E(x, t) = (E_{01} + E_{02})\,\exp\{i(k \cdot x - i\,\omega t)\}^{\circ}\,(E_{01}\,\varepsilon'_1 + E_{02}\varepsilon'_2)\,\exp\{i(k \cdot x - i\,\omega t)\}$$

where $\varepsilon'_1, \varepsilon'_2$ and $k$ unit vectors constitute an alternative set of orthogonal vectors.

If we write $E_{01} = |E_{01}|\exp(i\varphi_1)$ and $E_{02} = |E_{02}|\exp(i\varphi_2)$ so that

$$E(x, t) = [\varepsilon'_1\,|E_{01}| + \varepsilon'_2\,|E_{02}|\exp\{i(\varphi_2 - \varphi_1)\}]\exp(ik \cdot x - i\omega t + i\varphi_1) \qquad \ldots(11.6)$$

The overall phase of the field $\varphi_1$, represents a constant rotation in the complex plane. We can drop it and write the field as

$$E(x, t) = [\varepsilon'_1 |E_{01}| + \varepsilon'_2 \, \varepsilon' E_{02}| \exp \{i(\varphi_2 - \varphi_1)\}] \exp (ik \cdot x - i\,\omega t +). \quad \ldots(11.7)$$

To understand the behaviour of the actual electric field, we should be interested in the real part of the above equation of the field. We find from the above equn. (11.6) that when $(\varphi_2 - \varphi_1) = 0$ the wave is linearly polarized. The total real part of the above equation will then comprise of two parts.

$E_1 = |E_{01}| \cos (k \cdot x - \omega t)$ and $E_2 = |E_{02}| \cos (k \cdot x - \omega t)$ so that arctan $E_2/E_1$ remains constant as the field evolves in space and time. Under this condition the field vector oscillates along a constant direction making an angle arctan $E_2/E_1$ with respect to the $\varepsilon'_1$ direction. When $(\varphi_2 - \varphi_1) \neq 0$, the wave is elliptically polarized and rotates around $k$. Because of the phase difference between $E_{01}$ and $E_{02}$, the field component in the $\varepsilon'_1$ direction passes through its notes at different spatial locations and/or times compared with the other one. Though they do not change proportionately, the net effect is a rotation in the $\varepsilon'_1 - \varepsilon'_2$ plane. $E$ sweeps around once every $2\pi/\omega$ seconds, so that the angular frequency is maintained at $\omega$. It may also be seen that when as a special case $|E_{01}| = |E_{02}|$ and $(\varphi_2 - \varphi_1) = \pm\, \pi/2$, then the amplitude of the electric field $E$ is constant and $E_1(\text{real}) = |E_{01}| \cos(k \cdot x - \omega t)$ and $E_2(\text{real}) = |E_{01}| \sin(k \cdot x - \omega t)$. Under this condition $E$ then rotates around $k$ with a constant magnitude and angular frequency $\omega$. This is case of circular polarization. Under this condition the electric field is given by

$$E(x, t) = E_0(\varepsilon'_1 + i\, \varepsilon'_2) \exp \{i(k \cdot x) - \omega t\} \quad \ldots(11.8)$$

The reader may compare these results with the polarization discussed and described in the Chapter 10 on polarization of light.

## 11.2.2   Reflection and Refraction

Now, from (11.5) we can know that in the medium $m_1$

$$|k_i| = |k_r| = \sqrt{\mu_1 \varepsilon_1}\, \frac{\omega}{c} \qquad \ldots(11.9)$$

and for the refracted medium with $\mu_2$

$$|k_t| = \sqrt{\mu_1 \varepsilon_1}\, \frac{\omega}{c}. \qquad \ldots(11.10)$$

According to the way we have chosen the axes and as $k_i$, $k_r$ and $k_t$ are coplanar,

$$k_{ix}x = k_{rx}x + k_{ry}y$$

and so, $\qquad k_i x = k_r x \qquad$ and $\qquad k_{ty} = 0.$

From the above finding we can say $\sin \theta_i = \sin \theta_r$ because of equn. (11.9) which means that angle of reflection is equal to the angle of incidence and this is known as law of reflection.

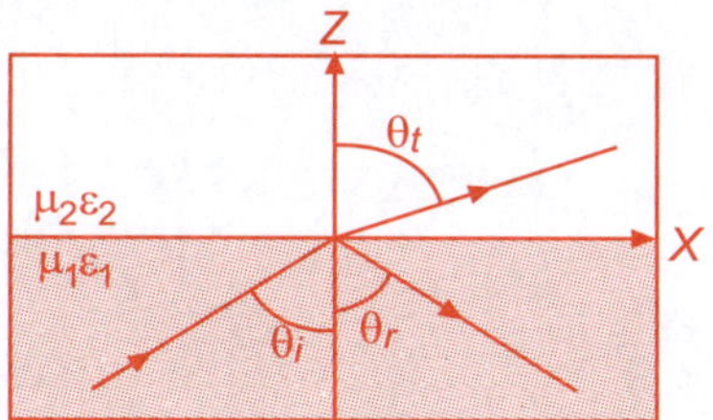


**Fig. 11.2** The directions of unit vectors $\varepsilon'_1$ and $\varepsilon'_2$ are respectively as on the plane of the figure and perpendicular to it and wave vectors $k_i$, $k_r$ and $k_t$ are respectively for incident, reflected and transmitted field and are coplanar

Now, again from the equns. (11.9) and (11.10), it implies that

$$|k_i|\sin\theta_i = |k_t|\sin\theta_t$$

so that

$$\frac{\sin\theta_i}{\sin\theta_t} = \sqrt{\frac{\mu_2\,\varepsilon_2}{\mu_1\,\varepsilon_1}}$$

This is known as Snell's law of refraction.

## REFERENCES


1.  S. B. Palmer and M.S. Rogalski- Advanced University Physics, Gordon and Breach Publishers, USA, 1996.

2.  F. Melia – Electrodynamics, The University of Chicago Press, Chicago, 2001.

# Statistical Mechanics

## 12.1 INTRODUCTION TO CLASSICAL STATISTICS

The classical mechanics founded on Newton's laws discusses certain principles such as conservation of energy and momentum. These principles are also applied to interacting particles. The classical statistical mechanics introduced in this chapter extended these principles to a system of many particles and is used to obtain collective or macroscopic properties of the system without even considering the detailed motions of individual particles. We use "particle' here in the broad sense, meaning a fundamental particles, such as electrons or aggregate of fundamental particle such as atoms, molecules etc., composing a given system. The necessity of this statistical approach can be realized from the simple fact that in one cubic centimeter of a gas at STP the number of molecules is $3 \times 10^{19}$ and to study its bulk properties like pressure and temperature, the approach to concentrate on each molecules individually to get the bulk properties is practically impossible and also unnecessary. As a reasonable alternative, statistical analysis of such many particle system, we make some reasonable estimate of the dynamical state of individual particles based on general properties of the particles. We make this estimate by introducing the concept of the probability of distribution of the particles among the different dynamical states in which they may be found. When we introduce the idea of probability, we take in to mind that particles do not move randomly or in a chaotic way without obeying any well defined laws.

## 12.2  STATISTICAL EQUILLIBRIUM AND BOLTZMANN'S PRINCIPLE

If we consider an isolated system which do not exchange matter or energy with other separate systems to be composed of a large number N of particles, say $n_1$, $n_2$, $n_3$, ….. etc., which are non interacting in a space of volume $V$. Then we can write

$$N = \sum_i n_i$$

and at a particular time particles are distributed among different energy states so that $n_1$ particles have energy $E_1$, $n_2$ particles have energy $E_2$; and so on. The total energy of the system is given by

$$U = \sum_i n_i E_i \, .$$

Now the parameters defined by $U$, $V$ and $N$ determine the macroscopic state and parameter $E_i$ determine the microstate of the $i$th particle.

Since all particles are assumed to be non interacting, it seems at first sight that this assumption is unrealistic. However, under special conditions we can use a technique called self-consistent field, in which each particle is considered subject to the average interactions of the others, with an average potential energy which depends only on its coordinates, so that we can assume the above two equations as valid. It is also reasonable to assume that for each macroscopic state of a system of particles there is a partition which is more favored than the other. In other words, "Given the physical conditions of the system of particles there is a most probable partition and when this partition is achieved, the system is said to be in statistical equilibrium".

The number of microstates compatible with a macroscopic state of the actual system is called "Thermodynamic Probability", $W$.

A given macroscopic state defined by parameters $U$, $V$, $N$ will be represented by phase points in a volume $\Delta\Omega$ in phase space. This phase volume is an increasing function of $U$ and also depends on $N$ and $V$. The thermodynamic probability is proportional to the phase volume $\Delta\Omega(U,V,N)$, so that its magnitude $\Omega(U, V, N)$ and the nature of its dependence on $U$, $V$, and $N$ will provide us with a bridge between statistical mechanics and thermodynamics. As $N$ is typically very large, the analysis will be carried out in the thermodynamical limit, where $N \rightarrow \infty$ and $V \rightarrow \infty$, so that $\dfrac{N}{V}$ and $\dfrac{U}{N}$ remain definite and a continuous distribution of energies $E_i$ is obtained. We find that the extensive variables, such as energy and entropy which refer to system as a whole become in this limit proportional to both $V$ and $N$ whereas the temperature and pressure remain independent of these parameters.

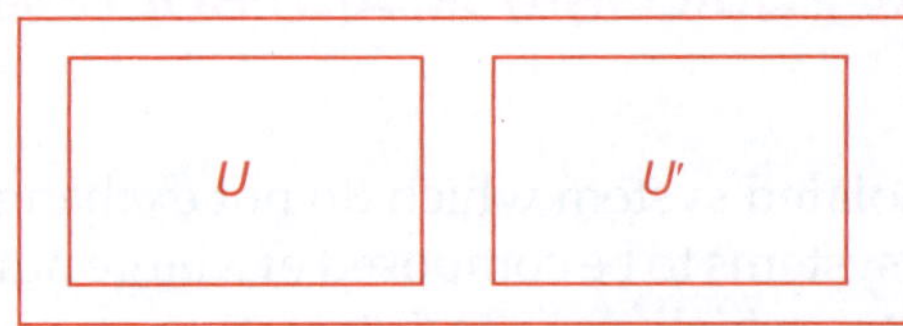


**Fig 12.1** Thermal contacts between two subsystems

A direct relation between the thermodynamic probability $W(U, V, N)$ and an extensive thermodynamic function can be derived by diving a system in to subsystems which are separately in states of thermodynamic equilibrium specified by $U, V, N$ and $W(U\ V\ N)$ and $U', V', N'$ and $W'(U'\ V'\ N')$ respectively. In the above figure the subsystems are separated by rigid walls so that $V, V'$ and $N, N'$ remain constants and the exchange of energy within the subsystems is allowed so that

$$U_0 = U + U' = \text{constant.} \qquad \qquad \dots(12.1)$$

Now, any one microstate of $W(U)$ can combine with any microstate $W(U')$ to yield a microstate of the system $W(U, U')$ and the rule for combining the microstates reads

$$W_0(U, U') = W(U)\,W'(U') = W(U)W'(U_0 - U) = W_0(U, U_0 - U). \dots(12.2)$$

The thermodynamic equilibrium of the system will be attained at the value of U which maximizes the thermodynamic probability $W_0\,(U, U_0 - U)$ and this state of equilibrium will be attained after the lapse of maximum time interval by the system. Now let $\bar{U}$ and $\bar{U}' = U_0 - \bar{U}$ be the definite values of $U$ and $U'$ that maximize the function (12.2) under the condition (12.1). Therefore,

$$d[W(U)\,W'(U')] = 0,\ dU + dU' = 0$$

This gives

$$\left\{\frac{\partial}{\partial U}\ln W(U)\right\}_{U = \bar{u}} = \left\{\frac{\partial}{\partial U'}\ln W'(U')\right\}_{U = \bar{u}'} \qquad \dots(12.3)$$

Introducing a parameter $\beta = \left\{\dfrac{\partial}{\partial U}\ln W(U, V, N)\right\}_{V, N}$ $\qquad \dots(12.4)$

so that at equilibrium $\beta = \beta'$.

The physical significance of this $\beta$ is clear when we take that in Boltzmann's principle, the thermodynamic entropy $S(U, V, N)$ is an extensive function directly related to the thermodynamic probability $W(U, V, N)$.

Now, recalling from the first law of thermodynamics, $dQ = dU + pdV$, which is $dU = TdS - pdV$ and these result $T = \left(\dfrac{\partial U}{\partial S}\right)_V$ and finally this can be written for this system as

$$\left\{\frac{\partial S(U, V, N)}{\partial U}\right\}_{V, N} = \frac{1}{T} \qquad \qquad \dots(12.5)$$

Therefore, at the condition of equilibrium the above system behaves as

$$T = T' \text{ or } \left\{\frac{\partial S(U)}{\partial U}\right\}_{E = \bar{E}} = \left\{\frac{\partial S'(U')}{\partial U'}\right\}_{E' = \bar{E}'} \quad \ldots(12.6)$$

Now, finding the correspondence between thermodynamic approach (12.6) and statistical approach (12.3), we get after using equns. (12.5) and (12.4)

$$\frac{1}{\beta T} = \left\{\frac{\partial S(U)/\partial U}{\partial \ln W(U)/\partial U}\right\}_{V,N} = \frac{\Delta S}{\Delta(\ln W)} = \frac{S - S_0}{\ln W - \ln W_0} = k_\beta . \quad \ldots(12.7)$$

Where the additive constant $S_0$ gives the law of increase of entropy must give the minimum entropy associated with the state of perfect order. This is the macro state realized through a single microstate, that is $W_0 = 1$. The equn. (12.7) can then be rewritten in the form

$$S = k_\beta \ln W + S_0.$$

At absolute zero the entropy must vanish if third law of thermodynamics is to hold and if we assume that thermodynamic probability is one at absolute zero, the $S_0$ is equal to zero, this gives:

$$S = k_\beta \ln W. \qquad \ldots(12.8)$$

This is known as Boltzmann's Relation which can also be written in the form:

$$W = e^{S/k}{}_\beta.$$

This shows that increase of entropy of a system in thermodynamic equilibrium is associated with increase in the number of associated microstates, W in phase space. The absolute value of entropy in terms of the number W of accessible microstates can then be taken as a measure of disorder of the system. The constant $k$, known as Boltzmann's constant allows us to measure the statistical parameter $b$ from (12.7) as

$$\beta = \frac{1}{kT} . \qquad \ldots(12.9)$$

The key problem of statistical mechanics is to find the most probable partition of an isolated system (distribution law), given its composition. Once the most probable partition has been found, the next problem is to devise methods for deriving macroscopic properties from it after adopting different assumptions so that the distribution law yield finally a result which is compatible to the experimental observations.

## 12.3 MAXWELL – BOLTZMANN DISTRIBUTION LAW

Classical statistical mechanics was developed in the last part of nineteenth century and the beginning of twentieth century as a result of the work of Ludwig Boltzmann (1844-1906), James C. Maxwell (1831-1879) and

J. W. Gibbs (1839-1903). The classical statistical mechanics has a broad applicability especially when it is applied for the study of many properties of gases. The two other statistics namely Fermi-Dirac and Bose-Einstein, belong to quantum statistics, which are to be discussed in the following sections. The classical statistics, however, can be considered as a limiting value of the two quantum statistics.

Let us consider a system composed of large number of identical yet distinguishable particles. The consideration that the particles seem to posses two contradictory characteristics that is they are distinguishable in one hand and identical on the other can be explained as, the particles are identical because they posses same structure and composition and they are distinguishable at the same time because we can distinguish or tell the difference between one particle and another identical particle. Though there exists an apparent lack of logic, the results we get are sufficiently simple and justify a preliminary discussion on the subject. Let us represent a particular partition $n_1$, $n_2$, $n_3$ ....etc., distributed in energy states $E_1$, $E_2$, $E_3$ .... etc.

To start with let us assume that all energy states equally accessible i.e. all energy states have the same probability of being occupied by particles. This can be summarized as that the probability of a particular partition is proportional to the number of different ways in which the particles can be arranged among available energy states to produce the partition. Let there be particles which are distinguishable are named as $p$, $q$, $r$ and $s$ and in a particular distribution or partition if $p$, $r$, $s$ exist in the energy states respectively as $E_1$, $E_3$ and $E_4$ will be different if all the three exist in energy states instead of $E_1$, $E_3$ and $E_4$, they exist in respectively states as $E_4$, $E_1$ and $E_3$ .This is because the particles are distinguishable. Now, there are $N$ number of total particles, then the first energy state can be filled by any one of the N particles to make one particular partition i.e. $N$ different ways. Now when first particle is selected, the same state can be filled up by a second particle in $(N - 1)$ ways and so the third by $(N - 2)$ ways. All these three particles existing in the first energy state can then be selected to fill the energy state in $N(N - 1)(N - 2)$ ways.

$$N(N - 1)(N - 2) = \frac{N!}{(N-3)!}. \qquad \qquad ...(12.10)$$

Now, which particular particle is selected first and which second does not make any different partition as each particle is identical and the partition is dependent on the number of particles selected and not which one is selected. Therefore, we must divide the expression (12.10) by 3! resulting in:

$\dfrac{N!}{3!\,(N-3)!}$ . The general expression for the total number of distinguish-able

different ways of placing $n_1$ particles in $E_1$ state is

$$\frac{N!}{n_1!(N-n_1)!} \qquad \qquad ...(12.11)$$

Now, if there will be $n_2$ numbers of particles to be selected in $E_2$ energy state when $n_1$ is already selected in $E_1$ then this can be done in $\dfrac{(N-n_1)}{n_2\,(N-n_1-n_2)!}$ and similarly for the third energy state $E_3$, $n_3$ number of particles can be selected when $n_1$ and $n_2$ are already selected in:

$$\frac{(N-n_1-n_2)}{n_3!(N-n_1-n_2-n_3)!} \quad \text{ways.}$$

Now, if this process is continued the total different ways to fill up $E_1$, $E_2$, $E_3$ ... states by $n_1, n_2, n_3,...$ particles, which is the thermodynamic probability $W$ can be obtained by multiplying each such terms. The final value of $W$ obtained on multiplication will cancel similar terms in numerator and denominator resulting in

$$W = \frac{N!}{n_1!\,n_2!\,n_3!....}. \qquad \qquad ...(12.12)$$

Now, here we have assumed that each energy state has equal intrinsic probability of being occupied by particles, but if it is not i.e. meaning that the states have different intrinsic probabilities, so that the number of particles $n_1, n_2...$ etc. are to decided by that then if for $E_i$ state has such intrinsic probability as $g_i$, then first particle out of $n_i$ particles can be selected in $g_i$ ways and $n_i$ particles $g_i \times g_i \times g_i \times ....$ to $n_i$ terms and that is $g_i^{ni}$ number of ways. Therefore, the total probability of a given partition is given by

$$W = \frac{N!\,g_1^{n1}\,g_2^{n2}\,g_3^{n3}\,\cdots}{n_1!\,n_2!\,n_3!...}. \qquad \qquad ...(12.13)$$

Now, if we remove the distinguishability and assume that all particles are identical and indistinguishable that is, there is no difference if any one of the number of particles accommodated in one energy state in a particular partition is selected before the others of the same state. In this condition all $N!$ permutations among the particles occupying the different states give the same partition. This means that we have to divide (12.3) by $N!$, resulting in

$$W = \frac{g_1^{n1}\,g_2^{n2}\,g_3^{n3}\,\cdots}{n_1!\,n_2!\,n_3!...} = \prod_{i=1}^{N}\frac{g_i^{n_i}}{n_i!}. \qquad \qquad ...(12.14)$$

This is the probability of a distribution in Maxwell – Boltzmann statistics.

Now, the most probable or equilibrium partition corresponds to that for maximum of probability $W$. It is mathematically easier to find the maximum of $\ln W$ instead of finding the maximum of $W$.

Now, taking log of equn. (12.4) we get

$$\ln W = n_1 \ln g_1 + n_2 \ln g_2 + n_3 \ln g_3 + \ldots - \ln n_1! - \ln n_2! - \ln n_3! \quad \ldots(12.15)$$

Now in the above equation we have $\ln n_1$, $\ln n_2$, $\ln n_3$ etc. are in the form $\ln x!$, which is $\ln x! = \ln 1 + \ln 2 + \ln 3 + \ldots + \ln (x-1) + \ln x$ and when $x$ is an integer $\ln x! = \sum_1^x \ln x$ and when $x$ is very large we can replace the summation by an integral without much error and so we can write $\ln x! = \int_1^x \ln x\, dx$ then integrating by parts (replacing $\ln x = u$ and $dv = x$) we get

$\ln x! = x \ln x - x + 1$. Neglecting 1 compared to large $x$ we get that what is called Sterling's approximation : $\ln x! = x \ln x - x$.

Now, applying this Sterling's approximation in the equn. (12.15)

$$\ln W = n_1 \ln g_1 + n_2 \ln g_2 + n_3 \ln g_3 + \ldots, - (n_1 \ln n_1 - n_1) - (n_2 \ln n_2 - n_2)$$
$$- (n_3 \ln n_3 - n_3) - \ldots$$
$$= -n_1 \ln n_1/g_1 - n_2 \ln n_2/g_2 - n_3 \ln n_3/g_3 - \ldots + (n_1 + n_2 + n_3 + \ldots)$$

Using $N \sum_i n_i$ we get

$$\ln W = N - \sum_i n_i \ln n_i / g_i. \qquad \ldots(12.16)$$

Now, differentiating this equation and as for a closed system $dN = 0$, we get

$$d(\ln W) = -\sum_i dn_i \ln n_i / g_i - \sum_i n_i\, d(\ln n_i / g_i)$$
$$= -\sum_i (dn_i) \ln n_i / g_i - \sum_i n_i (dn_i) / n_i$$

as intrinsic probability $g_i$ is constant. Therefore,

$$d(\ln W) = -\sum_i (dn_i) \ln n_i / g_i - \sum_i (dn_i). \text{ Now as } dN = 0,$$

$$\sum_i dn_i = 0. \qquad \ldots(12.17)$$

Therefore, $\quad -d(\ln W) = \sum_i \ln n_i / g_i\, dn_i = 0. \qquad \ldots(12.18)$

as in the equilibrium condition the probability of partition attains its maximum.

As again for the closed system the total energy $U$ is constant then as

$$dU = \sum_i E_i\, dn_i = 0. \qquad \ldots(12.19)$$

Now, to compensate for two conditions given by equns. (12.17) and (12.19), we adopt the Lagrange's method of undertermined multipliers, $\alpha$

and $\beta$ and multiplying the equn. (12.17) by $\alpha$ and (12.19) by $\beta$ and adding to equn. (12.18), we obtain

$$\sum_i (\ln n_i / g_i + \alpha + \beta E_i)\, dn_i = 0.$$

These $\alpha$ and $\beta$ are two arbitrary coefficients. The equilibrium condition of distribution is then obtained if

$$\ln n_i/g_i + \alpha + \beta E_i = 0$$

or, $\qquad\qquad\qquad n_i = g_i e^{-\alpha - \beta E_i}.$ $\qquad\qquad$ ...(12.20)

Now, as we know that $N = n_1 + n_2 + n_3 + \ldots\ldots$

$$= \left(g_1 e^{-\alpha - \beta E_1} + g_2 e^{-\alpha - \beta E_2} + g_3 e^{-\alpha - \beta E_3} + \ldots.\right)$$

$$= e^{-\alpha}\left(g_1 e^{-\beta E_i} + g_2 e^{-\beta E_2} + g_3 e^{-\beta E_3} + \ldots..\right)$$

$$= e^{-\alpha}\left(\sum_i g_i e^{-\beta E_i}\right) = e^{-a}Z.$$

Where, $\qquad\qquad\qquad Z = \sum_i g_i e^{-\beta E_i}$ $\qquad\qquad$ ...(12.21)

called the partition function. Writing $e^{-a} = N/Z$ the equn. (12.20) can be written as

$$n_1 = \frac{N}{Z} g_i e^{-\beta E_1} = \frac{N}{Z} g_i e^{-E_i/kT}.$$ $\qquad$ ...(12.22)

This expression (12.21) is called Maxwell-Boltzmann distribution law. As quantity $\alpha$ is related to partition function and $\beta$ is related to energy of the system.

### 12.3.1 Application of Boltzmann Statistics in Ideal Gas

It is now to find the system of particles in nature whose collective behaviour resembles the prediction of Maxwell-Boltzmann distribution law. It is found that most of the gases can be described according to Maxwell-Boltzmann statistics over a wide range of temperature. We consider that the gas considered is composed of monatomic molecules and do not possess any potential energy and possess only translational kinetic energy. Then

$E_i$ of equn. (12.21) is given by

$$E_i = \frac{1}{2m} p_i^2 = \frac{1}{2} m v_i^2$$ $\qquad$ ... (12.23)

Now, as the kinetic energy of an ideal gas occupying a large volume may be considered as not being quantized but as being a continuous spectrum, then the summation sign as given above: $Z = \sum_i g_i e^{-\beta E_i}$ can be replaced by an integral in the form as $b = 1/kT$:

$$Z = \int_0^\infty e^{-E/kT} g(E)\, dE. \qquad \qquad ...(12.24)$$

Where, $g(E)dE$ replaces $g_i$ and represents the number of molecular states in the energy range between $E$ and $E + dE$. This number arises from the different orientations of the momentum $p$ for a given energy.

Now, let us consider the number of energy states or levels in a small energy range $dE$ for a particle in a very large potential box. The momentum of the particle in $x$, $y$ and $z$-directions in a three dimensional potential well be given as $p_x$, $p_y$ and $p_z$ and all these three momentums are quantized so that for potential well or box being cube in shape can be given as

$$p_x = \frac{h n_1}{2a}, \; p_y = \frac{h n_2}{2a} \; \text{and} \; p_z = \frac{h n_3}{2a},$$

where $n_1$, $n_2$ and $n_3$ are integers

Now, the energy:

$$E = \frac{1}{2m} p^2 = \frac{1}{2m}\left(p_x{}^2 + p_y{}^2 + p_z{}^2\right)$$

$$= \frac{h^2}{8m a^2}\left(n_1{}^2 + n_2{}^2 + n_3{}^2\right).$$

Now as $n_1$, $n_2$ and $n_3$ are all positive, the number of energy states between zero and $E$, $N(E)$ approximately will be equal to the volume of the octant which is $1/8$ the of the sphere with radius

$$r = (n_1{}^2 + n_2{}^2 + n_3{}^2)^{1/2}.$$

Therefore,

$$N(E) = \frac{1}{8}\left(\frac{4}{3}\pi r^3\right) = \frac{\pi}{6}(n_1 + n_2 + n_3)^{3/2} = \frac{\pi}{6}\left(\frac{8m a^2}{h^2} E\right)^{3/2}.$$

As $a^3 = V$, the volume then

$$= \frac{\pi V}{6}\left(\frac{8mE}{h^2}\right)^{3/2} = \frac{8\pi V}{3h^3}\left(2m^3\right)^{1/2} E^{3/2}$$

$$dN(E) = \frac{4\pi V (2m^3)^{1/2}}{h^3} E^{1/2} dE. \qquad \qquad ...(12.25)$$

It is convenient to write $dN(E) = g(E)\,dE$ and replacing $g(E)$ in equn. (12.24), we get

$$Z = \frac{4\pi V (2m^3)^{1/2}}{h^3} \int_0^\infty E^{1/2} e^{-E/kT} \, dE. \quad \ldots(12.26)$$

The value of the integral in the right side of equn. (12.25) can be evaluated as $\frac{1}{2}\sqrt{\pi(kT)^3}$. Therefore,

$$Z = \frac{V(2\pi m kT)^{3/2}}{h^3}. \quad \ldots(12.27)$$

This gives the partition function of an ideal monatomic gas as the function of temperature and the volume of the gas. Taking natural logarithm we get $\ln Z = C + 3/2\,(\ln kT)$ where $C$ is the constant which includes all remaining constant quantities.

Now, from equn. (12.21) as $\beta = 1/kT$, $d\beta = -\,dT/kT^2$ and as the total energy of the closed system

$$U = n_1 E_1 + n_2 E_2 + n_3 E_3 + \ldots\ldots$$

$$= \frac{N}{Z}\left(\sum_i g_i E_i \, e^{-\beta E_i}\right)$$

and using the definition of partition function (12.21), we may write $U$ in the alternative form as:

$$U = \frac{-\dfrac{N}{Z}\dfrac{d}{d\beta}\left(\sum_i g_i e^{-\beta E_i}\right)}{}$$

$$= -\frac{N}{Z}\frac{dZ}{d\beta} = -N\frac{d}{d\beta}(\ln Z). \quad \ldots(12.28)$$

and so, $\qquad E_{average} = \dfrac{U}{N} = -\dfrac{d}{d\beta}(\ln Z).$

Now, writing the Maxwell-Boltzmann distribution law (12.22) in terms of temperature $T$ and since $\beta = 1/kT$; where $k$ is Boltzmann constant and then $d\beta = -\dfrac{dT}{kT^2}$, writing the expression of energy $U$ as

$$U = kNT^2 \frac{d}{dT}(\ln Z). \quad \ldots(12.29)$$

The average energy per particle $U/N$, is then given by:

$$E_{ave} = kT^2 \frac{d}{dT}(\ln Z). \quad \ldots(12.30)$$

Now, substituting the value of $\ln Z = C + 3/2 \ln kT$ in equn. (12.30) we get the value of the average energy of the molecules as

$$E_{ave} = \frac{3}{2} kT. \qquad \qquad …(12.31)$$

And as $$U = NE_{ave} = \frac{3}{2} NkT. \qquad \qquad …(12.32)$$

### 12.3.2 Observation on Boltzmann Distribution Law

Considering the equn. (12.22), since the exponential $e^{-E_i/kT}$ is a decreasing function of $E_i/kT$, the larger the ratio $E_i/kT$, the smaller the value of occupation number $n_i$. Therefore, at a given temperature, the larger the energy $E_i$, the smaller the value of $n_i$. Therefore, the occupation of states available to the particles decreases as their energy increases. Conversely at low temperature only the lowest energy levels are occupied. As $T \to 0$, all the available particles tend to occupy the lowest energy state. This is what is shown in Fig. 12.2 below.

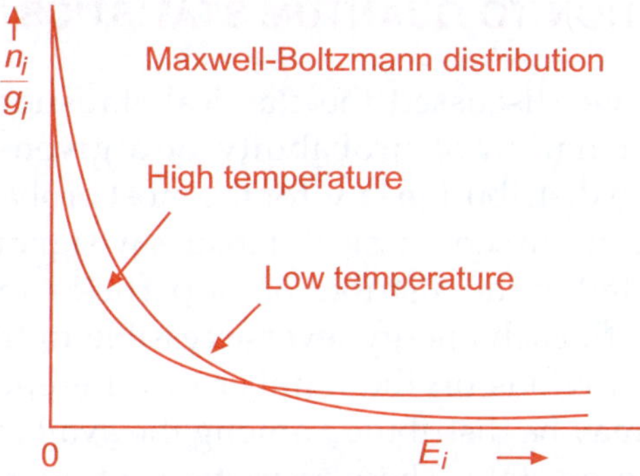


**Fig 12.2** Maxwell-Boltzmann distribution at high and low temperatures

Now, for an ideal gas we have seen that from equn. (12.22) after replacing $g_i$ by $g(E)\,dE$ and taking differential and after replacing $g(E)\,dE$

$$dn = \frac{N}{Z} e^{-E/kT} g(E)dE = \frac{N}{Z} \frac{4\pi V (2m^3)^{1/2}}{h^3} E^{1/2} e^{-E/kT} dE.$$

Where $dn$ is the number of molecules with energy between $E$ and $E + dE$. After introducing the value of $Z$ from (12.27) we have,

$$\frac{dn}{dE} = \frac{2\pi N}{(\pi kT)^{3/2}} E^{1/2} e^{-E/kT}. \qquad \qquad …(12.33)$$

This is Maxwell formula for energy distribution of molecules in an ideal gas. A plot of $dn/dE$ for two different temperatures are given in Fig. 12.3 below. The number of molecules per energy level i.e. $dn/dE$ shows the difference in packing of energy levels at two temperatures. The difference is marked in the figure. It is interesting to note that the expression (12.33) is independent of mass of the molecules.

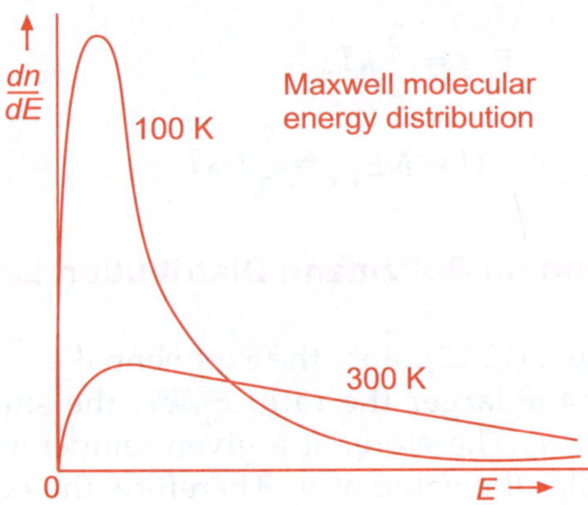


**Fig 12.3** Maxwell formula for energy distribution. While higher energy levels at high temperature is more densely packed by molecules, same level remain almost empty at lower temperature

## 12.4 INTRODUCTION TO QUANTUM STATISTICS

In the last section we discussed the classical statistics characterized by the method of calculation of probability of a given partition and by Maxwell-Boltzmann distribution law, for the most probable or equilibrium partition. When we discussed classical statistics we ignored any symmetry considerations related to the distribution of particles among the different states associated with each energy level accessible to the particles. There may be certain restrictions on the number of different ways in which a group of particles may be distributed among the available wave functions associated with energy state. These restrictions of quantum origin, affect the probability of a given partition. The theory in which these symmetry considerations are taken into account is called 'quantum statistics'.

There are two kinds of such statistics one where in the particles obey exclusion principle and hence described by antisymmetric wave functions. This kind is named as "Fermi–Dirac statistics". The particles which obey this statistics are known as 'Fermions'. In the other type of statistics, where the particles do not obey the exclusion principle and are described by symmetric wave functions, is known as Bose-Einstein statistics, The particles which obey this statistics are called 'Bosons'.

### 12.4.1 Fermi-Dirac Distribution Law

We now find the probability of partition of a system of particles where the particles are identical and indistinguishable and in addition they obey the exclusion principle. Under this principle no two particles can be in the same dynamical state and the wave function of the total system is antisymmetric. It has been stated that the particles which satisfy these conditions are known as 'Fermions' after Enrico Fermi (1901-1954) who first discussed this system. It has been found experimentally that all fermions have spin 1/2. Now in classical statistics, $g_i$ has been introduced as intrinsic probability, where as in the quantum statistics it is redefined as the different quantum states corresponding to a given energy *i.e.* the degeneracy of energy states. To each quantum state there corresponds a particle wave function. These wave functions are determined by each of the possible arrangements of quantum numbers corresponding to a given energy level. Explicitly, for particles of spin 1/2, not subject to magnetic forces, each particle may be in each energy state with spin up or down $(m_s = \pm 1/2\,)^*$ and so $g_i = 2$. This $g_i$ is the maximum number of particles (fermions) that can be accommodated in an energy level without violating the Pauli's exclusion principle. Therefore $n_i$, the number of particles in an energy level $E_i$ can not exceed the $g_i$ in any particular partition i.e. $n_i \le g_i$.

Now, to fill the energy level $E_i$ with available $n_i$ number of fermions, we can place the first particle in any of the $g_i$ states available and thus this can be done in $g_i$ different ways. Once one of such $g_i$ state is filled up, the second particle in $(g_i - 1)$ way and the third in $(g_i - 2)$ ways. This continues until all of the $n_i$ particles are placed in $g_i$ states with energy $E_i$. Therefore, this distribution in one particular partition can be done in $g_i(g_i - 1)(g_i - 2)(g_i - 3) \ldots\ldots(g_i - n_i + 1)$ ways.

This can be written in the form:

$$\frac{g_i\,!}{(g_i - n_i)!} \qquad\qquad \ldots(12.34)$$

* This will be discussed in one later chapter.

Now, in addition to the exclusion principle so far applied if now we introduce the concept that the particle are indistinguishable, then it is not possible to differentiate which of the particle from $n_i$ is chosen for the particular $g_i$. Therefore, it is not possible to distinguish the partition distribution, if the $n_i$ fermions are reshuffled among $g_i$ energy states in the level of $E_i$. The resulting conclusion is then the equn. (12.34) is to be divided by $n_i!$. Then the resulting distribution is

$$\frac{g_i\,!}{n_i!\,(g_i - n_i)!} \qquad\qquad \ldots(12.35)$$

We now can find the total number of distinguishable different ways of obtaining the partition $n_1, n_2, n_3 \ldots$ among energy levels $E_1, E_2, E_3, \ldots$ by multiplying all expressions like (12.35) for each energy levels available and thus obtain the partition probability as:

$$W = \frac{g_1\,!}{n_1!\,(g_1 - n_1)!}\,\frac{g_2\,!}{n_2!\,(g_2 - n_2)!}\,\frac{g_3\,!}{n_3!\,(g_3 - n_3)!}\cdots$$

$$= \prod_i \frac{g_i!}{n_i!\,(g_i - n_i)!} \qquad\qquad \ldots(12.36)$$

Taking the natural logarithm of this, we get after using the Sterling's approximation

$$\ln W = \sum_i \{g_i \ln g_i - n_i \ln n_i - (g_i - n_i)\ln(g_i - n_i)\}$$

Similar to our discussion on Maxwell- Boltzmann statistics at equilibrium, the probability becomes maximum and that is, $d \ln W = 0$.

Then,

$$-d(\ln W) = \sum_i \{\ln n_i - \ln(g_i - n_i)\}\, dn_i = 0. \qquad \ldots(12.37)$$

Considering once again the two conditions as discussed before for a closed system at equilibrium as $\sum_i dn_i = 0$ and $\sum_i E_i\, dn_i = 0$ and multiplying them by $\alpha$ and $\beta$ and adding with (12.37), we get

$$\ln n_i - \ln(g_i - n_i) + \alpha + \beta E_i = 0$$

or,

$$\frac{n_i}{g_i - n_i} = e^{-\alpha - \beta E_i}$$

i.e.

$$n_i = \frac{g_i}{e^{\alpha + \beta E_i} + 1}. \qquad\qquad \ldots(12.38)$$

This is Fermi–Dirac distribution Law.

The parameter $\beta$ here plays the same role that it played for Maxwell – Boltzmann distribution law and that was $\beta = 1/kT$. The quantity $\alpha$ is determined from the condition that $\sum_i n_i = N$ and in most cases it is found to be negative. It is however more convenient to introduce a new quantity, $E_F$ which is related with $\alpha$ by the relation

$$E_F = -\alpha kT$$

Then the equn. (12.38) can be written as

$$n_i = \frac{g_i}{e^{(E_i - E_F)/kT} + 1}. \qquad\qquad \ldots(12.39)$$

This quantity $E_F$ has a positive value in most cases and possesses tremendous importance. This is known as 'Fermi Level of Energy'. It represents a particular level of energy which is related to $E_i$, the energy corresponding to $i$th level as:

$$\lim_{T \to 0} e^{(E_i - E_F)/kT} = \begin{cases} 0 \text{ for } (E_i - E_F) < 0 \\ \infty \text{ for } (E_i - E_F) > 0 \end{cases}. \qquad \ldots (12.40)$$

In the following Fig. 12.4, the distribution of the function $\dfrac{n_i}{g_i}$, which physically, can be interpreted as the mathematical probability of filling up $g_i$ energy degenerate states by $n_i$ number of fermions is shown with its variation at different temperatures (high and low). It is seen from the following Fig. 12.4 at absolute zero, unlike Maxwell-Boltzmann all particles do not accumulate at the ground energy level, rather they occupy increasingly higher states until $E_F$, the Fermi energy state. This is to obey exclusion principle. As temperature is increased the fermions start occupying energy states higher than $E_F$. The probability of occupation until Fermi energy state is reached is 100% (at $T = 0$), where as this probability of higher energy states remain zero but increases as temperature is increased. Hence the Fermi energy state can be defined as the highest energy state occupied by fermions at zero degree Kelvin. Only those fermions with energy close to $E_F$ can move into higher unoccupied states by absorbing energy in order to maintain the exclusion principle. The corresponding Fermi temperature is introduced as $\Phi_F = E_F/k$.

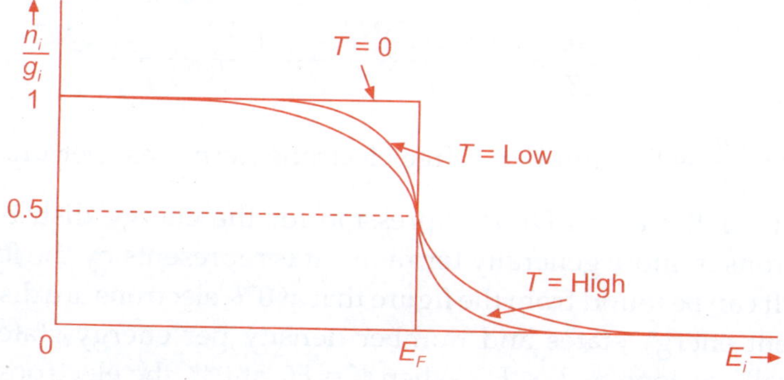


**Fig. 12.4** The variations of distribution function $\dfrac{n_i}{g_i}$ with energy $E_i$ at different temperatures

## 12.4.2 Electron Gas

We now consider an assembly of electrons in a metal and the electrons obey exclusion principle, they are fermions. These electrons in a metal exist in two bands, the lower is the Valance Band where all the energy states are filled up at any temperature and the higher one is Conduction Band. In conduction band the electrons exist in levels from the bottom of the conduction band to above. These electrons in metals are known as conducting or free electrons and they constitute current as they are free to move. As the energy spectrum in the conduction band may be considered as continuous because of large number states occupied by electrons obeying exclusion principle exist within the some energy limits in the band. Therefore, $g_i$ introduced in (12.34) as number of energy states in the energy level $E_i$, may be assumed to be expressed as $g(E)dE$ within the energy levels $E$ and $E + dE$. The number of electrons say $dn$ existing within this energy limits is given as (from 12.39):

$$dn = \frac{g(E)\,dE}{e^{(E - E_F)/kT} + 1} \qquad \text{...(12.41)}$$

where $E$ is measured from the bottom of the conduction band and $g(E)$ $dE$ is the number available states within that range or limit.

Now, recalling and using the equn. (12.25) we can write the expression for $g(E)dE$ the number of states, after multiplying the expression (12.25) by 2 to take into account two different states having same energy for two possible orientations of spin of electrons as

$$g(E)\,dE = \frac{8\pi V (2m^3)^{1/2}}{h^3} E^{1/2} dE \qquad \text{...(12.42)}$$

and substituting this equn. in (12.41), we get

$$\frac{dn}{dE} = \frac{8\pi V (2m^3)^{1/2}}{h^3} \frac{E^{1/2}}{e^{(E - E_F)/kT} + 1}. \qquad \text{...(12.43)}$$

Where $\dfrac{dn}{dE}$ is the number of free electrons (fermions) per unit energy range. This is the Fermi-Dirac expression for the energy distribution of free electrons or more generally fermions. It is represents by the following Fig. 12.5. It can be found from the figure that at $0°K$ electrons are distributed in different energy states and number density per energy states varies parabolically as long as $E < E_F$. When $E = E_F$ at $0°K$ the electrons occupy the highest energy state i.e. $E_F$ and when $E > E_F$ vanishes to zero (at $0°K$).

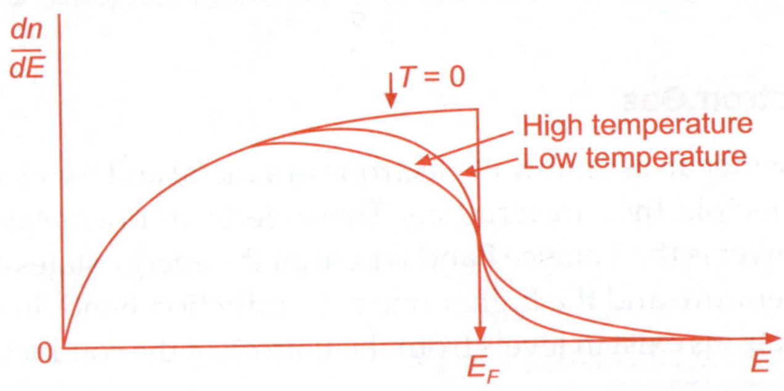


**Fig. 12.5** Energy distribution of fermions at different temperatures

As temperature is increased, electrons move from Fermi energy state and nearby lower energy states to higher energy levels. This effect continues with the increase of temperature. If we integrate the equn. (12. 43) over all energies we get the total number $N$ of the electrons in the conduction band. As the temperature is very low, we can assume that $\dfrac{dn}{dE} = g(E)$

$$N = \frac{8\pi V(2m^3)^{1/2}}{h^3} \int_0^{E_F} E^{1/2} dE$$

$$= \frac{16\pi V(2m^3)^{1/2}}{3h^3} E_F^{3/2}$$

$$E_F = \frac{h^2}{8m}\left(\frac{3N}{\pi V}\right)^{2/3} . \qquad \qquad ...(12.44)$$

This is the expression for the Fermi energy of the electrons in a metal and it can be calculated after knowing the value of $N/V$ of conduction electrons per unit volume.

### 12.4.3  Total and Minimum Energy of a Fermi Gas

As explained before the total energy $U$ of a gas classical or quantum is given as

$$U = \int E\,dn = \int E\frac{dn}{dE}\,dE \quad \text{and replacing the value of } \frac{dn}{dE} \text{ at low temperature}$$

so that $\dfrac{dn}{dE}$ can be approximated as $g(E)$, we get from (12.43) and (12.44)

$$U = \frac{8\pi V(2m^3)^{1/2}}{h^3} \int_0^{E_F} E^{3/2}\,dE = \frac{16\pi V(2m^3)^{1/2}}{5h^3} E_F^{5/2} . \qquad ...(12.45)$$

Replacing the value of $E_F$ from (12.45) we get

$U = \dfrac{3}{5} N E_F$. This is the minimum energy of a system of $N$ fermions. The average energy per particle is

$$E_{ave} = \frac{U}{N} = \frac{3}{5} E_F . \qquad \qquad ...(12.46)$$

This Fermi-Dirac statistics applied to the quantum particle like electrons obeying exclusion principle is applicable for the electrons in metals. The various characteristics of metals, semiconductors can be explained on the basis and concept of Fermi energy level.

These will be discussed in one later chapter.

## 12.5 BOSE-EINSTEIN DISTRIBUTION LAW

Now, let us consider the system where the particles like Fermi-Dirac are identical and indistinguishable but unlike Fermi-Dirac, they do not obey the exclusion principle. This results into the situation in which we can accommodate any number of such particles in an energy state. This statistical distribution law was first investigated by Prof. S. N. Bose and the statistics goes by the name Bose-Einstein statistics. The particles which obey this statistics are known after Prof. Bose as bosons. It has been found that all particles with integral spin (0 or 1) are bosons and the wave function describing such system of particles are symmetric. Like Fermi-Dirac statistics, each energy level with energy $E_i$ are divided into $g_i$ number degenerate states which accommodates $n_i$ number of particles without following exclusion principle. We under this statistical distribution can then accommodate any number of bosons in a particular degenerate state resulting into symmetric wave function. The helium nucleus and mesons are examples of Bose-Einstein statistics. This distribution can be pictorially represented by the following classical example showing and arrangement of particles in different divisions. Let us suppose that there are $n_i$ identical particles in a row and they are to be distributed in $g_i$ number of partitions (say, boxes), without limit to number of such particles that can be put within any partition or box. This is similar to finding $n_i$ number of bosons distributed in $g_i$ number of quantum states. The total number of particles and divisions is equal to the number of permutations of $(n_i + g_i - 1)$ objects in a row and this $(n_i + g_i - 1)$ !

Now, as all particles are identical and indistinguishable all permutations differ only on the number of particles in each partition or divisions and as the permutation of divisions yield same physical state, the permutations $(n_i + g_i - 1)!$ is to be divided by $n_i!$ and $(g_i - 1)!$ The total number of distinguishable arrangements of the $n_i$ particles in $g_i$ states therefore yields as

$$\frac{(n_i + g_i - 1)!}{n_i!\,(g_i - 1)!}. \qquad \qquad ...(12.47)$$

Now, by multiplying all expressions for different values of $i$, we get the different ways of partition of particles $n_1, n_2, n_3, ....$ among the energy levels $E_1, E_2, E_3, ....$ and the partition probability results then

$$W = \frac{(n_1 + g_1 - 1)!}{n_1!\,(g_1 - 1)!} \, \frac{(n_2 + g_2 - 1)!}{n_2!\,(g_2 - 1)!} \, \frac{(n_3 + g_3 - 1)!}{n_3!\,(g_3 - 1)!} \, ...$$

$$= \prod_i \frac{(n_i + g_i - 1)!}{n_i!\,(g_i - 1)!}. \qquad \qquad \dots(12.48)$$

We now find the most probable partition by finding the maximum of $\ln W$ keeping in mind that $\sum_i n_i = N$ and $\sum_i n_i E_i = U$.

$$\ln W = \sum_i \left[ (n_i + g_i - 1) \ln (n_i + g_i - 1) - n_i \ln n_i - (g_i - 1) \ln (g_i - 1) \right]$$

We now set $d \ln W = 0$ to find the maximum of $W$

$$-d \ln W = \sum_i \left[ -\ln (n_i + g_i - 1) + \ln n_i \right] dn_i = 0 \qquad \dots(12.49)$$

Now, to combine the expressions obtained from $\sum_i n_i = N$ and $\sum_i n_i E_i = U$ at equilibrium i.e. $\sum_i dn_i = 0$ and $\sum_i E_i dn_i = 0$ we multiply them by $\alpha$ and $\beta$ respectively and add with (12.49).

$-\ln (n_i + g_i - 1) + \ln n_i + \alpha + \beta E_i = 0$ as $n_i + g_i$ is very large compared to 1, therefore, neglecting 1 in the first term of the above expression we get

$$\ln \frac{n_i}{n_i + g_i} = -\alpha - \beta E_i$$

or, 
$$\frac{n_i}{n_i + g_i} = e^{-\alpha - \beta E_i}$$

and putting as $\beta = 1/kT$

or, 
$$n_i = \frac{g_i}{e^{\alpha + E_i/kT} - 1}. \qquad \qquad \dots(12.50)$$

This is Bose-Einstein distribution law. The distribution of bosons in different energy levels at different temperatures is shown by the following (Fig. 12.6)

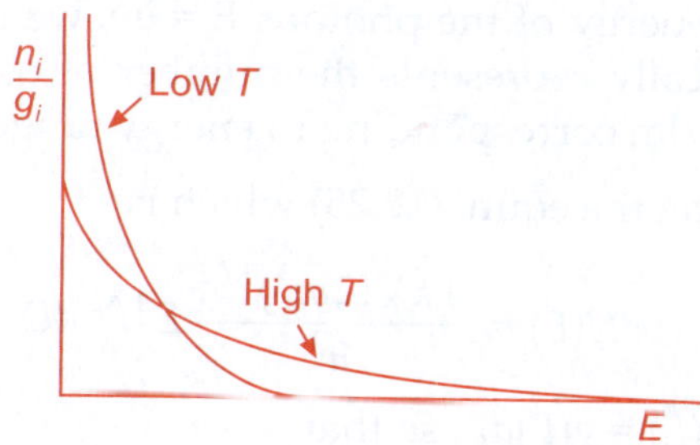


**Fig. 12.6** The distributions function $\frac{n_i}{g_i}$ at two temperatures in bose-einstein statistics

## 12.5.1 THE PHOTON GAS

The most important application of Bose-Einstein statistics lies in the analysis of Black Body radiation which is represented by electromagnetic radiation trapped in an enclosure (cavity) and is in thermal equilibrium with the atoms of the wall of the cavity. The atoms of the wall will be continuously absorbing the energy and emitting it so as to obtain a stage of equilibrium. This electromagnetic radiation is composed of 'photons' having energy $hv = hc/\lambda$, where $n$ is the frequency and $l$ is the wavelength. As any number of photons can have same energy for example a monochromatic beam of light (electromagnetic radiation) can have any intensity. The wavelength is related to the energy and intensity to the number of photons. Therefore, the photons can be considered as bosons and obey Bose-Einstein statistics. In black body radiation the number of photons is not constant as they are continuously absorbed or emitted by the atoms of the wall of the cavity.

Therefore, the condition so far considered valid i.e. $\sum_i dn_i = 0$ either for

Maxwell-Boltzmann (Ideal Gas) or Fermi-Dirac statistics ( Electron gas) is to be  dropped out and so we can set a = 0 and the equn. (12.50) is to be changed into

$$n_i = \frac{g_i}{e^{E_i/kT} - 1}.$$

In addition, the energy spectrum of the photons may be treated as continuous if the cavity is large compared to the average wavelengths of the radiation as in this case the energy difference between successive allowed energy values is extremely small. Under this condition we can write the above expression by replacing $g_i$ by $g(E)\, dE$ and the equn. (12.50) can be written as

$$dn = \frac{g(E)dE}{e^{E/kT} - 1}. \qquad \qquad \dots(12.51)$$

Now as the energy of the photons $E = hv$, we can write $g(E)dE = g(v)\, dv$, which physically represents the number of oscillatory modes in the frequency range d n corresponding to energy range $dE$.

Now, recalling the equn. (12.25) which is

$$dN(E) = \frac{4\pi V(2m^3)^{1/2}}{h^3} E^{1/2}\, dE$$

we write $dN(E) = g(E)dE$, so that

$$g(E) = \frac{dN}{dE} = \frac{4\pi V (2m^3)^{1/2}}{h^3} E^{1/2}.$$

Now, considering that the energy is related to momentum $p$ by $E = p^2/2m$ and also defining $g(p)$ so that $dN = g(p)dp = g(E)dE$, we get

$$dN = g(p)dp = g(E)dE$$

and subsequently,

$$\frac{dN}{dp} = g(p) = g(E)\frac{dE}{dp} = \frac{4\pi V}{h^3} p^2.$$

Applying this to the number of modes of "Longitudinal waves" trapped in a cavity of volume $V$. In such cases it is more convenient to use frequency $v$. Recalling that $p = h/\lambda$ and $v = c/\lambda$, where c is the phase velocity of the waves, we define $g(v)$ by $g(v)\, dv = g(p)dp$ and get

$$g(v) = g(p)\frac{dp}{dv} = \frac{4\pi V}{c^3} v^2. \qquad ...(12.52)$$

The above equn. (12.52) gives the number of oscillatory modes trapped in a cavity of volume $V$, but as an electromagnetic waves are transverse having two independent perpendicular directions of polarization, we have to multiply the above equn. (12.52) by a factor of 2. Therefore, the number of states in the black body radiation with frequency between $v$ and $v + dv$ or energy between $E$ and $E + dE$ is given by

$$g(E)\, dE = g(v)dv = \frac{8\pi V}{c^3} v^2 dv.$$

Therefore, we may write equn. (12.51) as

$$dn = \frac{8\pi V}{c^3} \frac{v^2 dv}{e^{hv/kT} - 1}. \qquad ...(12.53)$$

Now, the energy corresponding to $dn$ photons in the frequency range of $dv$ is $(hv)\, dn$ and the energy per unit volume is then $(hv)\, dn/V$ and defining the energy density distribution in black body radiation by $\varepsilon(v) = \dfrac{hv}{V}\dfrac{dn}{dV}$, we get

The energy per unit volume corresponding to radiation with frequency between $v$ and $v + dv$ from equn. (12.53) given as

$$\varepsilon(v) = \frac{8\pi h v^3}{c^3} \frac{1}{e^{hv/kT} - 1} \qquad ...(12.54)$$

This is the famous Planck radiation equation. The agreement of this equation with experimental observations is a strong support of the idea that electromagnetic radiations are composed of photons and the photons obey Bose-Einstein statistics. It may be stated here that the problem of explaining the black body emission spectrum Planck introduced the concept that when radiation interacts with matter, it is absorbed or emitted in energy quanta equal to $h\nu$. However, the flaw in Planck's proposition was that he assumed that the atoms of wall of cavity behave as an oscillators with energy $E = nh\nu$, instead of $E = (n + 1/2)h\nu$ to include zero-point energy and he also applied Maxwell-Boltzmann statistics. Though otherwise Planck's derivation of the law found in agreement with experimental observation at elevated temperatures, the result was found to deviate if zero-point energy is considered. This discrepancy was convincingly removed after the introduction of Bose-Einstein statistics.

### 12.5.2 Comparison between the Statistics

Let us now have a comparison between Maxwell-Boltzmann, Fermi-Dirac and Bose-Einstein statistics. These three statistics lead to the distribution laws which are derived respectively in equns. (12.20), (12.38) and (12.50) which may be recalled here.

1. Maxwell–Boltzmann  :  $n_i = g_i e^{-\alpha - E_i/kT}$

2. Fermi-Dirac  :  $n_i = \dfrac{g_i}{e^{\alpha + \beta E_i} + 1}$

3. Bose-Einstein  :  $n_i = \dfrac{g_i}{e^{\alpha + E_i/kT} - 1}$ .

Now, these three expressions can be written in one abbreviated form as:

$$\frac{g_i}{n_i} + \delta = e^{\alpha + E_i/kT} .$$

Here $\delta = 0$ for Maxwell Boltzmann, $-1$ for Fermi-Dirac and $+1$ for Bose-Einstein statistics.

It is evident that if $\dfrac{g_i}{n_i} \gg 1$ i.e. $\dfrac{n_i}{g_i} \ll 1$ which means that number of particles $n_i$ in the energy state $g_i$ is much less than the total number of available states $g_i$, then these three statistics yield almost same results. This is true at high temperatures because $\alpha$ increases with temperature. Therefore it may be concluded that except for low temperatures, we may ignore most quantum statistical effects and classical Maxwell-Boltzmann statistics can be safely applied.

## 12.6 BLACK BODY RADIATION

We see a body and its colour when light from the body reaches our eyes. The light may be emitted by the body, if it is light emitting source or be scattered by it. When light is scattered out of the spectrum of light received by it, it obeys a law known as Kirchhoff's law of radiation. According to this law a body scattered only those radiation which it would absorb when it acts as an absorber. A red coloured body do not absorb red colour from the white spectrum received by it but absorb all the rest colours.

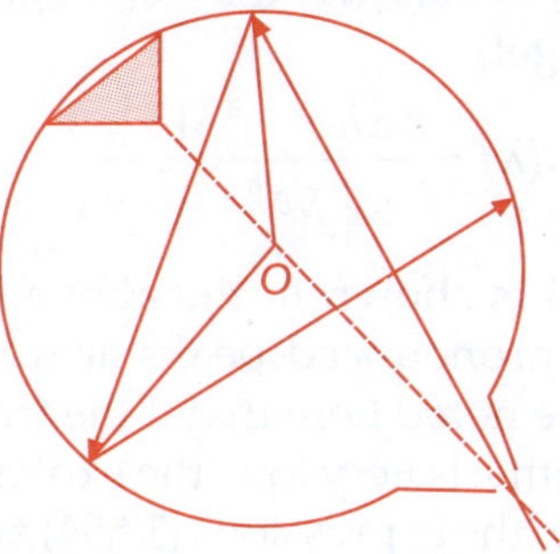

Fig. 12.7 Fery's black body

Therefore, when heated it can emit all colours except red because it did not absorb it. Now, if it absorbs all colours and does not reject any thing, it looks black and if this is true for all electromagnetic spectrum, the body is then named as Black body. When it heated such body would emit all wave lengths and is called white body. The figure shown above is an enclosure with insulated wall and having a narrow opening. The is a pyramid shaped structure just opposite to the opening. Any radiation entering through the opening will never be able to come out of the enclosure and suffer multiple reflections from the wall and thus may be assumed to be totally absorbed. Such enclosure devised by Fery is known as Fery's black body. When heated it emits all the radiations it absorbed and the experimentally observed distribution of intensity with frequency measured bears some peculiarities which led to the discovery of quantum ideas of radiation. The figure below (Fig. 12.8) shows the energy distribution spectrum of a black body radiation.

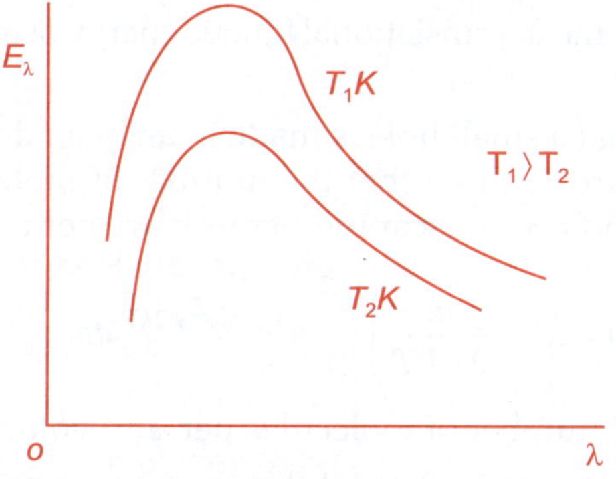


Fig. 12.8 Energy spectrum of black body radiation

Now, recalling the equn. (12.54) which is:

$$\varepsilon(\nu) = \frac{8\pi h\nu^3}{c^3} \frac{1}{e^{h\nu/kT} - 1}$$

and transforming it in terms of wavelength $\lambda$ by using $\nu = c/\lambda$ so that $d\nu = -(c/\lambda^2)\, d\lambda$ and $\varepsilon(\lambda) = -\varepsilon(\nu)d\nu/d\lambda = \varepsilon(\nu)\, c/\lambda^2$. Replacing these in the above equn. (12.54), we get:

$$\varepsilon(\lambda) = \frac{8\pi hc}{\lambda^5} \frac{1}{e^{hc/\lambda kT} - 1}. \qquad \ldots(12.55)$$

The variation of $\varepsilon(\lambda)$ is shown in the above Fig. 12.8 at two different temperatures. It shows pronounced peaks at wavelengths dependent on temperature. It should be noted here that if the frequency, $\nu$ is very high or conversely the wavelength $\lambda$ is very low, the expression $e^{h\nu/kT}$ would be very high and so the minus 1 in the expressions (12.54) and (12.55) may be ignored and the radiation is then governed by Wien's radiation law. Conversely, if frequency is very low i.e. wavelength is very large, then the expression $e^{h\nu/kT}$ could be neglected at high temperatures. The radiation law is then is called Rayleigh-Jeans law. While each of these two laws could explain the two different regions of the black body radiation energy spectrum, none of them developed classically could explain the total region of the spectrum. It was Planck introducing the concept of energy photons could successfully explain the radiation spectrum of black body. This Planck' radiation law was later more correctly derived by the application of Bose-Einstein statistics.

## REVIEW QUESTIONS

1. The possible particle energies of a system of particles are 0, $\varepsilon$, $2\varepsilon$, $3\varepsilon$, $n\varepsilon$,... show that the partition function of a with $g_i = 1$ is $Z = \left(1 - e^{-\varepsilon/kT}\right)^{-1}$.

2. Compute in the above problem the average energy per particle and also the limiting value of the average energy when $\varepsilon$ is much smaller than $kT$.

3. Compute the mean translational kinetic energy of an ideal gas molecule at 300°K.

4. Assuming that a small hole is made in an oven door containing a gas at temperature $T$, show that the number of molecules with velocity between $v$ and $v + dv$ escaping per unit are per unit time is given by:

$$dn = \pi N \left(\frac{m}{2\pi kT}\right)^{3/2} v^3 e^{-mv^2/2kT}\, dv$$

   $N$ is the total number of molecules per unit volume in the oven.

5. Find the average velocity and the average energy of the electrons at 0°K in a metal having $10^{22}$ electrons per c.c.

# REFERENCES


1. G. Wannier-Statistical Physic, John Wiley, N.Y. 1966.

2. E. Schrödinger-Statistical Thermodynamics, Cambridge University Press, Cambridge, 1963.

3. R. Feynman-The Feynman Lectures on Physics (Vol. I), Addison Wesley, Reading, Mass, 1963.

4. F Reif-Berkeley Physics Course (Vol. 5), McGraw Hill, N.Y., 1965.

5. M. Alonso and E. J. Finn - Fundamental University Physics, (Vol. III), Addison-Wesley Pub, Reading, Mass, 1968.

# Quantum Mechanics

"Quantum Mechanics is certainly imposing. But an inner voice tells me that it is not yet the real thing"
*Albert Einstein.*

## 13.1 INTRODUCTION: QUANTUM PHYSICS

The motion of the bodies independent of mutual interactions can be described on the basis of general rules based on experimental evidences. These rules or principles are like conservation of momentum, angular momentum and energy. These rules are framed on the assumption that we can localize a body in space and measure its momentum without disturbing the body so that its position and momentum remain measurable. The mechanics based on such assumptions is called Classical Mechanics. The ability to localize a body and measure its momentum simultaneously is found to be possible for large bodies. There is no limit to its largeness but down to its size it is found that the classical mechanics fails to yield the results for the particles which are the constituents of matter. It is found that for electron like small "particles" if we try to locate the object in space, it is so much disturbed that the measurement becomes impossible. The outcome of this result of impossibility is known as the uncertainty principle. The quantization of energy is another novel idea which does not appear in classical mechanics. The interaction of radiation and matter by means of emission and absorption of photons is another new concept.

Therefore, the need of a totally new formalism known as "Quantum mechanics" and the revolutionary development of this mechanics in the present form is due to the contribution of Luis de-Broglie, Erwin Schrödinger, Werner Heisenberg, Paul Dirac and Max Born and others in the late 1920's. The theory of this new formalism is mathematically elaborate but its basic concept is simple. In this chapter we will develop the fundamental aspects of quantum mechanics, which are adequate to explain some characteristics of atomic, molecular and nuclear structures.

## 13.2. PHOTOELECTRIC AND COMPTON EFFECTS: THE CORPUSCULAR NATURE OF RADIATION

**The photoelectric Effect:** When light is incident upon certain metallic surfaces, a radiative energy is transferred to bound electrons, liberating them from surface atoms. The kinetic energy of an electron produced by photoemission is simply given by: $\dfrac{mv^2}{2} = \varepsilon - W$, where $\varepsilon$ the radiative photon energy and $W$ is is the minimum work needed to be done on the electron to get it free from the metallic surface and is known as 'work function'. The difference of this two is contributed to the liberated electron as its kinetic energy. This liberated electron move because of its negative charge move towards the anode of the electrodes used to apply the field. Now, if $V_0$ is the retarding potential necessary to just stop the ejected electron then we can write $eV_0 = \varepsilon - W$. This $V_0$ and its variation for frequency $\omega$ of the incident light show the following results, Fig. 13.1. Now as $\varepsilon = h\nu$

where $\nu = \dfrac{w}{2\pi}$ from Planck's theory, we can write: $eV_0 = h\nu - W = h(\nu - \nu_0)$

where $h\nu_0 = W = \dfrac{hw_0}{2\pi}$ and $v_0 = \dfrac{\omega_0}{2\pi}$ is known as threshold frequency of the incident radiation (light).

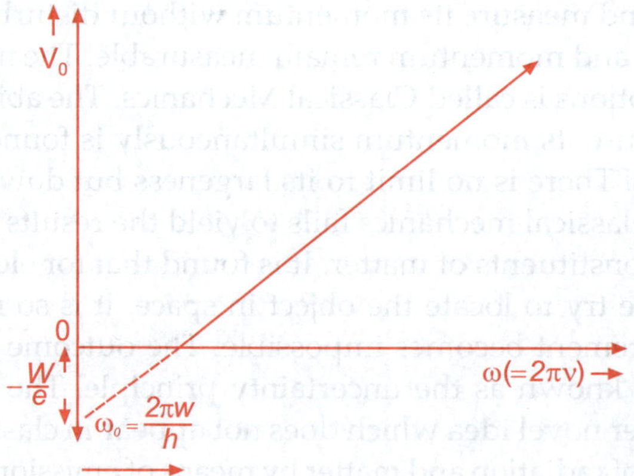


**Fig. 13.1** The variation of stopping potential with frequency of the incident light. $\omega_0$ is the threshold frequency

The slope of the above graph which is $\dfrac{h}{2\pi e}$ is constant (e is the charge of an electron).

This relation, experimentally confirmed gives the direct method for the determination of Planck's constant using the value of electronic charge.

**The Compton Effect:** The spectrum of X-ray scattered from carbon contains an undeflected beam of the same wavelength, $\lambda$ as the incident radiation and a deflected beam of longer wavelength. The effect was explained by Compton and based on the Einstein theory of photon. Compton suggested that the wavelength shift of the deflected beam of X-ray is independent of $\lambda$, the scattering material and varies only with the scattering angle $q$ according to:

$$\lambda' - \lambda = 0.0024(1 - \cos \theta).$$

Treating incident radiation of X-ray as photon having energy $e$ and which undergoes a collision with an electron of rest mass $m_e$, applying conservation of energy and momentum $p$ we get

$$\varepsilon + m_e c^2 = \varepsilon' + \varepsilon_e$$

$$\varepsilon + m_e c^2 = \varepsilon' + (p_e^2 c^2 + m_e^2 c^4)^{1/2} \qquad \text{...(13.1 a)}$$

and, $$\qquad\qquad p = p' + p_e \qquad\qquad\qquad\qquad \text{...(13.1 b)}$$

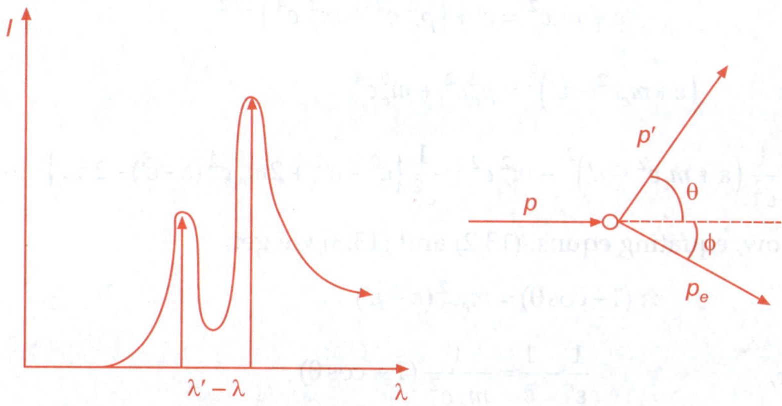


**Fig. 13.2** Compton effect: wavelength shift and momentum conservation

From (13.1 b) $p_e^2 = p^2 + p'^2 - 2\boldsymbol{p}\cdot\boldsymbol{p}' = p^2 + p'^2 - 2\boldsymbol{p}\cdot\boldsymbol{p}'\cos\theta.$

Now, from (13.1a) $\varepsilon = \dfrac{m_0 c^2}{\sqrt{1 - \dfrac{v^2}{c^2}}}$ and $p^2 = \dfrac{m_0^2 v^2}{1 - \dfrac{v^2}{c^2}}$ or $\dfrac{v^2}{c^2} = \dfrac{p^2}{p^2 + m_0^2 c^2}$

$$\left[ p^2 = \frac{m_0^2 v^2 c^2}{c^2 - v^2} = \frac{m_0^2 c^2}{\dfrac{c^2}{v^2} - 1} \text{ or } \frac{c^2}{v^2} - 1 = \frac{m_0^2 c^2}{p^2} \text{ or } \frac{c^2}{v^2} = \frac{m_0^2 c^2 + p^2}{p^2} \right]$$

$$\varepsilon = \frac{m_0 c^2}{\sqrt{1 - \dfrac{v^2}{c^2}}} = m_0 c^2 \gamma \quad i.e. \ \gamma = \frac{1}{\sqrt{1 - v^2\!/c^2}}.$$

Now, replacing $\dfrac{v}{c}$ we get $\gamma = \sqrt{1 + \dfrac{p^2}{m_0^{\,2} c^2}}$

Therefore,  $\varepsilon = m_0 \, c^2 \gamma = m_0 c^2 \sqrt{1 + \dfrac{p^2}{m_0 \, c^2}}$  or,  $\varepsilon^2 = p^2 c^2 + (m_0 c^2)^2$  and

writing for electron $\varepsilon_e = (p_e^2 \, c^2 + m_e^2 \, c^4)^{1/2}$ which is used in equn. (13.1a)

Now, recalling the momentum equation as above:

$$p_e^2 = p^2 + p'^2 - 2\, p \cdot p' = p^2 + p'^2 - 2\, p \cdot p' \cos\theta$$

and as $p = e/c$, we get

$$p_e^2 = p^2 + p'^2 - 2\, \boldsymbol{p.p'} = \frac{1}{c^2}(\varepsilon^2 + \varepsilon'^2 - 2\varepsilon\varepsilon' \cos\theta) \qquad\qquad \ldots(13.2)$$

Now, recalling equn. (13.1a) which is

$$\varepsilon + m_e c^2 = \varepsilon' + \left(p_e^2 \, c^2 + m_e^2 \, c^4\right)^{1/2}$$

or  $$\left(\varepsilon + m_e c^2 - \varepsilon'\right)^2 = p_e^2 c^2 + m_e^2 c^4$$

$$p_e^2 = \frac{1}{c^2}\left(\varepsilon + m_e c^2 - \varepsilon'\right)^2 - m_e^2 \, c^2 = \frac{1}{c^2}\left\{\varepsilon^2 + \varepsilon'^2 + 2 m_e \, c^2 \, (\varepsilon - \varepsilon') - 2\,\varepsilon\,\varepsilon'\right\} \cdot \;\ldots(13.3)$$

Now, equating equns. (13.2) and (13.3) we get

$$\varepsilon\varepsilon' (1 - \cos\theta) = m_e c^2 (\varepsilon - \varepsilon')$$

or,  $$\frac{1}{\varepsilon'} - \frac{1}{\varepsilon} = \frac{1}{m_e \, c^2} (1 - \cos\theta).$$

If for radiation the photon nature is correct then, $\varepsilon = h\nu = ch/\lambda$, the above equation can then be written as

$$\lambda' - \lambda = \frac{h}{m_e c} (1 - \cos\theta). \qquad\qquad \ldots(13.4)$$

This $\dfrac{h}{m_e c}$ is known as Compton's wavelength. It may be emphasized that the quantum nature of radiation was established from the observed fact that the change of wavelength of incident X-ray due to scattering by an electron depends only on the angle of scattering as per the equn. (13.4).

## 13.2.1  The Old Quantum Theory

The analysis of the radiation spectrum emitted by hydrogen atoms shows intense lines having definite frequencies which can be fitted to an empirical relation

$$\nu = R\left(\frac{1}{n^2} - \frac{1}{m^2}\right) \text{ where } n \text{ and } m \text{ are integers and } m > n \geq 1.$$

$R$ is a constant known as Rydberg constant and its value determined is $3.29 \times 10^{15}$ Hz.

For different values of the integers and the above expression fits give rise to different series. Similar series of formula have been found for the emission spectra of alkali metal atoms.

This result contradicts the classical picture of an electron orbiting round a nucleus and radiates electromagnetic waves leading to an unstable atom. Since the electron frequency of revolution will change smoothly, the emission spectra should be continuous and not that which is predicted by the above equation of series of discrete lines.

A successful interpretation of the discrete spectrum of one electron atoms was proposed on the basis of two assumptions known as "Bohr's Postulates".

1. **Postulate of stationary states:** "An electron moves only in certain permissible orbits which are stationary states, in the sense that no radiation is emitted. The orbital angular momentum of the electron in those stationary states equals an integer times of $h/2p$".

   **Conclusion:** The electron in a stable orbit is exempted from the requirement that accelerated charge must radiate.

2. **Postulate of discrete transitions:** "Emission and absorption of radiation occurs only when an electron makes a transition from one stationary state to another. The radiation has a definite frequency $\nu_{mn}$ gives by the condition: $h\nu_{mn} = \varepsilon_m - \varepsilon_n$".

   The stationary state for which $n = 1$ is known as "ground state" and the states for $n > 1$ are said to be excited states. The above condition states that radiation of frequency:

$$\nu_{nm} = Z^2 \frac{m_e e^4}{64\,\pi^3\,\varepsilon_0^2\,h^3}\left(\frac{1}{n^2} - \frac{1}{m^2}\right) \text{ is emitted when the electron drops from}$$

stationary state m to that of n .Where $m > n \geq 1$.

## 13.3  WAVE NATURE OF PARTICLE

The stable motion of electrons in the atom introduces integers, which can resemble classically to the wave motion phenomena such as normal modes of standing wave. A similar periodicity was assigned to electrons by Louis de Broglie, who derived the quantization condition of Bohr orbits by fitting a standing wave around each circumference. In other words he assumed in 1924 that electrons were accompanied by matter waves, which were he assumed that the electrons were accompanied by matter wave, which were regarded as localized with the particle, in contrast to the condition of classical waves.

The de-Broglie assumption was to assign the group velocity $v_g$ (Chapter 8) to a classical particle of momentum $p$. From equn. (8.29) which is:

$$v_g = \frac{\omega' - \omega}{k' - k} = \frac{d\omega}{dk} = \frac{p}{m}. \qquad \qquad ...(13.5)$$

Now, if free particle energy $\varepsilon$ obeys the Einstein relation: $\varepsilon = h\nu = \hbar\omega$

$$\left( \text{Where} = \hbar = \frac{h}{2\pi} \right)$$

i.e.     $\varepsilon = \hbar\omega = \dfrac{p^2}{2m}$ or $\omega = \dfrac{p^2}{2m\hbar}.$

The equn. (13.5) holds only for the choice: $p = \hbar k$. This is known as de-Broglie's hypothesis. The matter wavelength is then given by

$$\lambda = \frac{2\pi}{k} = \frac{h}{p}. \qquad \qquad ...(13.6)$$

This wavelength associated with electrons was experimentally measured by Davisson and Germer and this provided a strong reason to retain the de Broglie concept of matter waves and to modify the classical concept concerning the meaning of position and velocity of a particle. Davisson and Germer in their experiment demonstrated the formation of diffraction pattern obtained when a crystalline material was irradiated by electrons.

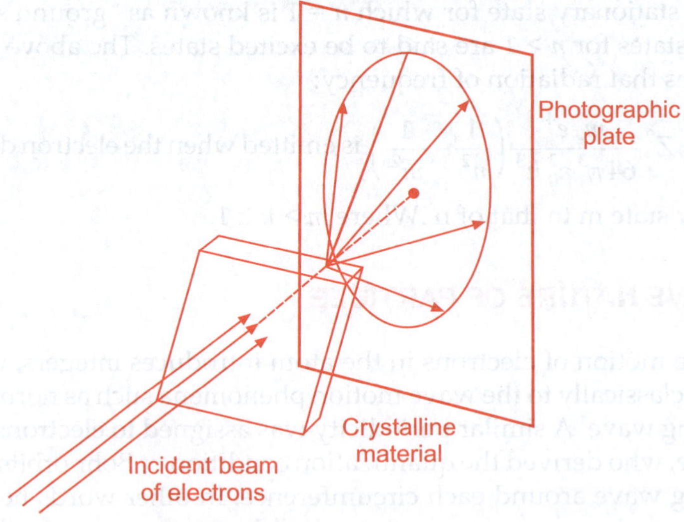


**Fig. 13.3**  Davisson–germer experiment of electron diffraction by crystalline material

Electrons accelerated by an electric potential $V$ gain an energy $eV$ and kinetic energy $p^2/2\,m_e = eV$. Therefore, $p = \sqrt{2\,m_e\,eV}$. Now, introducing the values of $e$, $m_e$ and $h$, we obtain the de-Broglie wavelength of such electrons $\lambda = h/\sqrt{2m_e eV} = 1.23 \times 10^{-9}/\sqrt{V}\ m.$

## 13.3.1 Uncertainty Principle

Let us now encounter a situation which can not be explained by classical mechanics and we start with an experiment given below. An electron stream is directed towards a slit say of opening $b$. According to de-Broglie hypothesis and demonstration of particle wave by Davisson– Germer, we will get a diffraction pattern on the screen. From the Fig. 13.4 it can be seen that uncertainty in the particle's momentum parallel to X-axis is determined by the angle $\theta$, corresponding to the central maximum. Since the electron after passing through the slit would most probably be moving within the angle $2\theta$. According to the diffraction by a single slit the angle $\theta$ is given by $\sin\theta = 1/b$. Then $\Delta p \sim p \sin\theta = (h/\lambda)\,(\lambda/b)$ is the uncertainty in the momentum parallel to X-axis. Now, if we want to determine the $x$-coordinate of the particle whether or not the particle passes through the slit of width $b$, then the precision of such measurement will be $\Delta x = b$. Now, combining these two uncertainties i.e. momentum and position we get

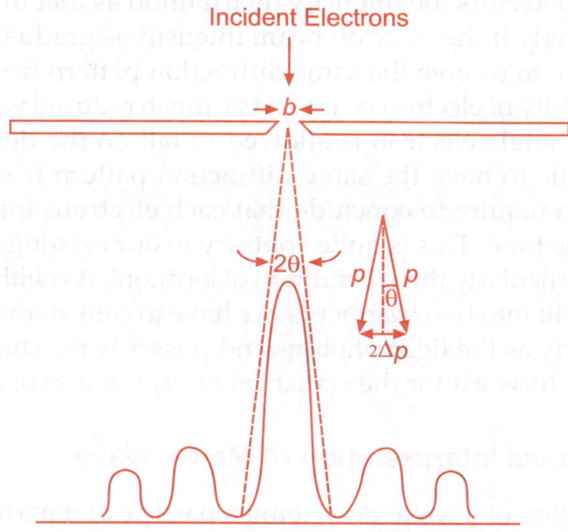


**Fig. 13.4**  Measurement of position and momentum of a particle passing through the slit

$$\Delta p \cdot \Delta x \sim h. \qquad \qquad \qquad \text{...(13.7)}$$

Now, in order to reduce the uncertainty in position determination the only alternative is to reduce the opening of the slit i.e. to reduce $b$ but in doing so the broadness of the diffraction central maximum will increase (Chapter 9). This broadening of the central maximum will increase then the uncertainty of measurement of momentum $p$ i.e. $\Delta p$. Conversely, in order to reduce the uncertainty in momentum measurement the central maximum must then be very narrow and to achieve this we must have a wider slit opening and this in return increase the uncertainty in position measurement i.e. $x$-coordinate of the particle. In most cases $x$ and $p$ are known with much less accuracy, so that we must write instead of equn. (13.7):

$$\Delta p \cdot \Delta x \geq h/2p. \qquad \qquad \ldots(13.8)$$

This result is known as Heisenberg's uncertainty principle and this can be stated in words as:

**"It is impossible to know simultaneously and with exactness both the position and the momentum of a particle".**

The above so long new and thrilling for the existing concept of matter, we may imagine an experiment. Let us now take a double slit and consider the double slit experiment as discussed in Chapter 9 (Section 9.2.1) with only difference is that we now replace light photons by a stream of electrons and on the screen position, there are a series of detectors of microscopic dimensions. The detectors show an interference pattern similar to that of Fig. 9.3. If we close one slit, we would get the diffraction pattern from single slit as Fig. 9.9 and Fig. 13.4. More explicitly when both slits are open we actually receive in the detectors the intensity distribution as that in Fig. 9.10. Now, more interestingly if the electron beam intensity is gradually reduced, we would continue to receive the same diffraction pattern from a double slit. When the intensity of electron beam is very much reduced we can essentially assume that a single electron is allowed to fall on the double slit and as we still continue to have the same diffraction pattern from a double slit, we would then require to conclude that each electrons interact with both slits at the same time. This is quite contrary to our existing classical idea of particle and particularly their 'condition of location'. As neither electrons nor photons can split into two fragments, we have to consider that the electrons behave similarly as the light photons and posses wave characteristics. We now can move forward for the statistical interpretation of matter wave.

### 13.3.2  Statistical Interpretation of Matter Wave

The interpretation of a wave describing a particle or a particle represented by a wave was a matter of discussion in the early years of the development of quantum mechanics. It was Max Born, who first put forward the statistical interpretation of the wave function. In the diffraction pattern of a light wave say from a single slit opening, it may be said that the probability of finding light photon at the peak of the central maximum of intensity distribution is maximum and it decreases with the decrease of intensity on both sides of the central maximum until again it increases at different positions on the screen giving rise the secondary maxima. Born introduces in analogy with the electromagnetic wave, a wave function $\psi$, which is a complex scalar function of the coordinates of the particle and time. The probability that a particle will follow a particular path is given by the intensity of the field which is the absolute square of this guiding field. In this case of scattering the intensity of the matter wave determines at every point the probability of finding an electron there. The square of the amplitude of the wave function $\psi$ is the intensity which determines the probability of finding a particle at a certain place. This $\psi$ which may be complex and the probability is real; we do not define $\psi^2$ as a measure of intensity but instead

$$| \psi^2 | = \psi \cdot \psi^*$$

where $\psi^*$ is the complex conjugate of complex function $\psi$.

Now, as the probability of finding a particle say $dW(x, y, z, t)$ within a certain volume is also proportional to the volume element $dV = dx.\, dy.\, dz$ at a time $t$, according to the statistical interpretation of matter wave, the following hypothesis may be adopted

$$dW(x, y, z, t) = | \psi(x, y, z, t) |^2\, dV.$$

Now, to get a quantity independent of volume, we introduce the spatial probability density

$$w(x, y, z, t) = \frac{dW}{dV} = \left| \psi(x,y,z,t) \right|^2 \qquad \ldots(13.9)$$

$$\int_{-\infty}^{\infty} \psi\psi^* dV = 1. \qquad \ldots(13.10)$$

This implies that the particle must be somewhere in the space and this normalization integral is independent of time otherwise we will not be able to compare the probabilities referring to different times. A state of the particle is bound when the wave function $y$ is square integrable and when it is not the particle is in the free state. It can be seen that wave function $\psi$ in the bound state where $E < 0$ is square-integrable and the bound states are localized within the potential well and can propagate only within the interior of the well. Free states are located above the potential well and are not bound.

## 13.4 INTERPRETATION OF QUANTUM MECHANICS AND SCHRÖDINGER EQUATION

We have by this time have probably established that in quantum mechanics the wave function $\psi(x)$ plays similar role of standing wave of amplitude $\xi(x)$ in classical mechanics. For one dimensional wave equation the amplitude of a standing wave of wavelength $\lambda$ satisfy the following differential equation: [Refer to equn. (8.6)]

$\dfrac{d^2\xi}{dx^2} + k^2\xi = 0$, where $k$ the wave number of the standing wave is given by $k = 2p/\lambda$. Now we may recall that in matter wave $p = \hbar k$ so that we can expect that wave function $\psi(x)$ to satisfy a similar equation

$$\frac{d^2\psi}{dx^2} + \frac{p^2}{\hbar^2}\psi = 0.$$

Now, the total energy of the system $E = p^2/2m + E_p(x)$, where $E_p$ is the potential energy and $p$ and $m$ are respectively as momentum and mass of the particle. Then we may write

$$p^2 = 2m[E - E_p(x)]$$

then, $\qquad \dfrac{d^2\psi}{dx^2} + \dfrac{2m}{\hbar^2}[E - E_p(x)]\psi = 0$ $\qquad\qquad$ ...(13.11)

This is known as Schrödinger's equation.

In case of free particle the potential energy is zero i.e. $E_p = 0$ and the Schrödinger's equation is then transformed in the form:

$$-\frac{\hbar^2}{2m}\frac{d^2\psi}{dx^2} = E\psi.$$

This may be written in the form:

$$\frac{d^2\psi}{dx^2} + \frac{2mE}{\hbar^2}\psi = 0.$$

For a free particle, $E = p^2/2m$ and we know $p = \hbar k$ where $k$ is the wave number, we then write

$$\frac{d^2\psi}{dx^2} + k^2\psi = 0. \qquad\qquad ...(13.12)$$

A slightly more elaborate derivation of this Schrödinger's equation in place of which that has been so far derived by assuming intuitively a correlation between matter wave and a classical stationary wave is given below. Students during their first reading may however omit this derivation.

In relativistic classical mechanics time coordinates and spatial coordinates are treated as four components of a four vector representing energy and momentum i.e.

$$x_v = (r,\, i\,c\,t),\ p_v = (p,\, iE/c), \text{ where } n = 1, 2, 3, 4.$$

By enlarging the operator representation of the three dimensional momentum to a four dimensional covariant vector operator, we get

$$\left(\hat{p},\, \frac{i}{c}\,\hat{E}\right) = -i\,\hbar\left\{\frac{\partial}{\partial x_v}\right\} = -i\,\hbar\left(\frac{\partial}{\partial x},\, \frac{\partial}{\partial y},\, \frac{\partial}{\partial z},\, \frac{\partial}{\partial(ict)}\right)$$

Now by comparison, the energy is replaced by the operator

$$\hat{E} = i\hbar\frac{\partial}{\partial t}.$$

In quantum mechanics we have three operators 1. Kinetic energy operator 2. Angular momentum operator and 3. Hamiltonian operator. These can be introduced before we proceed further:

### 1. Kinetic Energy Operator

In the relativistic case we have for the kinetic energy $T = p^2/2m$. With $\nabla^2 = \Delta$, we obtain the operator $T$ written as $\hat{T} = \dfrac{\hat{p}^2}{2m} = \dfrac{(-i\hbar)^2}{2m}\nabla^2 = -\dfrac{\hbar^2}{2m}\Delta$

### 2. Angular Momentum Operator

The classical definition of angular momentum of a particle is $L = r \times p$ and its quantum mechanical operator is given as:

$$\hat{L} = r \times (-i\hbar\nabla) = -i\hbar r \times \nabla$$

where $\hat{L}$ is a vector operator

### 3. Hamiltonian Operator

In analogy with Hamiltonian classical mechanics, let us define the Hamiltonian operator as the operator of total energy. In classical mechanics

$$H = T + V \quad \text{and} \quad H = \frac{p^2}{2m} + V(r)$$

where $T$ denotes the kinetic and $V$, the potential energy. Now, as $\hat{p} = -i\hbar\nabla$ and therefore, $\hat{H} = -\dfrac{\hbar^2}{2m}\Delta + \hat{V}(r)$.

Then we have two operators for the energy as both $\hat{E}$ and $\hat{H}$ describe the total energy and then be equated. This generates the Schrödinger's equation

$$\hat{E}\psi(r,t) = \hat{H}\psi(r,t).$$

Which is

$$i\hbar\frac{\partial}{\partial t}\psi(r,t) = \hat{H}\psi(r,t)$$

$$= \left[-\frac{\hbar^2}{2m}\Delta + V(r)\right]\psi(r) \qquad \ldots(13.13)$$

This is the non relativistic Schrödinger's equation.

To get more detailed picture of the quantum operators compared to classical definitions, the following table may be looked into.

**Table 13.1** Quantum Operators

| Quantity | Classical definition | Quantum operators |
|---|---|---|
| Position | $r$ | $r$ |
| Momentum | $p$ | $-i\hbar\,\nabla$ |
| Angular momentum | $r \times p$ | $-i\hbar\,r \times \nabla$ |
| Kinetic Energy | $p^2 / 2m$ | $-(\hbar^2 / 2m)\,\nabla^2$ |
| Total Energy Hamiltonian | $p^2 / 2m + E_p(r)$ | $-(\hbar^2 / 2m)\,\nabla^2 + E_p(r)$ |

## 13.5 SCHRÖDINGER'S PICTURE AND PARTICLE IN A POTENTIAL BOX

Let us now consider the electron in a piece of metal. The electron move freely within the metal but if we neglect the interaction of negative electrons with the positive ions on the sides and if the height of the potential barrier is large enough compared with the energy of the electron, then the electron will be moving within the metal but will not be able to come out from the surface automatically. This has resemblance with the gas molecules enclosed within an enclosure. The physical situation may be represented by the following figure which represents the simplified situation of potential energy situation and is known as potential box.

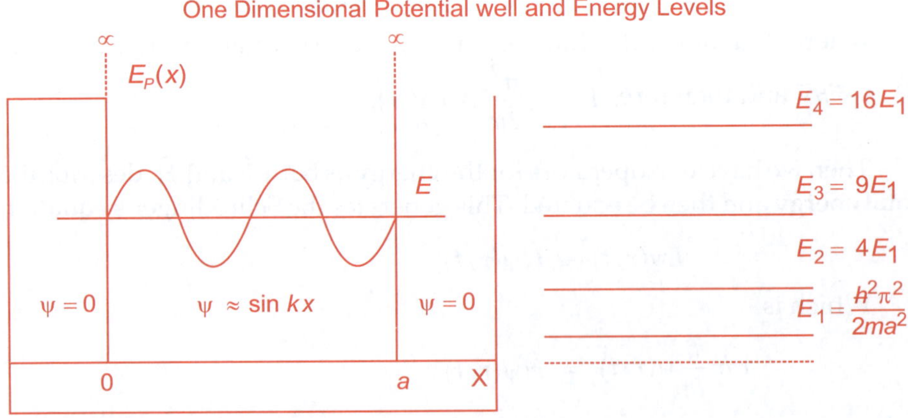


**Fig. 13.5** Particle in a potential box and energy levels

The electron within the potential well has only kinetic energy and no potential energy i.e. $E_p(x) = 0$ for $0 \langle x \langle a$. The electron then move freely within that region. There exists strong retarding force on the wall and can not be found either in the region $X \langle 0$ or $X \rangle 0$. The wave functions are also zero in these two regions i.e. $\psi = 0$ for $X \langle 0$ and also $\psi = 0$ for $X \rangle 0$.

Now, recalling the Schrödinger's equation for a free particle from equn. (13.12)

$$\frac{d^2\psi}{dx^2} + k^2\psi = 0.$$

As $k = \dfrac{2\pi}{\lambda}$ and so $k^2 = \dfrac{4\pi^2}{h^2}m^2v^2 = \dfrac{8\pi^2 m}{h^2}(\tfrac{1}{2}mv^2) = \dfrac{2m}{\hbar^2}E$ as $\hbar = \dfrac{h}{2\pi}$.

The particle will move back and forth between $x = 0$ and $x = a$ and forms standing wave.

The general solution of the above equation which is

$$\psi(x) = Ae^{ikx} + Be^{-ikx}. \qquad \qquad ...(13.14)$$

Where in the first part represents a particle moving in $+X$ direction and the second part the one moving in $-X$ direction. Now from the boundary condition mentioned above we get $\psi(x) = 0$ at $x = 0$ and also at $x = a$, then $y\,(x = 0) = A + B$ or, $A = -B$.

Therefore, $\psi(x) = A(e^{ikx} - e^{-ikx}) = 2i\,A\sin kx = C\sin kx.$

Here, $C = 2\,iA$. Now since at the boundary condition $x = a$, $y\,(x = a) = C\sin ka = 0$. But $C$ cannot be zero so, $\sin ka = 0$ or $ka = np$, where $n$ is an integer.

So, as $k = np/a$ and as $p = \hbar k = \dfrac{n\pi\hbar}{a}$. This gives the possible discrete values of momentum corresponding to different values on the integer $n$. Now, as the energy of the particle $E$ is given by

$$E = \frac{p^2}{2m} = \frac{\hbar^2 k^2}{2m} = \frac{n^2\pi^2\hbar^2}{2ma^2}. \qquad \qquad ...(13.15)$$

This equation is an important observation that is the energy levels of the particle trapped inside the one dimensional potential well and moving back and forth are discrete and there exists a minimum energy level which is not zero and equal to $E_1 = \dfrac{\pi^2\hbar^2}{2ma^2}$ for $n = 1$. Then we can say $E = E_1, 4E_1, 9E_1, \ldots$. These discrete energy levels are shown at the right side of Fig. 13.5. The particle can not then assume any arbitrary energy and neither can it attain any energy of its will as its energy is quantized. This quantization of energy of particle energy is observable whenever the Schrödinger's equation is applied and solved for potential energy which confines the particle to move in a limited region. The energy quantization is due to the fact that the wave function is determined by the potential energy and the boundary conditions of the physical problem and the solution exists only for certain values $E_1, E_2, E_3, \ldots E_n, \ldots$ of energy. This intuitive solution of Schrödinger's equation for particle in one dimensional potential well will not be valid if the potential field is of more complicated nature and then direct solution of the Schrödinger's equation would be necessary. For this simple case we can easily find resemblance with the vibration of stretched string fixed at two ends forming harmonics of different frequencies and energies as shown in the following (Fig. 13.6).

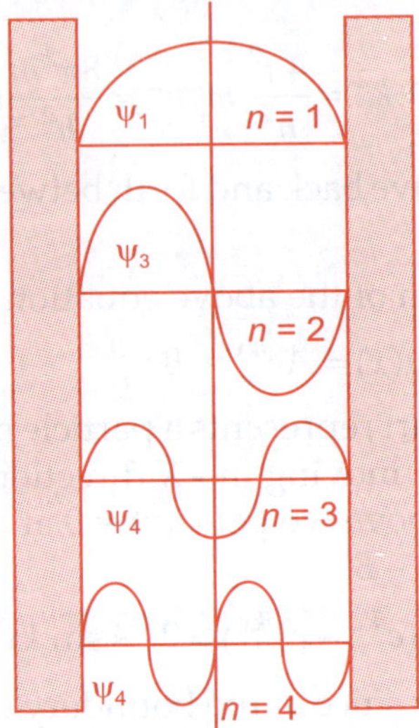


**Fig. 13.6** Vibration of stretched string fixed at the two ends and formation of different harmonics

### 13.5.1 Three Dimensional Box

Let us now extend the idea developed for one dimensional potential well to a three dimensional potential box of sides $a$, $b$ and $c$. Under this condition the momentum p will have three directions and given as $p_x$, $p_y$ and $p_z$ and they in relation with one dimensional case

$$p_x = \frac{\pi \hbar n_1}{a}, \quad p_y = \frac{\pi \hbar n_2}{b} \text{ and } p_z = \frac{\pi \hbar n_3}{c}.$$

Where, $n_1$, $n_2$ and $n_3$ are integers. The energy $E$ is given by:

$$E = \frac{1}{2m} p^2 = \frac{1}{2m}\left(p_x^2 + p_y^2 + p_z^2\right)$$

$$= \frac{\pi^2 \hbar^2}{2m}\left(\frac{n_1^2}{a^2} + \frac{n_2^2}{b^2} + \frac{n_3^2}{c^2}\right).$$

Now, as derived before for one dimensional potential well as $k = n\pi/a$, and wave function $y_n(x) = C \sin(npx/a)$, for three dimensional potential well the wave function is given by

$$\psi = C \sin\frac{n_1 \pi x}{a} \sin\frac{n_2 \pi y}{b} \sin\frac{n_3 \pi z}{c}.$$

Now, incase of cubical potential well where $a = b = c$, the energy $E$ is given by

$$E = \frac{\pi^2 \hbar^2}{2ma^2}\left(n_1^2 + n_2^2 + n_3^2\right) \qquad \ldots(13.16)$$

and the corresponding wave function is

$$\psi = C \sin\frac{n_1 \pi x}{a} \sin\frac{n_2 \pi y}{a} \sin\frac{n_3 \pi z}{a}. \qquad \ldots(13.17)$$

In the above expression $n_1$, $n_2$ and $n_3$ are as usual integers now for different values of these integers, let us examine one important case from the following table (Table 13.2)

**Table 13.2** Energy levels and degeneracy $\left(E_1 = \dfrac{\pi^2 \hbar^2}{2ma^2}\right)$

| Energy $(n_1^2 + n_2^2 + n_3^2)\,E_1$ | Combination of $n_1$, $n_2$ and $n_3$ | Degeneracy, g |
|---|---|---|
| $3\,E_1$ | $(1\,1\,1)$ | 1 |
| $6\,E_1$ | $(2\,1\,1),(1\,2\,1),(1\,1\,2)$ | 3 |
| $9\,E_1$ | $(2\,2\,1),(2\,1\,2),(1\,2\,2)$ | 3 |
| $11\,E_1$ | $(3\,1\,1),(1\,3\,1),(1\,1\,3)$ | 3 |
| $12\,E_1$ | $(2\,2\,2)$ | 1 |
| $14\,E_1$ | $(1\,2\,3),(1\,3\,2),(3\,1\,2),(3\,2\,1)$ | 4 |

We introduce here a term degeneracy which represents different energy states having the same energy. These different states result due to shuffling of the integers $n_1$, $n_2$ and $n_3$ but still keeping the same value of $n_1^2 + n_2^2 + n_3^2$ so that the energy remains same. The different energy states having same energy but still remaining as different states are known as "Degenerate States". Now, it can be shown that the number of states within some limiting positive values of $n_1$, $n_2$ and $n_3$ (the integers are only positive) corresponding to some maximum energy $E$, is approximately equal to the volume of an octant with radius $n_1^2 + n_2^2 + n_3^2$. As the volume of the octant is 1/8 th of the sphere of radius $r^2 = n_1^2 + n_2^2 + n_3^2$, the number of states $N(E)$. Therefore, using the equn. (13.16),

$$N(E) = \frac{1}{8}\left(\frac{4}{3}\pi r^3\right) = \frac{1}{8}\left(\frac{4}{3}\pi[n_1^2 + n_2^2 + n_3^2]^{3/2}\right) = \frac{\pi}{6}\left(\frac{2ma^2 E}{\pi^2 \hbar^2}\right)^{3/2}$$

$$= \frac{\pi V}{6}\left(\frac{2mE}{\pi^2 \hbar^2}\right)^{3/2} = \frac{8\pi V}{3h^3}(2m^3)^{1/2} E^{3/2} \text{ using } a^3 = V \text{ and } \hbar = \frac{h}{2\pi}$$

($V$ = volume of the box).

Now, the number of states within energy range between $E$ and $E + dE$ can be obtained by differentiating the above expression and this results:

$$dN(E) = \frac{4\pi V(2m^3)^{1/2}}{h^3} E^{1/2}\, dE$$

Now, writing $dN(E) = g(E)dE$, we get:

$$g(E) = \frac{dN}{dE} = \frac{4\pi V(2m^3)^{1/2}}{h^3} E^{1/2}. \qquad \ldots(13.18)$$

This equation gives the number of possible states per unit energy interval at the energy $E$ or may be said as density of states. It is found from the above equn. (13.18) that this density of state varies parabolically with energy $E$ and this variation is shown in the following (Fig. 13.7).

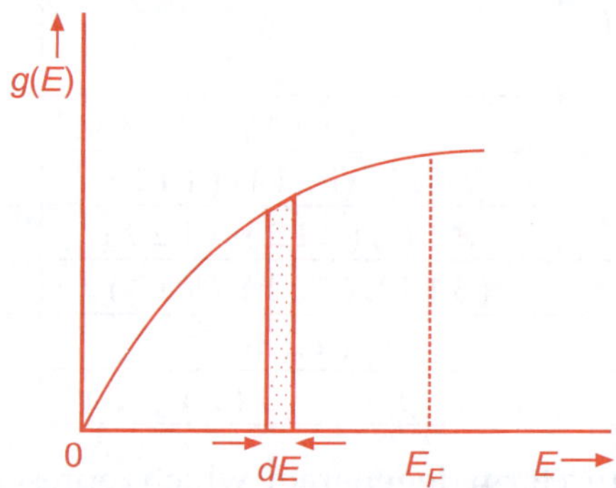


**Fig. 13.7** Variation of density of states $g(E)$ with energy $E$

Now, if we recall the Fermi-Dirac statistical distribution and the equn. (12.43) which was as : $\dfrac{dn}{dE} = \dfrac{8\pi V(2m^3)^{1/2}}{h^3} \dfrac{E^{1/2}}{e^{(E - E_F)/kT} + 1}$ and the Fig. 12.5,

it should be understood that while equn. (12.43) represents number of Fermions per energy states, the equn. (13.18) above represents the number of available energy states including degenerate states per energy. This difference must be kept in mind and as the variation of both parameters are function of $E^{1/2}$, the figures are both of the shape of parabolas.

## 13.5.2 Potential Barrier Penetration

We have seen above that the wave function while being hurdled by a potential barrier gets reflected from the wall and form standing wave.

This can be more easily conceived even from classical analogy (Vibration of stretched string) but the fact that a wave function may extend beyond the classical limits of motion and gives rise to an important and quantum phenomena known as "potential barrier penetration". Let us consider in the following Fig. 13.8, the potential barrier of height $E_0$ higher than the energy $E$ i.e.

$E \langle E_0$. Classical mechanics requires that for this case the particle would be reflected from the wall at $x = 0$ but from quantum mechanics, the wave function $\psi(x)$ has solutions in the regions as shown (1), (2) and also (3) and are shown in following second (Fig. 13.9).

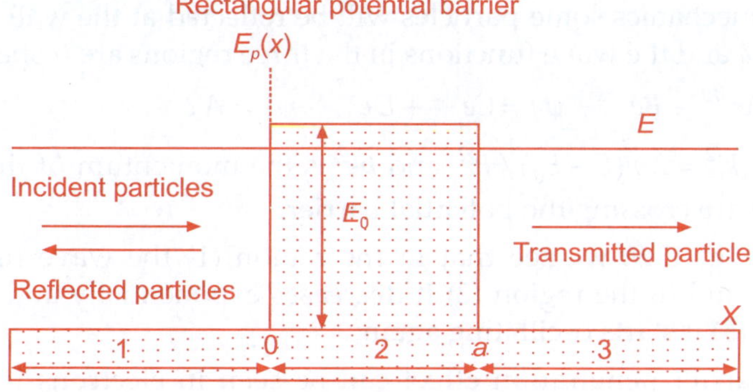


**Fig. 13.8** Rectangular potential barrier with height $E_0$ and width $a$

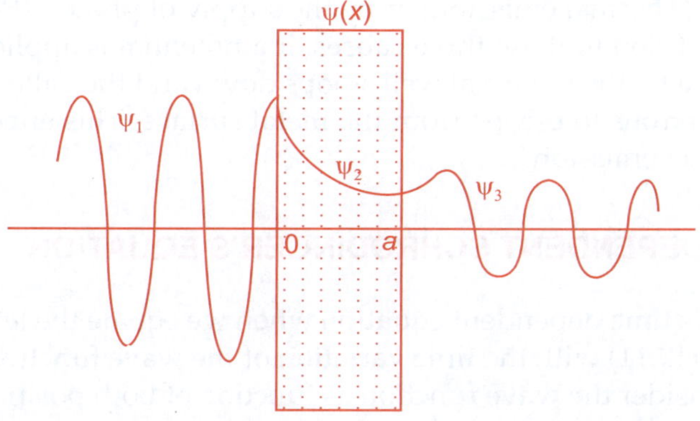


**Fig. 13.9** Wave functions and matter wave for an energy less than potential barrier

The wave function has components in the three regions as:

$$\psi_1 = Ae^{ikx} + Be^{-kx}, \quad \psi_2 = Ce^{\alpha x} + De^{-\alpha x}, \quad \psi_3 = A'e^{ikx}$$

where $k$ and $a$ are same and so as $k^2 = 2\,mE/\hbar^2$ so also $\alpha^2 - \dfrac{2m(E_0 - E)}{\hbar^2}$.

The wave function $\psi_1$ contains both the incident and reflected particles and when it is within the barrier the wave function still contains the positive exponential and therefore, $\psi_2$ is not zero even at $x = a$ and continue to exist in the third region (3) and is represented by $\psi_3$, for the transmitted particles. The transmitted particles have the same energy as the incident particles but the wave function has reduced amplitude different from A and so denoted by A'. Since $\psi_3$ is not zero there exists a definite probability of finding particle in the transmitted region (3). Therefore, in contradiction to classical mechanics, quantum mechanics confirms that it is possible for a particle to go through the potential barrier even if its kinetic energy is less than the height of the potential barrier.

When $E \rangle E_0$, classical mechanics says that all particles should cross the potential barrier and reach the other side of the barrier. However, in quantum mechanics some particles will be reflected at the wall $x = 0$ and also at $x = a$ and the wave functions in the three regions are respectively as

$$\psi_1 = Ae^{ikx} + Be^{-kx}, \quad \psi_2 = Ce^{k'x} + De^{-k'x}, \quad \psi_3 = A'e^{ikx}$$

where $k'^2 = 2m(E - E_0)/\hbar^2$ and $\hbar k'$ is the momentum of the particle while they are crossing the potential barrier.

From Fig. 13.9, we see that in the region (1) the wave function is oscillatory and in the region (2) it decreases exponentially and finally in the region (3) it starts oscillating again.

This barrier penetration effect can be seen in electrons in a metal. Usually the electrons from the Fermi energy level must be supplied with the minimum energy required to get free from the potential well, either in the form of heat (Thermal emission) or by the supply of photon (Photoelectric effect). In addition to these three causes, if a potential is applied across to ends of the metal the potential well stoops down and thus allows the most energetic electrons to escape from the metal surface. This effect is known as "Field effect emission".

## 13.6  TIME DEPENDENT SCHRÖDINGER'S EQUATION

Schrödinger's time dependent equation when we equate the left hand side of the equn. (13.11) with the time variation of the wave function, $\psi$. This is when we consider the wave function as function of both position i.e. space $x$ or $r$ and time i.e.

$\psi(x, t)$ in place of $\psi(x)$.

$$-\frac{h^2}{2m}\frac{\partial^2 \psi}{\partial x^2} + E_p(x)\psi = i\hbar\frac{\partial \psi}{\partial t}. \qquad \ldots(13.19)$$

The derivation of this Schrödinger's time dependent equation can be

followed from equn. (13.13) which is $i\hbar\frac{\partial}{\partial t}\psi(r, t) = \hat{H}\psi(r, t)$.

where Hamiltonian operator (quantum) $\hat{H}$ of total energy, $\hat{H} = -(\hbar^2/2m)\nabla^2 + E_p(r)$.

This equn. (13.19), the time dependent Schrödinger's equation is a fundamental law of nature. We may point out here that this Schrödinger's time dependent equation is of the first order in time derivative and second order in position or space derivative. If we try to obtain the a solution of the above equn. (13.19), we may assume that such solution has time and position variables separated and may have a trial solution as:

$$\psi(x,t) = \psi(x)e^{-iEt/\hbar} . \qquad \qquad \ldots(13.20)$$

We must now prove that $E$ is the energy and $\psi(x)$ is the amplitude satisfying the equn. (13.11) and for this:

$$\frac{\partial \psi}{\partial t} = -\frac{iE}{\hbar}\psi(x)e^{-iEt/\hbar} \text{ and } \frac{\partial^2 \psi}{\partial x^2} = \frac{d^2\psi}{dx^2}e^{-iEt/\hbar}.$$

Now, if we introduce these values in equn. (13.19) and simplify, we get the equn. (13.11). Therefore, we say that while $E$ is the total energy of the system and the matter wave given in the equn. (13.20) oscillates with angular frequency

$$\omega = E/\hbar \qquad \text{or} \qquad E = h\omega = h\nu.$$

This equn. (13.20) is a typical expression of standing wave because of separate space and time parts but unlike the other classical standing wave examples like vibration of string or air column, we can not express the time part in sine or cosine terms but instead the matter wave have always have complex time derivative like $e^{-i\omega t} = e^{-iEt}$.

## REVIEW QUESTIONS

1. What is the de-Broglie wavelength of thermal neutrons at a temperature of 25°C?

2. The position of an electron is determined with an uncertainty of $0.1\ A°$. Find the uncertainty in its momentum.

3. Explain the wave nature of particle and the Uncertainty Principle giving emphasis on their physical interpretation.

4. Solve the one dimensional Schrödinger equation for a finite potential well described by

$$V(x) \begin{cases} -V_0 & \text{if } |x| \le a \\ 0 & \text{if } |x| > a. \end{cases}$$

5. Calculate the zero point energy of a neutron which is confined within a nucleus which has a size $10^{-15}\ m$.

## REFERENCES


1. J. Slater-Quantum Theory of Matter, McGraw Hill, New York, 91951).

2. R. Hall-The Philosophical basis of Bohr's interpretation of Quantum Mechanics, Am. J. Phys. 33, 624 (1965).

3. R. Feynman–Lectures on Physics, (Vol-III), Addison Wesley, Reading, Mass, 1963.

4. E. H. Wichmann–Berkeley Physics Course, (Vol-4), McGraw Hill, New York, 1967.

5. W. Greiner-Quantum Mechanics: An Introduction, (4th Edn), Springer-Verlag, Germany, 2001.

6. S. N. Ghoshal–Introductory Quantum Mechanics. Calcutta Book House, Kolkata (2005–2006).

# 14

# Solid State Physics

## 14.1 STRUCTURE OF SOLIDS: AN OVERVIEW

When the atoms or molecules the basic constituents of matter, come closer to each other they fall within a potential trap and loose their freedom of movement. The only movement which will be allowed is their oscillating motion within limiting amplitude which depends on the temperature. Such state of existence of the matter is called the solid state. The atoms while coming close form a bonding between them and according to the bonding; they are classified as either Covalent Solids or Ionic Solids. In covalent solids the atoms are bound together by localized directional bonds. The crystal lattice discussed later is determined by the orientation and nature of the directional bonds. The other state of bonding is the ionic bonding where a regular array of positive and negative ions resulting from the transfer one electron or more from one kind of atom to another. In solid state of matter the constituents may either exist bearing regularity determined by this bonding, in their arrangement in space or not. These two states of arrangements are known respectively as crystalline state or amorphous. The crystalline state of matter bears some special properties which are lost if they lose their regular arrangements and become amorphous. The study of this regularity of arrangements is named as "Crystallography" which is therefore, an important aspect of condensed matter physics.

The regularity of arrangements of the constituents when bears a symmetry form a 'pattern' and a pattern either two dimensional like printed cloths or three dimensional like crystal has two aspects (1) The constituents named as 'Motifs' (2) Mode of their arrangements known as symmetry. If we change either of these two the entire pattern is changed from one type of regularity to the other. Let us now start with the second aspect of a pattern *i.e.* mode of arrangements or symmetry.

The 'symmetry' the characteristic which is found in a pattern is the result of an 'operation' which when done on a motif or its site, the entire pattern returns to a situation of self coincidence.

The symmetry operations we have just talked about are classified as follows:

1. Mirror Plane of Symmetry $\qquad$ (*m*)

2. Centre of Symmetry $\qquad$ $(\overline{1})$ $\left\{ \begin{array}{c} \text{Hermann-Mauguin} \\ \text{Symbols} \end{array} \right\}$

3. Rotation axis of Symmetry $\qquad$ (1, 2, 3, 4, 6)

4. Rotation Inversion axis of Symmetry $\qquad$ $\left(\overline{3}, \overline{4}, \overline{6}\right)$

**Mirror plane of Symmetry:** If a single or more number of planes may be imagined to exist so that the motifs on one side appear to be the mirror reflection of the opposite side, then those planes are mirror planes of symmetry. m is the Hermann–Mauguin symbol.

**Centre of Symmetry:** If we can find a point within the pattern and if one motif is found to be on the opposite side at equal distance and inverted, it is called centre of symmetry. The symbol of this centre of symmetry is $(\overline{1})$.

**Rotation axis of Symmetry:** If we identify an axis through the pattern about which if the pattern is rotated through an angle $\theta$ and the pattern returns to its position of self coincidence then if, $\dfrac{360^{\circ}}{\theta} = n$, the pattern is said to have a n fold of rotation axis of symmetry. Any object which having a random external shape possesses 1-fold of rotation. A rectangle has 2-fold, an isosceles triangle has 3-fold, a square a 4-fold and a regular hexagon 6-fold axis of symmetry. 5-fold rotation is said to be absent in geometrical crystallography, as it fails to make a compact structure.

**Rotation Inversion Axis of Symmetry:** A pattern can be said to have rotatory inversion axis of symmetry if it can be transformed in to self coincidence by the combined effect of rotation and inversion. $\overline{1}$ which is 1-fold rotation and inversion is already considered in centre of symmetry and $\overline{2}$ which is 2-fold rotation and inversion is exactly equivalent to mirror plane of symmetry. $\overline{5}$ is not considered as 5-fold rotation is said to be absent in conventional geometrical crystallography.

All these ten number of symmetry operations are grouped as "Macroscopic Symmetry Elements" as they are manifested on the external shape of the crystal which is three dimensional pattern and they can be identified by simple observation of the symmetry present on the external faces of the crystal.

In real pattern or crystal the sites of the motifs must demonstrate the symmetry that exists in the pattern and these sites represented by geometric points are known as Lattice Points and the arrangements of these lattice points in space constitutes the "Space Lattice". The mode of repetition of a pattern is specified by this array of lattice points.

Now the above ten macroscopic symmetry elements can be combined without any repetition in 32 possible ways and every crystal must have macroscopic symmetry elements that can be described by one of the 32 combinations and there would be no others. They are known as "Point Groups". Now, if these 32 point groups are regrouped, based on the fact that each group has one symmetry element common between members of that group then we get seven crystal systems or crystal classes. They are:

**Table 14.1** Seven crystal systems

| Set | System or Class | Common feature (Symmetry elements) |
|---|---|---|
| A | Triclinic | 1 fold axis. |
| B | Monoclinic | One 2 fold axis. |
| C | Orthorhombic | Three mutually perpendicular 2 fold axes. |
| D | Rhombohedra | One 3 fold axis. |
| E | Tetragonal | One 4 fold axis. |
| F | Hexagonal | One 6 fold axis. |
| G | Cubic | Four 3 fold axes at 70° 32′ to each other. |

Now, these seven crystal systems are built up by joining lattice points, having non co-planar unit translations as edges and are representative of the lattice. A unit cell of the space lattice is completely specified by the unit translations $a$, $b$ and $c$ and the angles $\alpha$, $\beta$, and $\gamma$ between them. The relative values of these parameters required by minimum symmetry properties of each crystal system are as follows.

**Table 14.2** Unit cells of seven crystal systems

| Set | Crystal Class or Systems | Relation between $a, b$ and $c$ | Relation between $\alpha, \beta$ and $\gamma$ |
|---|---|---|---|
| A | Triclinic | $a \neq b \neq c$ | $\alpha \neq \beta \neq \gamma$ |
| B | Monoclinic | $a \neq b \neq c$ | $\alpha = \gamma = 90^0 \neq \beta$ |
| C | Orthorhombic | $a \neq b \neq c$ | $\alpha = \beta = \gamma = 90^\circ$ |
| D | Rhombohedral | $a = b = c$ | $\alpha = \beta = \gamma \neq 90^\circ$ |
| E | Tetragonal | $a = b \neq c$ | $\alpha = \beta = \gamma = 90^\circ$ |
| F | Hexagonal | $a = b \neq c$ | $\alpha = \beta = 90^0, \gamma = 120^\circ$ |
| G | Cubic | $a = b = c$ | $\alpha = \beta = \gamma = 90^\circ$ |

For any space lattice, the unit cell which is geometrically simple and which adequately displays the essential symmetry of the lattice may be primitive, base centered, face centered or body centered. By considering this aspect we can have only 14 different types of lattices without disturbing the unit cell characteristics and without any repetition. These 14 different types of lattices are called "Bravais Lattices".

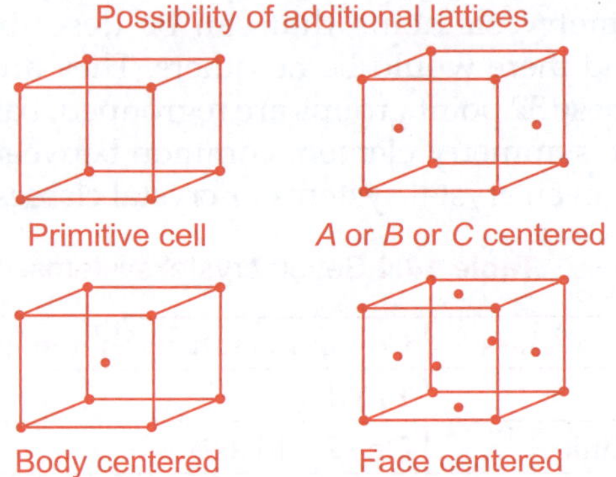


Additional lattices created with additional sites without changing the crystal system

The following table shows all such possible Bravais lattices:

**Table: 14.3** The possible bravais lattices

| SET | Crystal Class or System | Symbol or possible space Lattices | Total no. possible in the Class | Total no. of space (Bravais) Lattices |
|---|---|---|---|---|
| A | Triclinic | P | 1 | 1 |
| B | Monoclinic | P and C | 2 | 2 + 1 = 3 |
| C | Orthorhombic | P, C, I and F | 4 | 3 + 4 = 7 |
| D | Rhombohedral | P | 1 | 7 + 1 = 8 |
| E | Tetragonal | P and I | 2 | 8 + 2 = 10 |
| F | Hexagonal | P | 1 | 10 + 1 = 11 |
| G | Cubic | P, I and F | 3 | 11 + 3 = 14 |

The number of lattice points, $N$ (atoms or molecules in actual crystals) in a Unit Cell is given by:

$$N = 1 + (½) f + b,$$

where $f$ and $b$ stand for number of points on the centre of the faces and at the centre of the body of the unit cell.

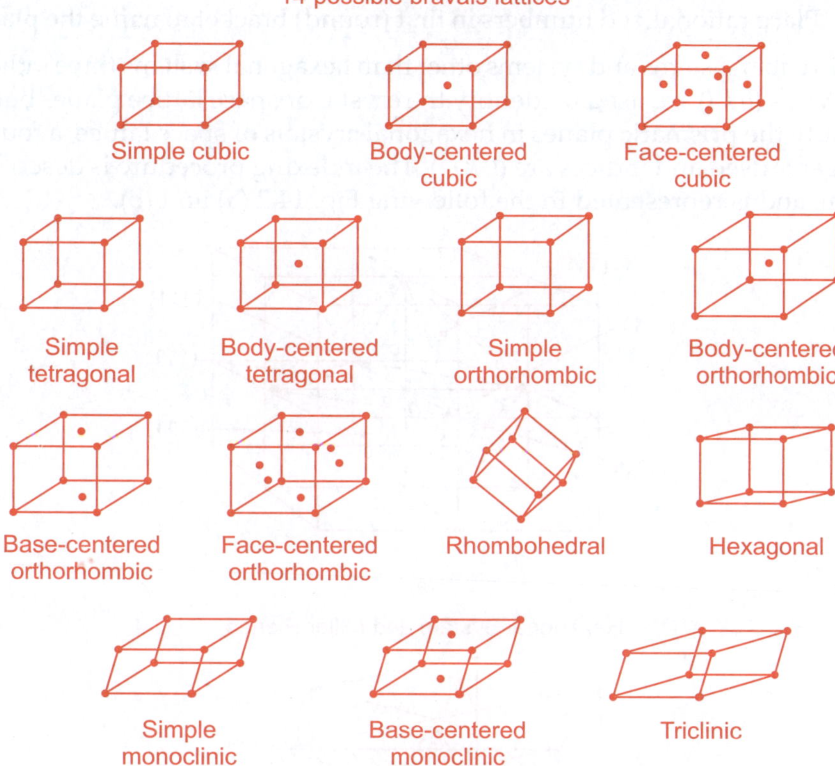


**Fig. 14.1** Fourteen bravais lattices

A primitive Cubic lattice unit cell has atoms at the corners but each one of them is shared by eight neighbouring unit cells and therefore the total contribution of corner atoms is equivalent to only 1. A body centered cubic unit cell has only $2(f=0)$ and a face centered cubic unit cell has 4, one due to eight corner points 3 due to centre points on each of six faces. These face centered points are shared by two neighbouring unit cells.

Now, it is essential to know different planes and the planes are designated by integers known as "Miller Indices".

The Miller indices of a plane or of a direction in a crystal or in a space lattice are that set of three numbers enclosed in the appropriate brackets that identifies the particular plane or particular direction and distinguishes it from others.

To assign the Miller indices we must follow the steps as:

1. Measure the intercepts that the plane makes on the axes of the lattice.

2. Divide the intercepts by the appropriate unit translations.

3. Invert the dividends.

4. Rationalize the inverted dividends.

5. Place rationalized numbers in first (round) bracket to name the plane.

In cubic systems and systems other than hexagonal systems three indices known as ($h\,k\,l$) are used to identify the crystal or space lattice planes but to identify the prismatic planes in hexagonal crystals of space lattice, a fourth integer is used and indices are ($h\,k\,i\,l$). The indexing procedure is described above and is represented in the following Fig. 14.2 (a) and (b).

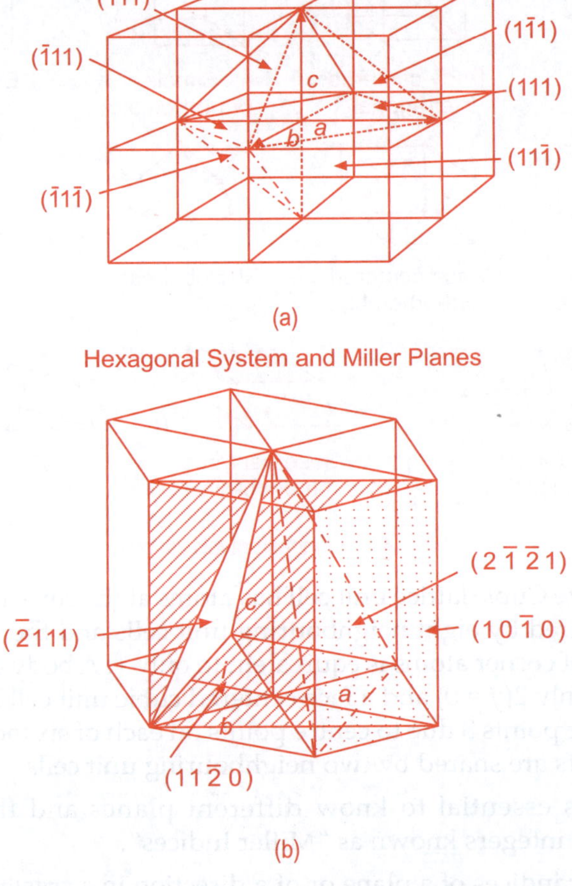


**Fig. 14.2** Miller indices of some planes in (a) Cubic system (b) Hexagonal system

A set of planes ($h\,k\,l$) is noted by {$h\,k\,l$} and {$1\,1\,0$} set of planes include a family of six planes namely : (110), (011), ($\bar{1}10$), ($1\bar{1}0$), ($0\bar{1}1$), ($\bar{1}0\bar{1}$).

The crystallographic directions within the space lattice can also be designated by Miller indices and they are done as follows:

1. Determine the coordinates of any point on the direction.

2.  Divide the coordinates of the point by the respective unit translations.

3.  Rationalize the dividends.

4.  Place these rationalize dividends in square brackets.

The indices of the direction in a space lattice do not depend on the size or shape of the unit cell and so similar indices in lattice of any type indicate same direction.

Now, if we assume the atoms in any Bravais lattice are represented by solid sphere of radius $r$, we can evaluate the "packing fraction" $f$ defined as the ratio of the volume of the atoms in the unit cell to the volume of the unit cell. Therefore, for close packed *FCC* structures, there are four atoms of radius $r$ per unit cell of volume $a^3$ and as $(4r)^2 = 2a^2$ so that $r = a\sqrt{2}/4$ and

the packing fraction is $$f_{FCC} = \frac{4(4\pi/3)\left(\sqrt{2}\,a/4\right)^3}{a^3} = \frac{\sqrt{2}\,\pi}{6} = 0.74$$

Similarly, for another important *BCC* lattice packing fraction is 0.68.

The billiard ball model used to find the packing fraction is given Fig. 14.3.

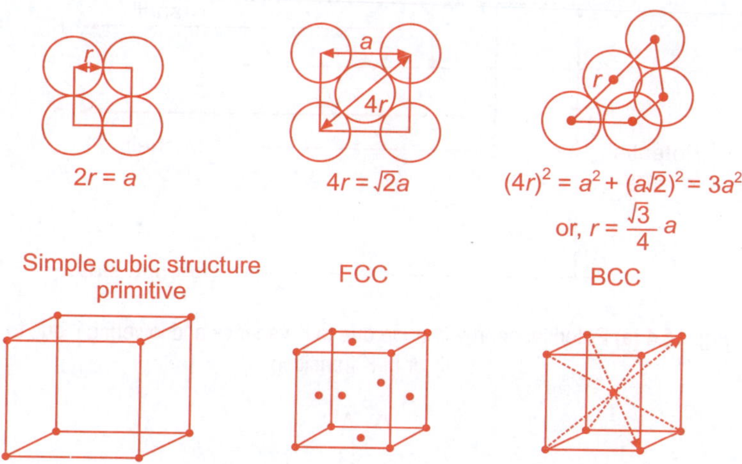


**Fig. 14.3** Cubic systems' packing of atoms

## 14.1.1 X-ray Diffraction of Crystalline Solids

X-rays was discovered by Roentgen, a German Physicist in the late nineteenth century. It is important in the structure analysis or rather indispensable as its wavelength is of the order of the inert atomic distances of solids satisfying the basic requirement of diffraction. There are two types of X-ray radiation and this classification is based on its physics of production. The bombarding electrons from the source are accelerated and when they enter the orbital electrons of the target atom, they are rapidly and continuously decelerated which results into the emission of invisible X-radiation or X-rays. Due to this reason of its production, this type X-ray is called 'General radiation'. Superposed on it at higher accelerating voltage is the second type of X-rays which are sharp peaks on the continuous general radiation and it is called 'Characteristic radiation'. They are called characteristic as the wavelengths of these peaks depend on the atomic number of the target atom. Physics behind the production of this characteristic radiation is that with the accelerated kinetic energy of the bombarding electrons, they dislodge the orbital electrons of the target atoms and the atom as a consequence of this loss of orbital electrons is raised to higher potential energy and to lower down this potential energy, the electrons from higher orbits jump to the vacant site of the orbit by releasing the photons having specific wavelength. The wavelength of the radiation is equal to the energy difference of the orbits. This phenomenon is shown in the following Fig. 14.4 (a). The net observed X-ray spectra is shown in Fig. 14.4 (b). This spectra is however characteristic of the target and so also the characteristic wavelengths.

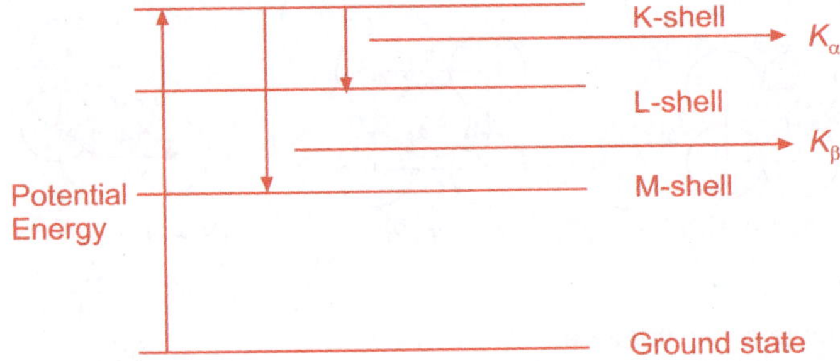


**Fig. 14.4 (a)** Potential energy change due to K-vacancy and lowering it due to L-K or M-K transition

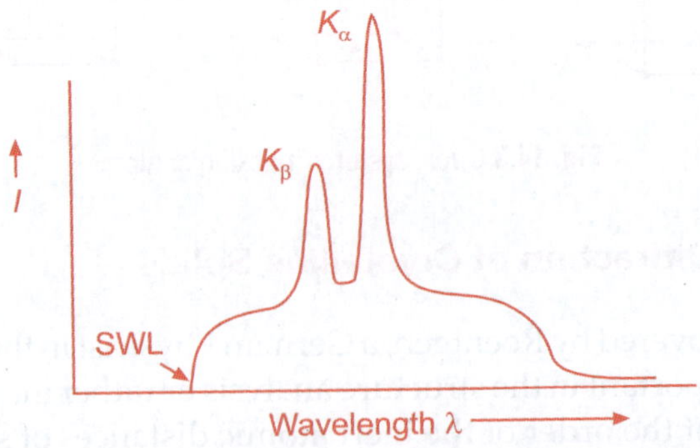


**Fig. 14.4 (b)** The X-ray spectra, showing the peaks of characteristic $K_\alpha$ and $K_\beta$ radiations and also short wavelength limit (SWL)

The short wavelength limit as indicated in the spectra is generated when the total energy of the bombarding electron is utilized in the production of radiation and this will be the $X$-ray radiation of shortest wavelength. These characteristic radiations have two components and as $K_\alpha$ radiation is of highest intensity, the $K_\beta$ is eliminated by absorbing this component through filters.

### 14.1.2 Laue Equations and Bragg's Law

Now, coming back to symmetry and ordering in crystals we may state here that these ordering of crystals are of two types 1. Short range order, which is effective within few atomic limits and 2. Long range order which is effective all through the bulk of crystal. A real crystal may possess both of these two types or the second one. When both of these are followed it is regarded as Single crystal and when the second one is followed all through but the short range is violated at some places, it is called Poly crystal. In both of stages of existence the structure of crystals can only be detected by either X-ray or by electron beams as their wavelengths are of the order of inter atomic distance and can cause of diffraction. The diffracted beam brings out the inside structural information of the crystalline substance. We will start this study within the limitations of this book with Laue equations. Let us consider an array of atoms within the crystal with atomic spacing as a and X-ray upon incidence is diffracted as shown in the following (Fig. 14.4).

In these figure $S_0$ and $S$ are the unit vectors defining the directions of incident and diffracted beams.

Now, when the path difference between incident and diffracted beams equals the integral multiple of the wave length $\lambda$, then the interference maxima condition will be satisfied.

i.e. $BC - AD = a\cos\theta - a\cos\varphi = n\lambda$, this is one dimensional Laue equation. When the other two directions are considered then the corresponding Laue equations are: $b\cos\theta'' - b\cos\varphi' = n\lambda$ and $c\cos\theta'' - c\cos\varphi'' = n\lambda$ in vector form these three Laue equations can be written as:

(a) $(S - S_0) = n\lambda$　　　(b) $(S - S_0) = n\lambda$　and　(c) $(S - S_0) = n\lambda$.

When these three conditions are simultaneously satisfied the entire diffraction phenomenon may be equivalent to a planer reflection and gives rise to Bragg's Law.

It can be shown that the wavelengths of X-ray for which the Laue conditions are satisfied are not the characteristic radiation but general radiation.

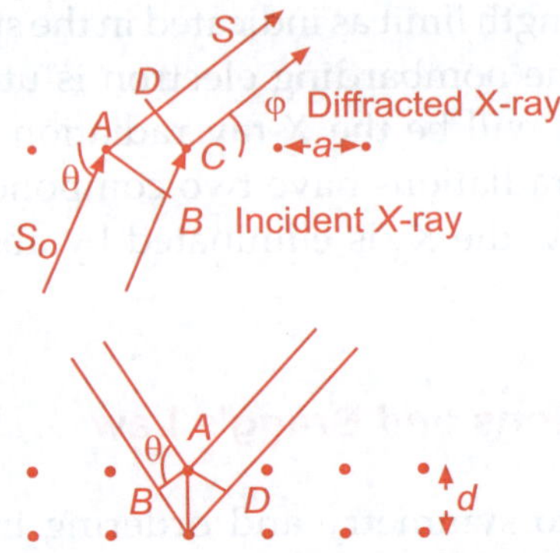


**Fig. 14.5** Laue and Bragg's Diffraction from a set of planes having indices ($h\,k\,l$)

The path difference between the incident and diffracted wave fronts: $AB$ and $AD$ will be $BC - CD = 2d\sin\theta$ and when this is equal to an integral multiple of the characteristic wavelength $\lambda$, then there will be 'reflection' and the reflected beams will interfere constructively i.e. $2d\sin q = n\lambda$. This is known as Bragg's Law and the angle $q$ is known as Bragg angle. The d is the inter planer spacing. As the values of the inter planar spacing *i.e.* d are different for different sets of planes the angle $q$ will also be dependant on the plane for a definite wave length of X-ray radiation. So, the Bragg's condition for reflection can be more generalized as:

$$2d_{hkl}\sin\theta_{hkl} = n\lambda$$

The X-ray is diffracted from the planes and brings about amongst many two very important information; one is the angle of diffraction and the other one is intensity of diffraction. The angle of diffraction gives us the Miller indices of the diffracting planes and the intensity from the diffracted planes gives us the clues to find the crystal structure of the crystals. However, the discussion of these aspects is beyond the scope of this book and the interested readers may consult the references given.

## 14.2 FREE ELECTRON APPROXIMATION

We will now consider a metal to consist of positive ions located on lattice sites, and bathed in a sea of conducting electrons in the conduction band and take a particular example where, $N$ valence electrons move in a structure containing $N$ monovalent ion cores. Then we can assume that these electrons move in a constant average potential energy and are free and can move independently. Also for simplicity we consider a linear lattice. Then we can apply the Schrödinger's Equation as described in Chapter 13, equn. (13.13), which was

$$i\hbar\frac{\partial}{\partial t}\psi(r,t) = \hat{H}\psi(r,t) = [-\frac{\hbar^2}{2m}\Delta + V(r)]\,\psi(r)]$$

We may then drop the potential term and write the approximate wave function of an electron having momentum $p = \hbar k$ is $\psi = e^{ikx}$ the $k$ can be either positive or negative to allow the motion of electrons in two opposite directions. Now, in a three dimensional lattice, we must have a wave function which can be given as

$$\psi(r) = e^{ik.r}. \qquad \qquad ...(14.1)$$

For both of the above two functions we must have $|\psi|^2 = 1$, which means that the electron has same probability of being found at any place in the lattice. This is however a crude approximation as the electrons will have greater probability for being found near the positive ions at the lattice sites but we can ignore this in the 'Free electron Model" as this gives some important insight in to the properties of many solids.

The plane wave solution of the above equn. (14.1) is:

$$\psi(r) = \frac{1}{\sqrt{V}} \, e^{ik \cdot r}.$$

Now, the normalized function with respect to volume $V$, energy eigenvalues of the electron described by wave functions as above and when we disregard the constant average potential energy, is

$$E = p^2/2\,m_e = \hbar^2 k^2 / 2m_e \qquad \qquad ...(14.2)$$

The variation of energy $E$ with $k$ is parabolic and is given in Fig. 14.5.

The momentum eigenvalues of the free electrons result from

$$P\psi(r) = \frac{\hbar}{i}\left\{\frac{1}{\sqrt{V}} \, e^{ik. \, r}\right\} = \hbar k \psi(r).$$

The free electron model allows all values of $k$ and therefore $E_k$, which means that the model does not provide any information about the width of the band, but we may estimate it in the following way.

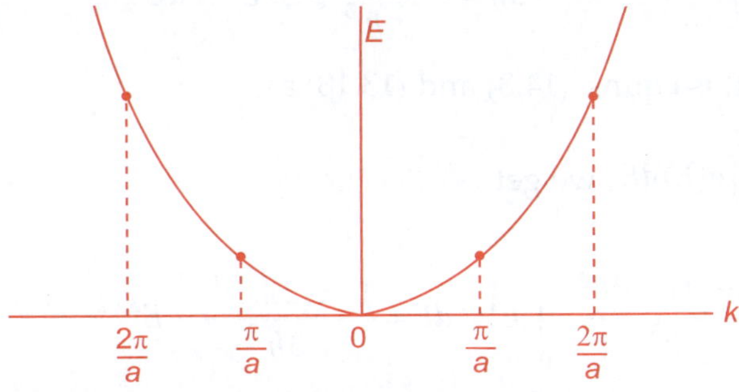


Fig. 14.6 Energy of free electron as a function of $k$ ($p = \hbar k$)

Consider a liner lattice of length $L$ composed of $N$ ions separated by a distance a, so that $L = Na$. At the boundary the wave function vanishes and to sustain a standing wave we have $n(\lambda/2) = L$ and for each value of na stationary state results.

The possible values of $n$ is then are 1, 2, 3, … $N$ and noting that $k = 2\pi/\lambda$, we get:

$$K = n\pi/L = n\pi / Na, n = 1, 2, 3, \ldots, N.$$

The difference between successive values of $k$ is $\pi/Na$ and for $N$ being very large $k$ may be safely assumed to be continuous though the values are quantized. Now, setting $n = N$ the maximum value of $k$ is $k_{max} = \pi/a$.

Thus the range of $k$ values allowed within the band is between $-\pi/a$ to $+\pi/a$ and the maximum energy of the band which is also the width of the band is then

$$E_{max} = \frac{\hbar^2 \pi^2}{2\, m_e\, a^2}.$$

Now, it is very important to know how the electrons distribute themselves in a band among the energies from zero up to $E_{max}$, we recall the equation derived in Chapter 13, equn. (13.18) and as $g(E)$ is equal to the number of energy levels within the range $E$ and $E + dE$,

$$g(E) = \frac{dN}{dE} = \frac{4\pi V(2m^3)^{1/2}}{h^3} E^{1/2}$$

and therefore $\quad dN(E) = \dfrac{4\pi V(2m^3)^{1/2}}{h^3} E^{1/2} dE.$

Considering the spin up and down of the electrons in each levels, the total number of electrons per unit volume with energy between $E$ and $E + dE$ in the band is

$$dn = \frac{8\pi(2m_e^3)^{1/2}}{h^3} E^{1/2}\, dE. \qquad \ldots(14.3)$$

Using this equns. (14.3) and (13.18) as

$$n = \int_0^E g(E)\, dE \text{, we get}$$

$$n = \frac{8\pi(2m_e^3)^{1/2}}{h^3} \int_0^E E^{1/2} dE = \frac{16\pi(2m_e^3)^{1/2}}{3h^3} E^{3/2}. \qquad \ldots(14.4)$$

This quantity $g(E)$ and the distribution of electrons in the energy sates in the conduction band are shown in the following Fig. 14.6.

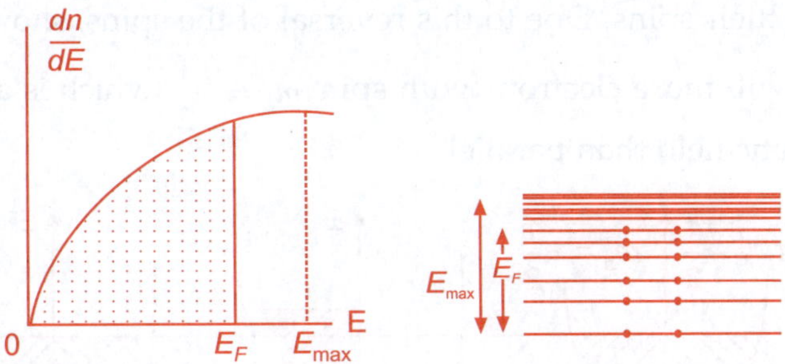

**Fig. 14.7** Density of states for free electrons in a solid and their distributions

## 14.3 ELECTRON SPIN PARAMAGNETISM

Now, consider the effect of magnetic field on these free electrons. The magnetic field is represented by the magnetic induction $B_0 = \mu_0 H$, where $H$ is the intensity of external magnetic field and $B_0$ is the field induced within the solid. At absolute zero and in the absence of any field the energy band for free electrons is represented by the plot of energy $E$ and $N(E)$ discussed in the last chapter and given by

$$N(E) = \frac{8\pi V}{3h^3}(2m^3)^{1/2}E^{3/2}.$$

The electrons with spin parallel to Z-axis say $N_p(E)$ and those anti parallel spin, $N_a(E)$ are separated onto two halves of the horizontal axis. This is given in the Fig. 14.7 a. If now the direction of the applied field is taken as the Z-axis, the energy $E$ of the spinning electrons is raised or lowered, depending on the spin orbit interaction. As the electron is spinning, it will posses an intrinsic spin magnetic moment and the spin magnetic moment $\mu_S$ is proportional to the spin angular momentum $S$ by the relation:

$$\mu_S = -g_S \frac{e}{2m_e} S.$$

The eigenvalue of its spin component by:

$$-\mu_S \cdot B_0 = g_S \frac{e}{2m_e} S \cdot B_0 = 2\frac{e}{2m_e} m_S \hbar B_0 = \pm \mu_B B_0.$$

The result is instantaneous displacement of two halves of the energy $b$ and as in Fig. 14.7 b. This is a non equilibrium configuration and as a consequence the electrons with $m_S = +\frac{1}{2}$, which have higher energy than those with $m_S = +\frac{1}{2}$ fall into the empty states and occupy the states by reversing their spins. Due to this reversal of the spins, shown in Fig. 14.7 (c), there will more electrons with spin $m_S = -\dfrac{1}{2}$ which is anti-parallel to the magnetic field than parallel.

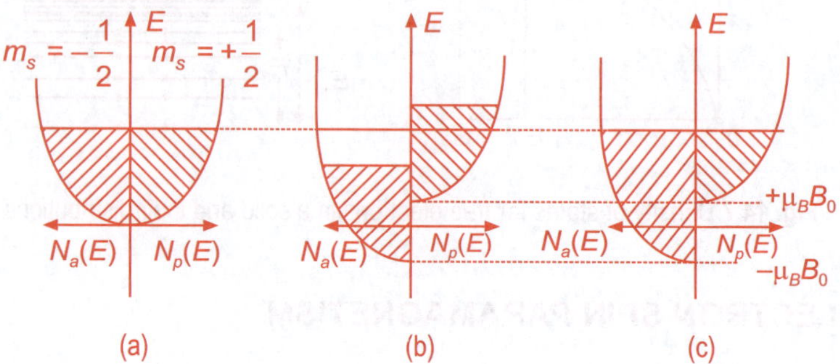


Fig. 14.8 Energy bands for free electrons in applied magnetic field

Due to this reversal of the electron spin direction in the direction anti parallel to the applied field, the net spin magnetic moment increases in the parallel direction and magnetization as a result is induced in the solid as long as the applied external field continues. This phenomenon of induced magnetism in the direction of applied field is known as "paramagnetism". This will be discussed in details in a later Chapter 16 along with other types of magnetism.

## 14.4 ELECTRICAL CONDUCTION

The electrons flow in a conductor in a conductor under the influence of an external electric field and this can be associated with a change of the free electron distribution say $f(\mathbf{k})$ in $\mathbf{k}$ space. Let us consider a simple situation when a small electric field $V$ is applied in $x$-direction and electron flow under its influence. The force applied by this field on the electron is given by using $\mathbf{p} = \hbar\mathbf{k} - eV = \hbar\dfrac{dk_x}{dt}$.

The rate of change of the distribution function $f(\mathbf{k})$ in $x$-direction is associated with the electron drift and is expressed as:

$$\left(\frac{df}{dt}\right)_d = \left(\frac{df}{dk_x}\right)\left(\frac{dk_x}{dt}\right) = -\frac{eV}{\hbar}\left(\frac{df}{dk_x}\right).$$

For small deviation of the distribution function from equilibrium, we assume that in real crystal $f(\mathbf{k})$ returns to its thermal equilibrium form $f_0(\mathbf{k})$ exponentially with time due to electron lattice collisions as:

$$(f - f_0)_t = (f - f_0)_{t=0}\, e^{-t/\tau}$$

where $t$ is the relaxation time of the system. Alternatively we may say that the rate of change of $f(k)$ caused by collisions is proportional to the deviation of the function from equilibrium i.e.

$$\left(\frac{df}{dt}\right)_c = -\frac{f - f_0}{\tau}.$$

Together these can be equated as:

$$\left(\frac{df}{dt}\right)_d = \left(\frac{df}{dt}\right)_c = -\frac{f - f_0}{\tau} = -\frac{eV}{\hbar}\frac{df}{dk_x}$$

The distribution function can then be equated as:

$$f = f_0 + \frac{e\tau V}{\hbar}\frac{df}{dk_x} = f_0 + e\tau V\frac{1}{\hbar}\frac{df}{dE}\frac{dE}{dk_x}$$

As $E = p^2/2m_e$ and $p_x = \hbar k_x$ $E = \dfrac{\hbar^2 k_x^2}{2m_e}$ and $\dfrac{dE}{dk_x} = \dfrac{\hbar^2 k_x}{m_e}$  ...(14.5)

Replacing this value above we get:

$$f = f_0 + \frac{e\tau V \hbar k_x}{m_e}\frac{df}{dE} = f_0 + e\tau V v_x\frac{df}{dE}. \text{ Here, } p_x/m_e = \frac{\hbar k_x}{m_e} = v_x.$$

$vx$ is the velocity of the electron in the $x$-direction.

Now, providing the potential $V$ is small, the distribution function $f$ may be assumed to be same as Fermi distribution which is the first term of the above equation so that

$$df/dE = df_0/dE$$

and $f_0 + e\tau V v_x \dfrac{df_0}{dE}$. In $k$ space $\displaystyle\int_0^{k_F} f_0(k) = 4\pi^3\, N/v = 4\pi^3 n$

where $N$ is the total number of electrons in volume $v$ and $n$ is the electron density.

$$dn = dN/v = \frac{1}{4\pi^3} f(k)dk.$$

The current density $\quad j_x = -\displaystyle\int ev_x\, dn = \frac{N}{v}\frac{e^2 \tau_F}{m_e} V$

and as current density $j_x = \sigma V$ from classical free electron theory, we

get the electrical conductivity $s = \dfrac{ne^2 \tau_F}{m_e}.$  ...(14.6)

## 14.5 BLOCK THEOREM: PERIODIC POTENTIAL

So far in the free electron model, we have considered that the electrons are either not influenced by a potential field due to non existence of it or they are not effectively influenced as the potential remains constant. However, now to improve the free electron model, we incorporate the periodic structure of the lattice. Let us start with the possible variation of the wave function. The effect of the periodic lattice is to change the particle wave function $e^{ik.r}$ so that instead of having a constant amplitude, the wave function has a varying amplitude and the variation changes with the period of the lattice.

Therefore, we may write the wave function as:

$$\psi(r) = e^{ik \cdot r} u(r) \qquad \qquad \dots(14.7)$$

where $u(r)$ is the modulating amplitude which repeats itself from one unit cell to the other in the lattice having spacing a and in a linear lattice the above equn. (14.6) is modified as:

$$\psi(x) = e^{i\,kx} u(x) \qquad \qquad \dots (14.8)$$

where $u(x)$ satisfies the condition

$$u(x + a) = u(x). \qquad \qquad \dots(14.9)$$

The equns. (14.7) and (14.8) together constitute the "Bloch's Theorem". The fact that the function $u(x)$ obey the periodicity of the lattice can be seen for one dimensional lattice in the following way:

If we consider a linear lattice of period "a" such that the potential energy $E_p$ is related as:

$E_p(x) = E_p(x + a)$ is also periodic with period "$a$" then the probability distribution of the electrons must show also the same periodicity as the potential energy so that

$$\left|\psi(x)\right|^2 = \left|\psi(x+a)\right|^2.$$

The above equation implies that $\psi(x+a) = C\psi(x)$ and $|C|^2 = 1$. Thus, we may write $C = e^{ika}$ having $k$ as an arbitrary parameter. Therefore, we may write:

$\psi(x) = e^{-ika}\,\psi(x + a)$. Now multiplying both sides by $e^{-ikx}$ we get

$$e^{-ikx}\,\psi(x) = e^{-ik\,(x + a)}\,\psi(x + a)$$

The above equation shows that if we consider the function $u(x) = e^{-ikx}\,\psi(x)$ then $u(x) = u(x + a)$. This is Bloch theorem.

Now, recalling the Schrödinger's equation from (13.13) which is

$$i\hbar \frac{\partial}{\partial t}\psi(r, t) = \hat{H}\psi(r, t) = \left[-\frac{\hbar^2}{2m}\Delta + V(r)\right]\psi(r) \text{ and replacing } \psi(r) \text{ by}$$

$y(r) = e^{i\,k \cdot r}\,u(r)$ we get an equation for $u(r)$.

$$-\frac{\hbar^2}{2m}\{k \cdot \nabla - k^2\}\, u(r) + V(r)\, u(r) = E u(r).$$

Considering the complex conjugate equation we can obtain that

$$E(-k) = E(k).$$

This further shows that each energy band is symmetric about the origin of $k$ space.

The energy of the electron is not entirely kinetic as it is in the case free electron, because of the potential energy due to the lattice ions. The detailed expression for energy in terms of $k$ is complicated and depends upon the geometry of the lattice. The important observation is that the energy has a discontinuity or gap at certain values of $k$ which for a linear lattice of spacing a, are given by $k = n\pi/a$, where $n = \pm 1, \pm 2, \pm 3, \dots$ and it is shown in the following Fig. 14.8.

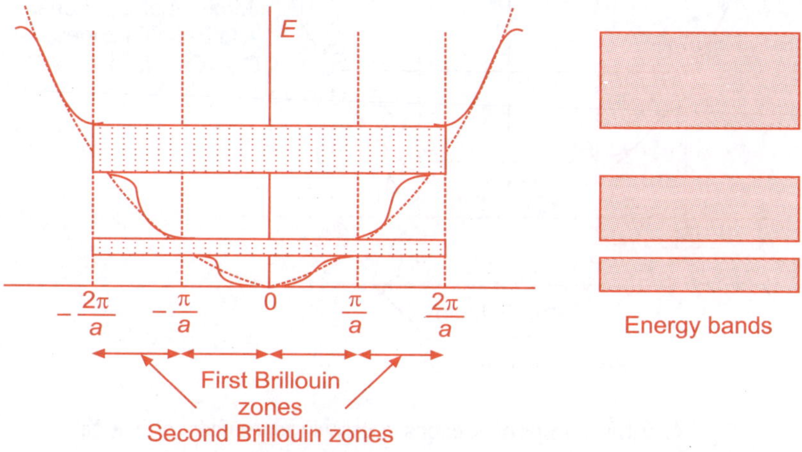


**Fig. 14.9** Brillouin zones in linear lattice

It is now be said that electrons can move freely within the lattice without encountering any resistance except when $k$ is close to the values $n\pi/a$. The motion of the electrons in the lattice can be considered similar to the propagation of electromagnetic wave in a crystal. It has been seen that the electromagnetic wave in the lattice gives to a reinforced scattering when Bragg's condition is satisfied as discussed before. For normal incidence the Bragg's law is written as: $2a = n\lambda$ for linear lattice. Now, writing $\lambda = 2\pi/k$, we get: $k = n\pi/a$ which is in agreement with the expression written before. Therefore, these values of $k$ are those at which the linear lattice blocks the motion of electrons in a given direction by forcing them to move in the opposite direction. The range of $k$ values between $-\pi/a$ to $+\pi/a$, constitute the first Brillouin zone and between $-2\pi/a$ to $+2\pi/a$, constitute second Brillouin zone and so on.

Now, when a force $F$ is applied, the $F = \hbar \dfrac{dk}{dt}$ and this can be considered as the equation of motion of electron in the lattice. The electron in the first Brillouin zone under the effect of the field suffers increase in its $k$ value and its velocity $v$ also suffers increase. When $k$ reaches a certain value close to $\pi/a$, the velocity begins to decrease i.e. electron decelerates even when the field continues to be applied. When $k = \pi/a$, the velocity becomes zero and the wave packet suffers a Bragg's reflection . the velocity of electron continuously decreases as it moves opposite to the direction of the applied field and continues until velocity becomes zero at $k = 0$ and then again it turns back and continues to move in the direction of the field and the cycle repeats itself as shown in Fig. 14.9.

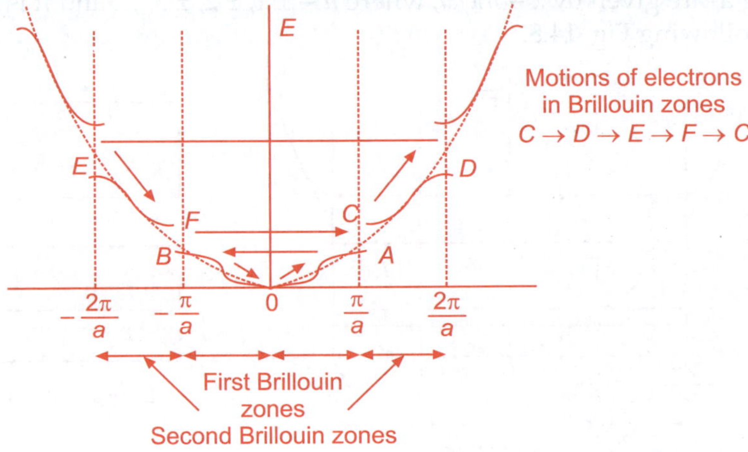


**Fig. 14.10** The motions of electrons in Brillouin zones under electric field

An important conclusion to be understood from the above kinematical description is that an external force cannot remove the electron from the Brillouin zone unless it is acted by photon and enough energy is gained. But the electrons and their motions discussed here is for the valence electrons in the metal atoms, which are rather free to move in the periodic lattice within which the potential variation is also periodic with the same periodicity as the lattice structure. For a complex structure, this potential variations are also complex and it is approximated here as a linear simple lattice. This approximation taken to develop Bloch' theorem is called "Weak binding approximation" and this approximation is valid for metallic bonding in metals and conductors.

It is now necessary to define the effective mass of electrons $m^*$, which is defined as $m^* = F/a$ where $F$ is the applied force due to electric field and a is the acceleration of the electrons due to this field and the lattice interaction. Therefore $m^*$ is different from $m_e$ and is also not a constant.

$$a = \frac{dv}{dt} = \frac{dv}{dk}\frac{dk}{dt}.$$

Now, as $E = \hbar\omega$ and group velocity $v_g = \frac{d\omega}{dk}$ so, $v = \frac{1}{\hbar}\frac{dE}{dk}$ and also as

$F\,v\,dt = (dE/dk)\,dk$ and using the above expression for $v$, we get

$$F = \hbar\frac{dk}{dt} \quad \text{and so} \quad m^* = \frac{\hbar^2}{d^2E/dk^2}. \qquad \ldots(14.10)$$

This should be noted here that when electrons are free and its energy is given by equn.(14.2) which was $E = p^2/2\,m_e = \hbar^2 k^2/2m_c$, we have $m^* = m_e$ and therefore $m^*$ is the parameter for the lattice and of the electrons lattice momentum $\hbar k$. It is interesting to point out here that in (Fig. 14.9) above $m^*$ is positive at the bottom of the of an energy band and negative at the top because at the bottom it moves in the direction of force resulting in the increase of velocity and at the top in the opposite to the force resulting velocity decrease.

Let us conclude the discussion on the electron motions in a periodic lattice by considering the density of states, which has already been introduced in free electron model. If we consider the variation of $g(E)$ which is $dn/dE$ with energy $E$ in following Fig. 14.10, it is found to be parabolic at the bottom of the band and parabolically increases but unlike Fig. 14.6, it does not continue it but again parabolically decreases at the top of the band. However, the shape of the variation of $g(E)$ given in the figure is only a qualitative representation. The correct representation of $g(E)$ variation depends on the structure of the lattice and the position of the band. The electron at the Fermi level faces some empty energy levels and can move up with the application of field and increase of temperature. If $E_F$ coincides with $E_{max}$ and so there will be no energy levels available and that condition electrons fail to get influenced by the applied field. The behaviour of the solids then changes from conductor to insulator.

The free electron theory under weak binding approximation discussed here is more classical than quantum mechanical as detailed quantum mechanical derivation and treatment is beyond the scope of this book. However semi quantum mechanical treatment for other types solids namely non metals whose valence electron are not as free as that in metals is discussed in the following section.

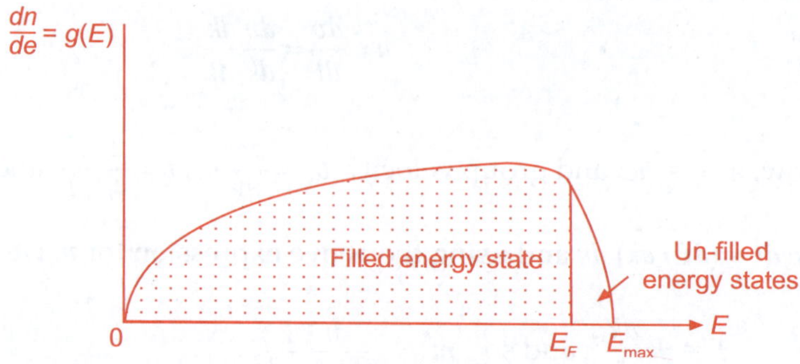



**Fig. 14.11** The density of states in a band

## 14.6 THE TIGHT BINDING APPROXIMATION

We also can demonstrate the existence of electronic energy band with gaps between energy bands, by solving one electron problem in the tight binding approximation, where we think electrons as bound to separate atoms in the solid. Such an approximation is valid for electrons in insulators. For an electron in linear lattice of spacing $a$ the wave function $y$ is given as :

$\psi = \sum_n e^{ikna}\phi(x - na)$, where $f$ is the atomic wave function of an electron in

a stationary state of an isolated atom and $n = 1, 2, \ldots, N$ identifies each of the atoms in the lattice. Thus, $f(x - na)$ is the wave function corresponding to the $n$th atom and $y$ is a linear combination of atomic wave functions with convenient phase factors. We will now see that the above equation satisfies the Bloch theorem. We can write the above equation can also be written as

$$\psi = e^{ikx} \sum_n e^{-ik(x - na)}\phi(x - na) \qquad \ldots(14.11)$$

which by comparison with $\psi = e^{ikx}$ results

$$u(x) = \sum_n e^{-ik(x - na)} \phi(x - na) \text{ and}$$

$$u(x + a) = \sum_n e^{-ik[x - (n-1)a]} \phi[x - (n-1)a]$$

The summation in both of the expression of $u(x)$ and $u(x + a)$ are identical if $N$, the number of atoms in the lattice is very large and thus as $u(x) = u(x + a)$, the Bloch's theorem is verified. The average energy of an electron described by the wave function (14.11) is given by

$$E_{ave} = \frac{\int \psi^* H\psi \, dx}{\int \psi\psi^* \, dx}$$

where $H$ the Hamiltonian operator given as before by $H = -\frac{\hbar^2}{2m_e}\frac{d^2}{dx^2} +$ $E_p(x)$ where $E_p(x)$ is the periodic potential energy of the electron in the lattice.

# REFERENCES


1. B. E. Warren X-ray Diffraction, Addison Wesley Pub. Co., 1969.

2. S. K. Chatterjee–Crystallography and world of Symmetry, Springer–Verlag, 2008.

3. M. M. Woolfson–An introduction to X-ray diffraction, Cambridge University Press, 1997.

4. F. C. Philips: An Introduction to Crystallography, Oliver and Boyd, Edinburgh, 1971.

5. L. S. Dent Glasser : Crystallography and its applications, Van Nostrand Reinhold Co., New York, 1977.

6. W. L. Bragg: The Crystalline state ,Vol .I: A general Survey,George Bell, London, 1933.

7. H. M. Rosenberg-The solid state, Oxford University Press, 1988.

8. R. E. Hummel–Electronic Properties of Materials, 2$^{nd}$ Ed., Springer-Verlag,1992.

# 15

# Semiconductor Physics

## 15.1 SEMICONDUCTOR PHYSICS

An interesting property of solids is their electrical conductivity. There are some solids which are Good conductors and Semiconductor at the other extreme lay some solids which are insulators. Amongst the conductors are metals like copper, aluminium and silver and amongst the insulators are most covalent and ionic solids including diamond and quartz. There is wide difference between the electrical conductivity within these two solid classes, for example, while for copper the conductivity at room temperature is $10^{20}$ times greater than that of quartz. Intermediate between these two classes is a third kind of solid named as Semiconductor which totally unlike conductors and insulators does not obey Ohm's law. It has been seen that the resistance increases with increase of temperature for conductors and insulators, for semiconductors it decreases. The semi conductors at room temperature show much poorer conductivity than conductors. Typical semiconducting materials are germanium and silicon.

When atoms of the substance come close enough to get transformed into solid state, not only they form normally crystalline structure but their electronic orbits overlap to form bands from discrete energy levels. Due to this possible situation the band theory of solids give a simple explanation of the marked different electrical behaviour of solids. The first analysis of this electrical behavior has been done in the last chapter (Chapter 14) by means of free electron model which was later refined by taking into account the periodic structure of the lattice. Let us consider a metal having band structure as shown in the following Fig. 15.1 which might correspond to the energy levels of sodium (Na) (Z = 11).

It can be seen from the above figure that the bands corresponding to 1$s$, 2$s$, 2$p$ are completely filled, whereas the band corresponding to 3$\varepsilon$ which can accommodate two electrons per atom but have only one is half filled and other half is emptydue to thermal excitation the electrons, these

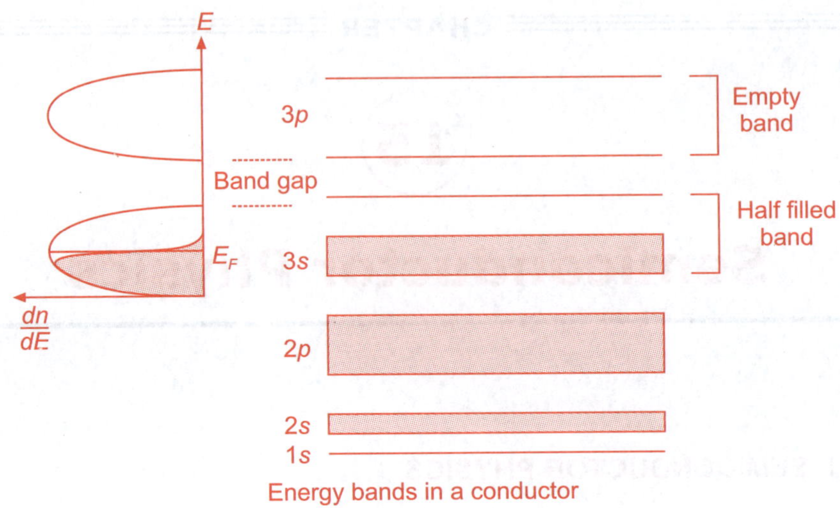



**Fig. 15.1**  Energy bands in sodium like metal

3s electrons from the Fermi level can increase their energy and go to higher energy level, depending on the thermal excitation. This is shown in the left half of the above Refer to Fig. 12.5. On the application of electric field these electrons can be accelerated and go to the nearby higher level or state of the same band without violating the Pauli's exclusion principle. Thus the electrons gaining momentum opposite to the direction of the field constitute electric current. Therefore, the substances having band structures similar to that of the above figure should be good conductors of electricity and also be good thermal conductors because of the same cause. To summarize, good conductors of electricity also called metals are those solids in which the upper most occupied band is not completely filled. Actual situation in conductors are slightly more complex for example in magnesium the 3S and 3p bands overlap eliminating the band gap and therefore the electrons in the 3s level have plenty of vacant states to move up. This is as shown in the following Fig. 15.2 is the usual state of affairs for all conductors. So, in magnesium the case is slightly different, if there is no overlapping and as 3S band is filled up and 3p band is empty, magnesium should an insulator. But because of overlapping, upper most electrons of the 3s band have the lowest energy states of the 3p band available and thus some 3s electrons move to occupy some low 3p levels until an equilibrium energy level for both bands is established. That is the reason for magnesium being a good conductor. Those substances whose atoms have complete shells but which in the solid state are conductors because of this overlapping of filled band and an empty band are often called "semimetals".

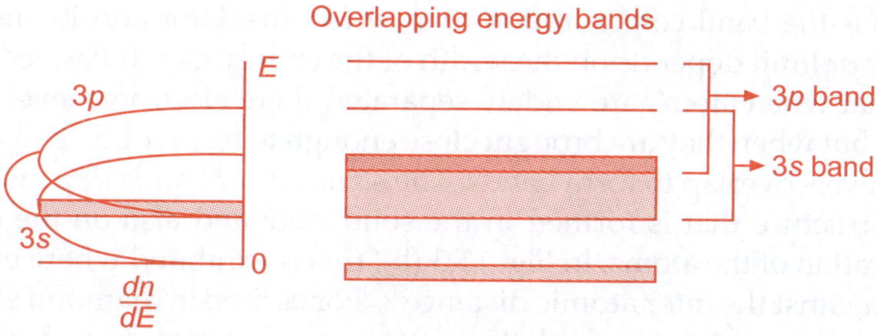



**Fig. 15.2** Overlapping of bands in the band structure

In transition metal group, such as Iron, the overlapping bands are 3*d*, 4*s*, and 4*p* and the number of electrons is insufficient to fill these bands. This property for Iron is a special and for this iron is ferromagnetic which will be discussed in a latter chapter. Similarly for rare earth group, the overlapping bands involved are 4 *f*, 5*d*, 6*s* and 6*p* and hence these elements in solid state are conductors. Now, let us refer to the figure above Fig. 15.3. In Fig. 15.3 a the valence band is totally filled up whereas the conduction band is empty and in between them there is a large energy gap.

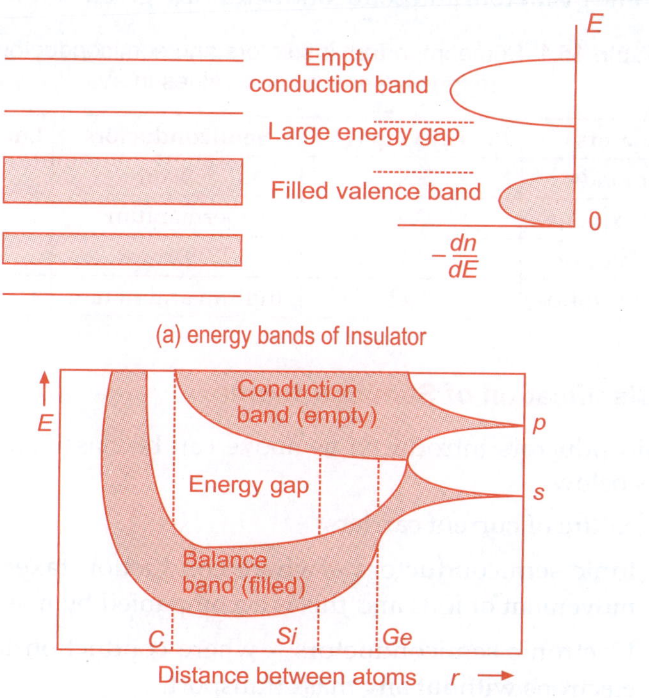



**Fig. 15.3** The energy band structures for an insulator (*A*) and for diamond silicon and germanium

This is the band configuration of a perfect insulator and its insulating property or limit depends on the width of the energy gap. It has been stated above that when atoms are widely separated there electronic levels do not over lap but when they are brought close enough to form solid, their electron energy levels overlap to form bands. The structure of bands depends on the crystal structure that is formed in the solid state and also on the electron configuration of the atoms. In Fig. 15.3 (b), this is exhibited where energy is plotted against the inter atomic distance $r$. For carbon in diamond structure when carbon atoms form solid, the energy gap is very wide and as a result diamond is a perfect insulator both for electricity and heat transmission. For silicon and germanium, the energy gaps are much narrow though other characteristic namely the filled valence band and empty conduction band remain same. This characteristic property for silicon and germanium allows the electrons from valence band to jump across the energy gap due to thermal agitation and enter into the conduction band and start conducting. This special property of silicon and germanium classifies them into an intermediate state other than conductors and insulators and that electrical state is called "Semiconductors". A comparative study of the value of energy gap for some insulators and semiconductors are given below in Table 15.1.

**Table 15.1** Some important insulators and semiconductors with their respective energy gap values in eV

| Insulators | Energy Gap (eV) | Semiconductors | Energy Gap (eV) |
|---|---|---|---|
| Diamond | 5.33 | Silicon | 1.14 |
| Zinc oxide | 3.2 | Germanium | 0.67 |
| Silver chloride | 3.2 | Tellurium | 0.33 |
| Cadmium sulfide | 2.42 | Indium antimonite | 0.23 |

## 15.1.1 Clssification of Semiconductors

The semiconductors introduced as above can be classified into different classes as below:

I :  On nature of current carriers

    (a) Ionic semiconductors – where conduction takes place by the movement of ions and this is accompanied by mass transport.

    (b) Electronic semiconductors – where conduction takes place by electrons without any mass transport.

II :  On composition

    (a) Elemental semiconductors – elements in Group-IV, C, Si, Ge, Sn.

    (b) Compound semiconductors – compounds of inter metallic Group III–V and Group II–VI.

III : According to structure

    (a) Amorphous – having local, short range order in quasi periodic structure.

    (b) Poly crystals – similar to single crystals in electrical behaviour but with significant lower conductivity.

    (c) Single crystals – most of the semiconductors are single crystals superior in all qualities.

In compound semiconductors, the semiconductors are formed from the elements in (i) Group III A with those of Group V A as:

| Group III A | Group V A | | | |
|---|---|---|---|---|
| $\downarrow$    $\rightarrow$ | N | P | As | Sb |
| B | | | | |
| Al | Al N | | | |
| Ga | GaN | **GaP** | **GaAs** | **GaS** |
| In | | **InP** | **InAs** | **InSb** |

Compounds in bold are stable and do not disintegrate with time. In this III-V category of semiconductors each atom on the average has 4 valence electrons and the bonding is of covalent nature. But as Group III elements are more electropositive than Group IV and Group V is more electro negative than Group IV, the cohesive force between the atoms has two terms one due to the contribution of covalent bonding and the other due to ionic contribution. As a result the cohesive force and the strength with which the valence electrons are bound to the atoms are higher for these crystals than for Group IV elements. Therefore, due to this along with melting point, the band gap $E_g$ in this type of III-V compound semi-conductor is higher than Group IV elements. The technical importance of these III-V semiconductors is that they give wider choice of band gap than the elemental semiconductors and some of them have their energy band gap within the visible region of light and so are applied as LED and In Sb and In As are used in galvanometric devices.

However, due to the difference of melting point and vapor pressure of the constituents, it is difficult to maintain the stoichometry of the compound.

**(ii) Group II–VI Compound semiconductors**

    These semiconductors are formed by combining the elements in Group II B and VI A.

| Group II B | $\downarrow$ | Group VI A $\rightarrow$ | | |
|---|---|---|---|---|
| | | O | S | Se | Te |
| Zn | | | **ZnS** | **ZnSe** | **ZnTe** |
| Cd | | | **CdS** | **CdSe** | **CdTe** |
| Hg | | | **HgS** | **HgSe** | **HgTe** |

The average number of valence electrons per atom is 4 and the bonding is a mixture of covalent and ionic type and large band gap compared to the other covalent semiconductors. They deserve a potential source of electro-luminescent devices.

**(iii) Group IV–VI Compound semiconductors**

These semiconductors are receiving wider attention because of their applications in infrared detectors. PbS, PbSe, PbTe and SnTe are among this category. They crystallize in simple NaCl structure, the bonding is ionic with some covalent characterization. They have small band gap and low effective mass of carriers.

## 15.1.2  Concept of Holes

From the characteristics of semiconductors as described above it can be said that semiconductors are insulators in which the energy gap between valence band and the conduction band is about one eV or less (1.1 eV in silicon and 0.7 eV in germanium) so that it is relatively easy to thermally excite electrons from valence band to the conduction band. The electrons transferred to the conduction band then face empty energy levels and can be said to act as conducting electrons as that in metals and other conductors. As long as these electrons were confined to the covalent bonding in semiconductors like germanium or silicon, they are said to exist in the valence band. With the increase in temperature, the thermal energy dislodges the electrons in the valence band and transports it to the conduction band. The bond they leave suffers a vacancy or absence of negatively charged electrons which may be considered as equivalent to the presence of equal amount of positive charge like positive electrons and exist in the valence band. These positive charge equivalents are named as "hole". With thermal excitation negative electrons break from the bonds and may fill up the holes and causing the transport of hole from one position to the other in the valence band. This effect increases with increase of temperature. For example in silicon the number of excited electrons is increased by a factor of $10^6$ when temperature is raised from 250°K to 450°K. We thus have electric conduction from the excited electrons in the conduction band and from holes in the valence band and this conductivity increases with temperature as number of excited electrons in the conduction band and holes in the valence band increase. This current due to electrons in the conduction band and positive holes in the valence band is an intrinsic property of the semiconductors and is called "intrinsic conductivity".

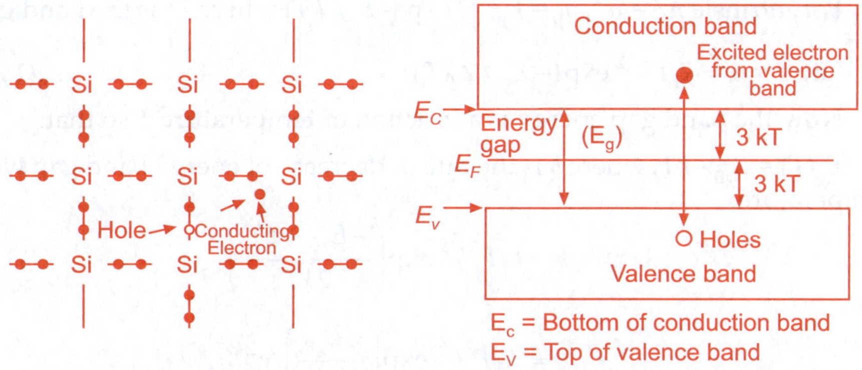



**Fig. 15.4** Conduction electrons and holes in the valence band

### 15.1.3 Carrier Density and Band-gap Determination in Intrinsic Semiconductor

Now, if $N_e$ represents the number of density of states in the conduction band and $N_V$ represents the density of states in valence band and $n_e$ and $n_p$ denote the number of conducting electrons and valence band holes, then:

$$n_e = N_e f(E) \text{ and}$$

$$n_p = N_V f(p)$$

where $f(E)$ is the Fermi-Dirac distribution of electrons in the conduction

band and $f(p) = 1 - f(E)$ writing:

$$f(E) = \exp\left[-\frac{(E_C - E_F)}{kT}\right]$$

and $f(p) = \exp\left[-\frac{(E_F - E_V)}{kT}\right]$ $E_C$, $E_V$ and $E_F$ are shown in Fig. (15.4).

$$n_e \cdot n_p = N_e N_V f(E) \cdot f(p)$$

$$= N_e N_V \exp\left[-\frac{(E_C - E_V)}{kT}\right] \qquad \dots(15.1)$$

Now, $N_e = 2\left(\frac{2\pi m_e^* kT}{h^2}\right)^{3/2}$ and $N_p = 2\left(\frac{2\pi m_p^* kT}{h^2}\right)^{3/2}$

Therefore, $N_e \cdot N_V = 32\left(\frac{\pi^2 m_e^* m_p^* k^2}{h^4}\right)^{3/2} xT^3 - k_1^2 T^3$

For intrinsic $n_i^2 = n_e \cdot n_p = k_1^2 T^3 \exp(-E_g/kT)$ where $E_g$ is the band gap

or,    $n_i = k_1 T^{3/2} \exp(-E_g/2kT)$.    ...(15.2)

Now the band gap energy is a function of temperature $T$ so that:

$E_g(T) = E_{g0} - bT$, where $b$ is the rate of decrease of energy band gap with temperature.

$$n_i = k_1 T^{3/2} \exp\left(\frac{-E_{g0} + bT}{2kT}\right)$$

$$= k_1 T^{3/2} \exp\left(\frac{-E_{g0}}{2kT}\right) \cdot e^{\frac{b}{2k}}.$$

Now writing    $k_2 = k_1 e^{b/2k}$

$$n_i = k_2 T^{3/2} \exp\left(\frac{-E_{g0}}{2kT}\right).$$    ...(15.3)

Now, putting the of values $m_e^*$, $m_p^*$, $b$ and $E_{g0}$, the value of $n_i$ can be determined as a function of temperature. For example:

For Ge: $n_i(T) = 1.76 \times 10^{16} T^{3/2} \exp(-4550/T)$ cm$^{-3}$   and

For Si: $n_i(T) = 3.88 \times 10^{16} T^{3/2} \exp(-7000/T)$ cm$^{-3}$.

As conductivity $\sigma$ is proportional to $n_i$ and therefore to $\exp\left(\dfrac{-E_{g0}}{2kT}\right)$

and as resistivity, $\rho = 1/\sigma$, the resistivity is given by: $\rho = \text{const.} \exp\left(\dfrac{E_g}{2kT}\right)$.

Now, taking log of both sides $\log_e \sigma = \log_e \text{const.} + E_g/2kT$.

Therefore, $\dfrac{\partial \log_e \rho}{\partial\left(\dfrac{1}{T}\right)} = \dfrac{E_g}{2k}$ and the slope of the plot of $\log_e r$ vs. $\dfrac{1}{T}$ gives

a measure of band gap, $E_g$.

### 15.1.4  Carrier Mobility: Intrinsic and Extrinsic (Impurity) Semiconductors

Now, if $n$ is the carrier per unit volume, $v$ the drift velocity, $F$ the electric field and $e$ the charge of the carrier, then current I is given by

$$I = n\, e\, v\, \alpha \text{ where } \alpha \text{ is the cross section}$$

The current density, $J = I/\alpha = n\, e\, v$.

The carrier mobility $\mu$ is defined as the drift velocity per unit field strength and so

$$\mu = \frac{v}{F} \quad v = \mu F \text{ and so } J = n\, e\, \mu F$$

The conductivity, $\sigma$ of the carriers is defined as current density per electric field i.e.

$$\sigma = J/F = ne\,\mu.$$

Now, this number density of carrier per unit volume are:

$n_e$ = number of electrons as carrier in the conduction band

$n_p$ = number of holes as carrier in the valence band and

$m_n$ = mobility of the electrons $\mu_p$ = mobility of holes

Then as

$\sigma = \sigma_n + \sigma_p$ where suffixes $n$ represents for conducting electrons and $p$ the holes in the valence band. Therefore,

$$\sigma = e(n_e\,\mu_n + n_p\,\mu_p)$$

For intrinsic semiconductors as discussed above

$$n_e = n_p = n_i$$

Therefore, $\quad \sigma = n_i\,e(\mu_n + \mu_p).$ ...(15.4 a)

And the current density $j$(a vector) can be written for intrinsic semiconductor as

$$j = (\sigma_n + \sigma_p)\,F = (n_n\,e\mu_n + n_p e\mu_p)F. \qquad ...(15.4\ b)$$

Now, recalling the Fermi distribution function from equn. (12.39) which was as

$$n_i = \frac{g_i}{e^{(E_i - E_F)/kT} + 1}.$$

Now the probability that $i$th level having energy $E_i$ will be occupied by conducting electrons is given by

$$f(E) = \frac{n_i}{g_i} = \frac{1}{1 + \exp\left[\dfrac{E_i - E_F}{kT}\right]}. \qquad ...(15.5)$$

Now, referring to the Fig. 15.4, we know that:

$E_C$ = Energy level at the bottom of the Conduction Band

$E_V$ = Energy level at the top of the Valence Band

and $\quad E_g$ = Forbidden energy gap

when $(E_i - E_F) > 3\,kT$, the Fermi-dirac distribution can be approximated by Maxwell Boltzmann classical distribution.

Therefore, $\displaystyle f(E) = \frac{n_i}{g_i} = \left\{1 + \exp\left[\frac{E_i - E_F}{kT}\right]\right\}^{-1} \approx \exp\left[-\frac{E_i - E_F}{kT}\right] = \exp\left[\frac{E_F - E_i}{kT}\right].$

Now, to increase conductivity in pure intrinsic semiconductors if impurity atoms either having 5 valence electrons (pentavalent) or alternatively if 3 valence electrons (trivalent) are doped, we get impurity or extrinsic semiconductors. They are respectively called as *n*-type or *p*-type. These extrinsic semiconductors are discussed later in this chapter.

Now, the number of excess carrier of electricity is negative electrons in *n*-type and in *p*-type it is positively charged holes due to impurity effect. Now, $n_e$ is the number of excess carrier in *n*-type and $n_p$ that in *p*-type semiconductors and $n_i$ is the number of charge carriers in any semiconductors, *n*-type or *p*-type and $E_F$ the Fermi energy level lies in the forbidden energy gap and $3\,kT$ energy away from conduction and valence band edge, (Fig. 15.4). Then we get:

$$n_e = n_i \exp\frac{E_F - E_i}{kT}. \qquad \ldots(15.6)$$

Now for holes
$$f(p) = 1 - f(E)$$

$$= 1 - \frac{1}{1 + \exp\left[\dfrac{E_i - E_F}{kT}\right]} = \frac{\exp\left[\dfrac{E_i - E_F}{kT}\right]}{1 + \exp\left[\dfrac{E_i - E_F}{kT}\right]}$$

$$= \frac{1}{1 + \exp\left[-\dfrac{(E_i - E_F)}{kT}\right]} \approx \exp\left[\dfrac{E_i - E_F}{kT}\right].$$

Therefore,
$$n_p = n_i \exp\left[\frac{E_i - E_F}{kT}\right] \qquad \ldots(15.7)$$

Now, multiplying equns. (15.3) and (15.4) we see that

$$n_e \cdot n_p = n_i^2. \qquad \ldots(15.8)$$

This is an important relation which shows that the product of number of two different types of carriers is equal to the square of the total number of intrinsic carriers.

### *Extrinsic (Impurity) Semiconductors*

Now the conductivity of pure intrinsic semiconductors is limited and can be enhanced by the addition of impurity atoms in the original sites of the covalent atoms of the semiconductor. This a bit special type of addition of impurity in the pure lattice is termed as "doping". Now two cases may arise, first suppose these impurity atoms have more electrons than those of the semiconductor for example in tetra valent Germanium or Silicon contributing four electrons per atom to the valence band, penta valent impurity like Phosphorus or Arsenic, each of which contribute five electrons per atom in the valence band is added. We then have one extra electron per impurity atom. This extra electron can not be accommodated in the valence band of the original lattice and so occupy some discrete energy levels just below the conduction band, (Figs. 15.5 and 15.6).

The separation between these discrete energy levels and the lower edge of conduction band is only a few tenths of 1 $eV$. This is a common feature for all such impurity atoms of same valency. These excess electrons from the energy levels can then be easily transported (excited) to the conduction band, (Fig. 15.6). Now these transported or excited electrons in the conduction band contribute to the electrical conductivity and another important feature which should be noted that these electrons excited from impurity atoms to the conduction band are in excess compared to holes created in the valence band due to intrinsic effect. Therefore, number of majority carriers is electrons and so these impurity semiconductors are called $n$–type or negative type. The discrete energy levels from which these electrons are excited to the conduction band are called "Donor levels".

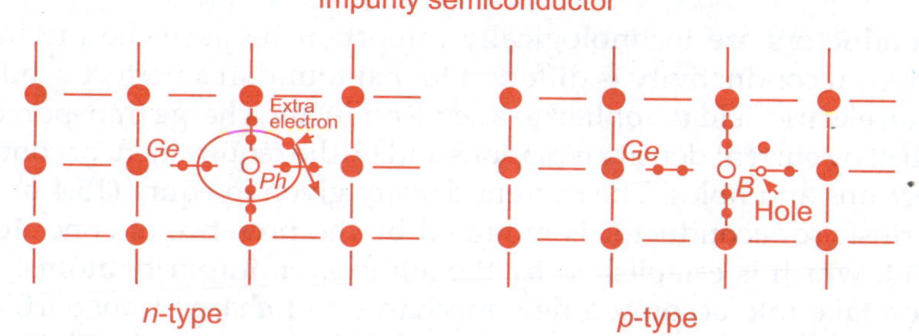


**Fig. 15.5** Doping of penta valent phosphorus in $n$-type and trivalent boron in tetra valent germanium host lattice

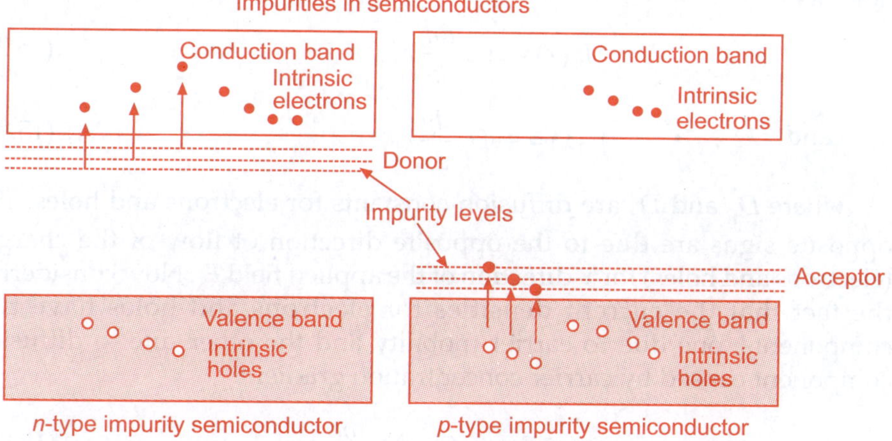


**Fig. 15.6** Donor levels in $n$-type and acceptor levels in $p$-type semiconductors

Conversely, if the impurity atom has fewer electrons than the host atoms like germanium or silicon the impurity atoms may be either Boron or Aluminum, each of which contributes three electrons. In this situation the impurity atoms introduce vacant discrete energy levels close to the top of Valence band Figs. 15.5 and 15.6. It is then easier for more energetic electrons in the valence band to get excited in these levels (named as "Acceptor levels") and thus creating holes in the valence band without creating their corresponding electrons in the conduction band. Therefore majority carriers are holes in this type of semiconductor and are called $p$-type impurity semiconductor.

The effect of insertion of the impurities either to form $n$-type or $p$-type in increasing the conductivity is so enormous that only one single impurity atom per million host semiconductor atoms makes significant change in conductivity. We will discuss the various important applications of these impurity semiconductors in latter sections.

## 15.2 CARRIER TRANSPORT PHENOMENA

Semiconductors are technologically important because the mechanism for electrical conductivity is different to that found in a perfect conductor. When an electric field is applied to a semiconductor, charge transport can be described by current density equations and by the requirement of continuity for electrons and holes. The current density given in equn. (15.4 b) above for intrinsic semiconductor is modified by the presence of concentration gradient, which is established by the addition of impurity atoms. We in addition take into account a new mechanism of charge transport, which is not significant in metals, consisting of diffusion of carriers by thermal motion and occurring whenever a concentration gradient exists. The current density caused by diffusion can be respectively written for electrons and holes as:

$$j_n(x) = eD_n \frac{dn}{dx}, \qquad \qquad ...(15.9a)$$

and
$$j_p(x) = -eD_p \frac{dp}{dx}. \qquad \qquad ...(15.9b)$$

where $D_n$ and $D_p$ are diffusion constants for electrons and holes. The opposite signs are due to the opposite direction of flow of the charges (electrons and holes) for $x$-direction of the applied field $F_x$. Now considering the fact that the current densities for electrons and holes have two components, one due to carrier mobility and the other due to diffusion component caused by carrier concentration gradient:

$$j_n(x) = en\mu_n F + eD_n \frac{dn}{dx}, \qquad \qquad ...(15.9\ c)$$

$$j_p(x) = ep\mu_p F - eD_p \frac{dp}{dx}. \qquad \qquad ...(15.9\ d)$$

Now, under no electric field there will be no current flow and the both $j_e$ and $j_p$ separately equal to zero and then the above equation reduces to:

$$n\mu_n F = -D_n \frac{dn}{dx}.$$  ...(15.10)

This results in the implication that the diffusion causes an internal electric field which produces an electrostatic potential given by

$$F = -\frac{dV(x)}{dx}.$$

The energy distribution of carriers however, follows the Maxwell-Boltzmann distribution and can be given as: $n = N_C \exp\left[\dfrac{eV(x)}{kT}\right]$, where $N_C$ is the carrier electrons or holes.

Then
$$\frac{dn}{dx} = N_C \frac{e}{kT} \frac{dV(x)}{dx} \exp\left[\frac{eV(x)}{kT}\right] = n \frac{e}{kT} \frac{dV(x)}{dx}$$

Hence equn.(15.10) becomes $-n\mu_n \dfrac{dV(x)}{dx} = -D_n\, n \dfrac{e}{kT} \dfrac{dV(x)}{dx}$

This yield
$$D_n = \frac{kT}{e}\mu_n$$  ...(15.11a)

Similarly for holes we get

$$D_p = \frac{kT}{e}\mu_p$$  ...(15.11b)

Now, these two relations are known as Einstein Relations which on substitution in (15.9 c) and (15.9 d) give to:

$$j_n(x) = en\mu_n F + kT\mu_n \frac{dn}{dx}$$

and
$$j_p(x) = ep\mu_p F - kT\mu_p \frac{dp}{dx}$$

In the absence of the electric field $F$, $j_n = 0 = kTm_n \dfrac{d}{dx}\left\{ne^{(E_F - E_{Fn})/kT}\right\}$

Where $n$, the number of electron carrier is replaced by using Fermi distribution and $E_F$ and $E_{Fn}$ represent Fermi energy levels for carrier in general and the electrons respectively and on differentiation we get

$\mu_n n \dfrac{dE_F}{dx} = 0$ and similarly also the current density for holes and we can

conclude that as $\dfrac{dE_F}{dx} = 0$ in general, the Fermi energy is constant throughout an inhomogeneous semiconductor in thermal equilibrium.

## 15.3 THE JUNCTION DIODES, TRANSISTOR AND APPLICATIONS

### 15.3.1 Junction Diodes

It may be once again stated that $p$ and $n$ type semiconductors respectively have positive holes and negative electrons as majority carriers. Now, two such semi conducting materials are joined together. Then, there will be transport of holes and electrons through the junction due to thermal diffusion to the other side. In doing so, the holes and the electrons from p and n side accumulate at the junction and prevent further flow. This results an opposing potential difference which acts as a barrier and is known as barrier potential. The minority carrier electrons from $p$-type on entering $n$–type through junction (which is not affected by barrier) combine with minority carrier holes in the $n$-type and make it electrically negative and cause a flow of small current $I_1$ from from left ($p$-type) to right ($n$-type). Conversely the minority carriers, holes from $n$–type diffuse through the junction and combine with minority carrier electrons in the $p$-type and form

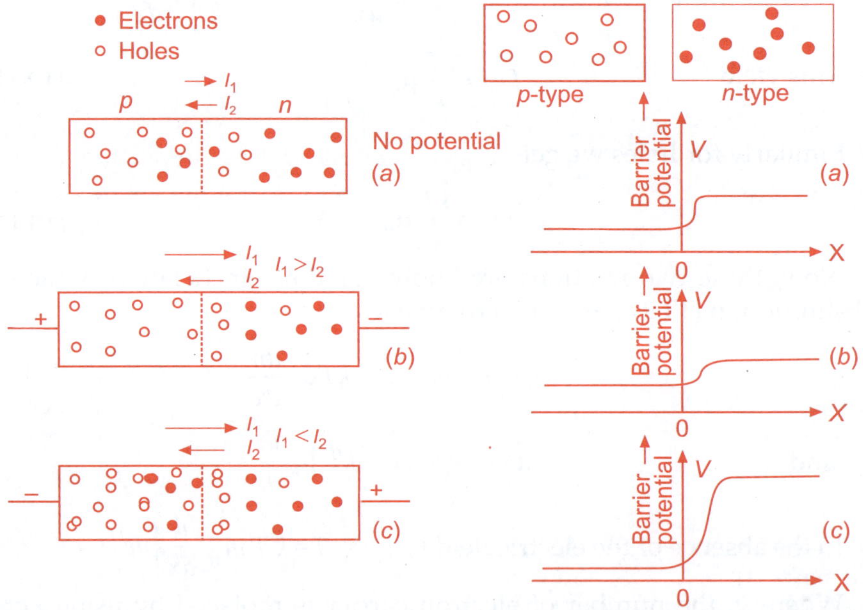


**Fig. 15.7**  The $p$-$n$ junction diode

an opposite current $I_2$ from right to left. At equilibrium both of these two currents are equal i.e. $I_1 = I_2$ Fig. 15.7 (a). Now, if a potential difference is applied with $p$-side connected to positive polarity and $n$-side to the negative polarity of the potential source, then majority carrier holes from $p$-side can overcome the barrier and reach the $n$-side and combine with electrons. This constitutes a comparatively large flow of current from left to right i.e. $I_1$. But on the other hand, the flow of electrons from $n$-type to the $p$-type will be limited due to potential barrier and constitutes $I_2$ from right to left which is less than $I_1$ i.e. $I_1 > I_2$, Fig. 15.7 (b). If now the polarity is reversed, the current $I_2$ due to minority holes from right side to left will remain same but $I_1$ is decreased due to potential barrier i.e. $I_1 < I_2$, Fig. 15.7 (c). The polarity shown in (*b*) for which $I_1 > I_2$ is called "Forward bias" and the polarity shown in (*c*) for which $I_1 < I_2$ is called "Reverse bias". The variation net current $I = I_1 - I_2$ with potential difference $V$ is shown in the following Fig. 15.8.

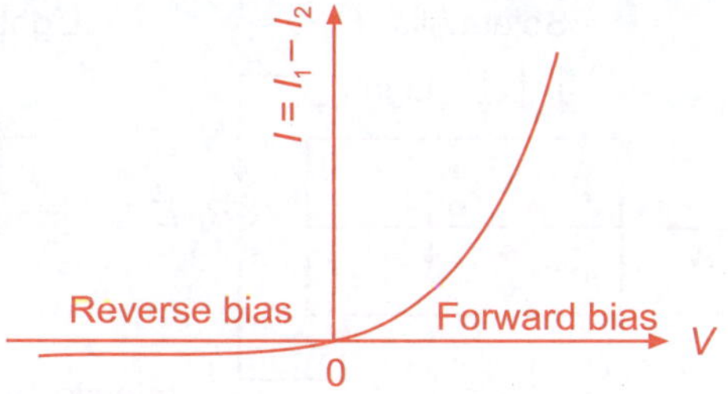


**Fig. 15.8** Current as function of voltage

Now, when the input voltage is oscillating between forward and reverse biased conditions, the out put current will be obtained only for the forward bias condition and the potential across a resistor will be direct instead of alternating in polarity. This action of diode is known as rectification of alternating signal into direct voltage.

Now, again if the voltage in reverse bias condition is increased the bonds in the lattice break up to produce further electron – hole pairs. This effect is prominent if the depletion region is narrow and the diode is heavily doped. This is known as Zener Diode. The main application of this diode is to make reverse bias potential constant as this Zener break down occurs at a particular voltage, Fig. 15.9 (a). and the reverse bias voltage across the diode $V_Z$ does not change. In Fig. 15.9 (b) the Solar Cell is represented. Solar cells are $p$-$n$ junction diodes which transform the sunlight to electricity with large conversion efficiency. Light photons on incidence produce electron-hole pairs in $p$ and $n$-regions. These electron-hole pairs produced near the junction can reach the depletion region of width say $W$ by thermal diffusion. The electrons in the $p$-side can slide down the barrier potential and reach the $n$-side. Conversely, the holes from $n$–side to the $p$-side. These effects cause $p$-side to be more reach in hole concentration and $n$-side with electron concentration and thus develop a potential difference. When joined by a resistance, $R_L$ a current I flows and the strength of this current will be proportional to the intensity of the light and will continue as long as there is light. However, instead of connecting a resistance the potential difference created by sun light during day time is used in charging a battery which acts as potential source during night. In Fig. 15.9 (c) Light Emitting Diode is represented. The diode is connected in forward bias; this accelerates the holes to move from $p$-side to $n$-side and electrons from $n$-side to $p$-side through the depletion region and if the transported electrons from conduction bands of $p$-side and also of $n$-side jump the energy gap and meet the holes of the valence band, an energy equivalent to the energy gap, $E_g$ is emitted. In this LED the released energy is photons within the visible region i.e. $n = E_g$. The different LED materials emit different coloured lights depending on $E_g$.

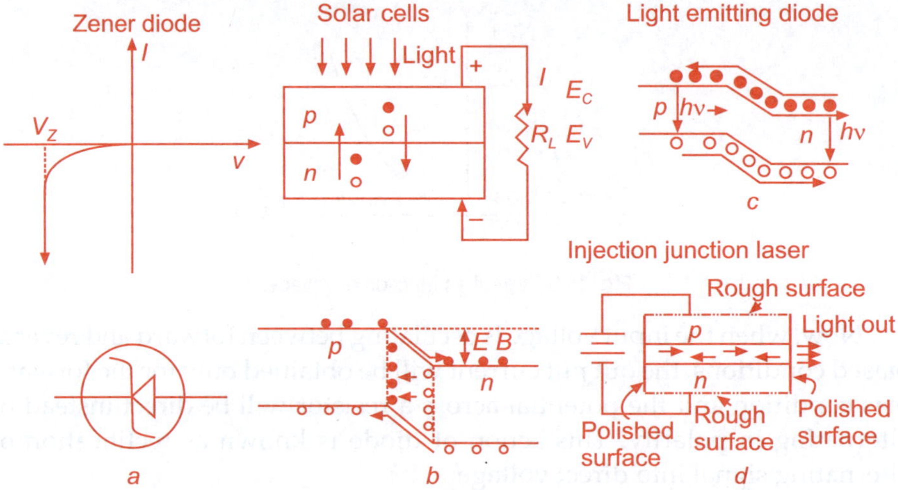


**Fig. 15.9** Uses and applications of *p-n* junction diode

*Ga As* however is used as a source of infrared and $Ga\ As_{1-x}\ P_x$ an alloy semiconductor emit different colours depending on the values of $x$. As when $x = 0.65$, it emits orange; when $x = 0.85$, it emits yellow and when $x = 1$, it emits green.

While LED emits visible light (power consumption $\approx$ milli watts) by spontaneous emission, Injection junction Laser (solid state laser) described in Fig. 15.9 (d), emits radiation due to stimulated action. This laser diode is basically LED type forward biased diodes but the difference is that the emitted radiation due to the combination of electrons with holes across the energy gap itself triggers or stimulates emission. The photon of energy emitted $h\,n$ is again equal to energy gap $E_g$ and it then is capable to stimulate emission by forcing electrons of conduction band to jump the energy gap and meet with holes in the valence band and emit radiation. This effect is similar to forced oscillation in classical oscillators.

Another application of junction diodes needs to be mentioned. Photo diodes in reverse bias, when illuminated with light extra electron-hole pairs are generated and as result minority carrier concentration increases significantly near the junction and increases the reverse bias current. This reverse bias current is proportional to light fluxes or intensity and is used extensively in decoding sound track and various light operated switches.

Another important application of diodes is the Tunnel Diode, the detailed explanation of this tunneling phenomenon though needs quantum mechanics (Chapter 13), it can however be stated qualitatively. We know that under forward bias condition the conduction band $E_C$ of $n$-region is lifted up, decreasing the energy level difference between $E_C$ of $n$-region and $p$-region of the junction diode. In reverse bias condition exactly the opposite situation occurs, the $E_V$ of the $p$-region is lifted up as $E_C$ of the $n$-region is lowered. Thus, the Fermi level energy levels of $p$ and n regions which were equal at zero potential condition, now after biasing become different. The $E_F$ of $n$-region moves higher than $E_F$ of $p$-region in forward bias and an opposite situation occurs during reverse bias. Now if in addition, the band gap is narrow, the electrons from either $n$-region (in forward bias) or from $p$-region cross over the gap and enter the $p$-region or $n$-region to equalize the difference in Fermi level. This is explained in the following Fig. 15.10.

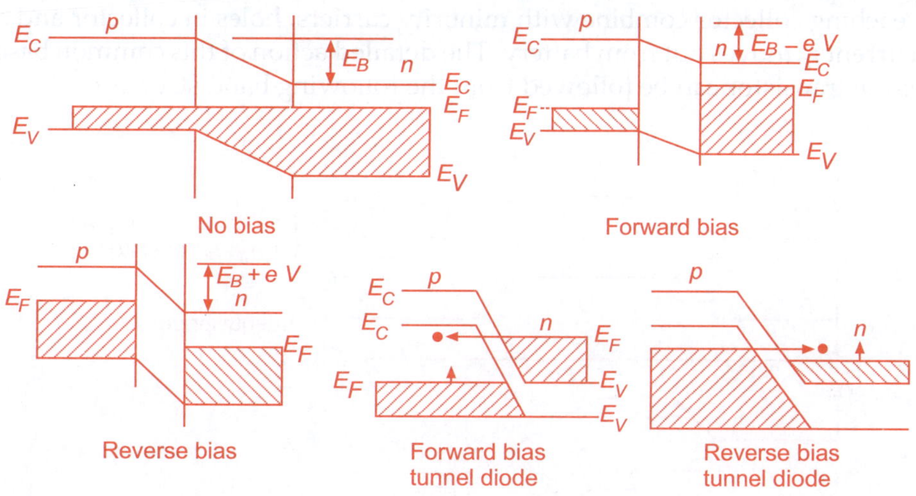


**Fig. 15.10** Diodes under forward and reverse biasing and tunnel diode

## 15.3.2 Transistors

The junction transistor consists of two $n$-$p$ junctions joined together in the two configurations, either in $n$-$p$-$n$ or $p$-$n$-$p$. $n$-$p$-$n$ is shown in the following Fig. 15.11.

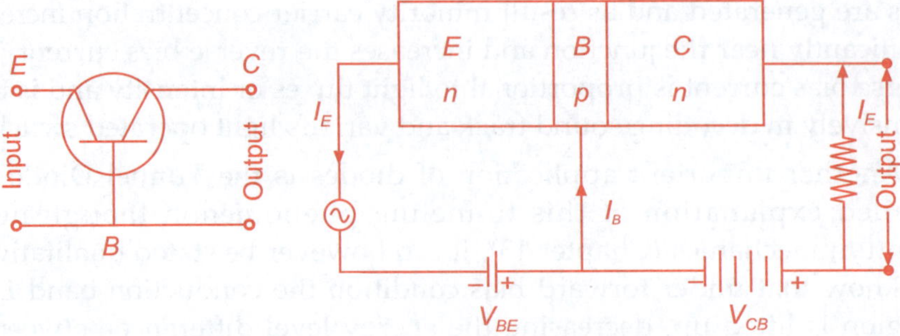

**Fig. 15.11** Junction transistor (common base)

Let us take *n-p-n* transistor. The emitters (E, *n*-type) is heavily doped and due to direct bias the majority carriers, electrons cross over to base (B, *p*-type) and constitute an emitter current $I_E$. These excess electrons can pass through base (B), which is thin and enter into the collector (C). The collector is comparatively larger in size than emitter and so contains a large number of minority carrier, holes. The electrons combine with majority carriers, holes in the collector base junction. For every electron thus combining with hole in B, a free electron moves out from battery to the base ($I_B$). The electrons avoiding such combination cross through base and reaching collector combine with minority carriers, holes in collector and a current $I_C$ moves out from battery. The detailed action of this common base *n-p-n* transistor can be followed from the following band structure.

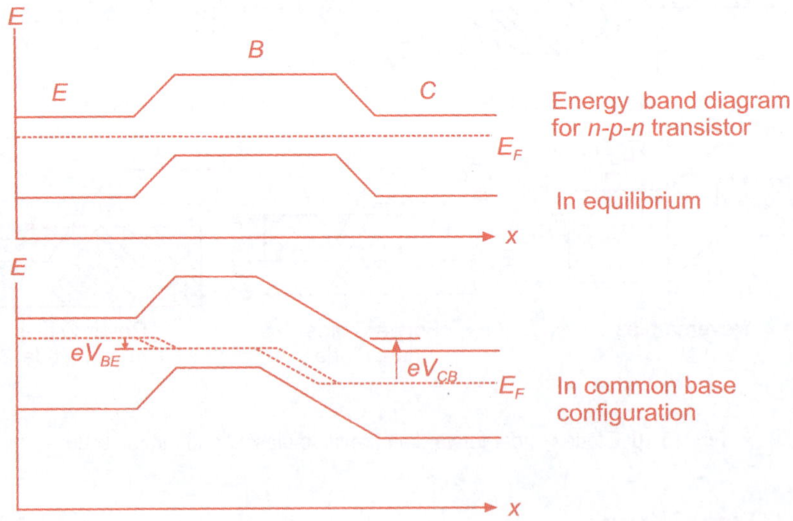



**Fig. 15.12** The energy band diagram of *n-p-n* transistor

Figure 15.12 (a) shows the arrangement of Fermi level in emitter, base and collector, which remain same in all the three sections when no potential is applied. Now, on applying the direct bias emitter-base potential $V_{BE}$, the Fermi level of emitter ($n$-region is moved up and correspondingly in base ($p$-region) is moved down. This initiates flow of electrons from $E$ to $B$. These electrons on combination with holes on the $E$-$B$ junction, produces the base current $I_B$. The excess electrons move past the base and as base–collector potential is reversed biased, the Fermi level of base ($p$-region) will go up and collector ($n$-region) will come down, resulting in the flow of electrons from base to collector. These electrons on combination with minority carrier holes in collector ($n$-type) produce current $I_C$ and potential difference $V_{BC}$. From Fig. 15.11, we can write $I_E = I_{eE} + I_{hE}$, where $I_{eE}$ and $I_{hE}$ respectively mean the electron and hole components of the emitter current $I_E$, then the emitter efficiency $g$ is defined as

$$\gamma = \frac{I_{eE}}{I_E} = \frac{I_{eE}}{I_{eE} + I_{hE}} = \frac{1}{1 + I_{hE}/I_{eE}}.$$

The base transport efficiency is defined by:

$\delta = \dfrac{I_{eC}}{I_{eE}}$ , where $I_{eC}$ stands for the electron component of the collector current. From the above two expressions we get

$$\gamma = \frac{I_{eC}}{\delta I_E} \quad \text{and so,} \quad I_{eC} = \gamma \delta I_E.$$

But as, $\qquad\qquad I_C = I_{eC} + I_{hC}$

Therefore, $I_C = \gamma \delta I_E + I_{hC}$ and from Fig. 15.11 we get $I_E = I_C + I_B$

Introducing "Current gain", $\alpha$ is related as $\alpha = \gamma \delta$.



Another configuration for $n$-$p$-$n$ transistor is common emitter type shown above. This configuration is useful for achieving larger current gain. The static characteristics of this configuration is same as the common base type but the only difference is that here the input current is $I_B$ and not $I_E$ as in common base configuration. The current gain which is given in common base as $\alpha = \gamma \delta$ is given here as:

**Current gain:** $\beta = \dfrac{\alpha}{1-\alpha} = \dfrac{1}{1/\alpha - 1}$, which is much greater than common

base and it is also important to know that $\alpha = \gamma\,\delta$ is to be as near to unity as possible. Now to summarize, the major use of these transistors either in *n-p-n* or *p-n-p* configuration is amplification of any input signal put in the emitter-base circuit due to current gain and are extensively used in practice.

## REFERENCES


1.  G. Burns - Solid State Physics, Oxford University Press, Oxford, 1989.

2.  H. P. Myer – Introduction to Solid State Physics, Taylor and Francis, London, 1991.

3.  S. M. Sze – Semiconductor Devices, Physics and technology, John Wiley and Sons, New York, 1985.

4.  K. E. Lonngren- Introduction to Physical Electronics, Allyn and Beacon, Inc, Boston, 1988.

# Magnetism - II

## 16.1 SOLID STATE MAGNETISM: AN INTRODUCTION

When a solid is put in a magnetic field of strength $H$, the magnetization induced with it and say of strength $M$ then they are related as:

$$M = \chi H$$

where $c$ is introduced as magnetic susceptibility. This is a constant for isotropic substance and is a tensor in an anisotropic substance and in that case $M$ and $H$ are not in the same direction. The magnetic induction $B$ is defined as:

$$B = \mu_0 (M + H) = \mu_0 H(1 + c) = \mu_0 \cdot \mu_r H$$

where $m_0$ and $m_r$ are respectively known as magnetic permeability in free (empty) space and relative permeability. Again, $\mu = \mu_0 \cdot \mu_r = \mu_0 (1 + \chi)$ is the magnetic permeability of the medium. This should be noted here that $\mu_0$, which is the free space permeability is constant and its value is given as: $\mu_0 = 4p \times 10^{-7}$ henry/$\mu$.

In an isotropic medium, $B$, $H$ and $H$ vectors and m is a scalar.

On the basis of the magnetic susceptibility $\chi$ the solids are classified into (a) Diamagnetic (b) Paramagnetic and (c) Ferromagnetic.

If $\chi \langle 0$ and has a very low value and $|\chi| \langle\langle 1$, then the slope of $M$ vs. $H$ is negative, constant and independent of the variation of $H$, the solid is Diamagnetic.

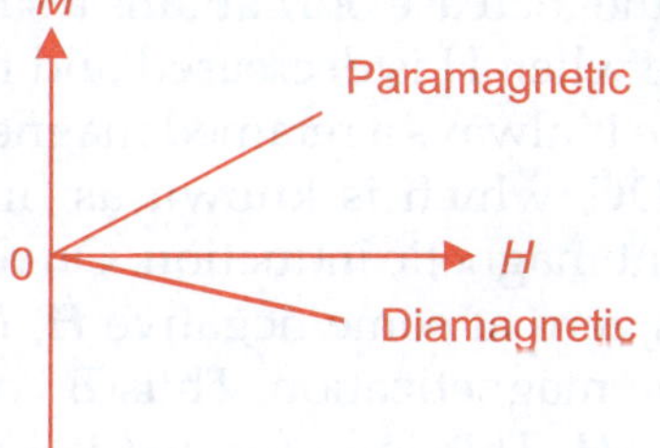


Some typical diamagnetic substances with their $\chi$ values

$Bi = -15 \times 10^{-5}$
$Cu = -0.9 \times 10^{-5}$
$Ge = -0.8 \times 10^{-5}$
$Si = -0.3 \times 10^{-5}$

But if c is positive i.e. $\chi > 0$, though $|\chi|$ remains as $\langle\langle\, 1$, then in **M** vs. **H** curve the slope becomes positive and c is independent of **H** but depends on temperature. The solid is then named as Paramagnetic. The values of $\chi$

For some typical paramagnetic substances at room temperature is given below;

$$CaO = 580 \times 10^{-5}, \qquad Fe\,Cl_2 = 360 \times 10^{-5},$$
$$NiSO_4 = 120 \times 10^{-5}, \qquad Pt = 26 \times 10^{-5}.$$

If now c is positive and has usually very high value i.e. $\chi \gg 1$, then the solid is named as Ferromagnetic. Apart from first transition group of materials like Fe (Iron Z = 26), Co(Cobalt Z = 27) and Ni (Nickel Z = 28) some rare earth metals like Gadolinium (Z = 64), Holmium (Z = 67), Erbium (Z = 68) and also some alloys exhibit ferromagnetism. For ferromagnetic the magnetization **M** first increases rapidly with **H** and then it is saturated. As $\chi$ is obtained from the slope, for ferromagnetic the $\chi$ first increases rapidly with **H** until it attains a maximum and then decreases to zero as **H** attains saturation. Another phenomenon which is typical property of ferromagnetic is the 'Hysteresis'. This is demonstrated in the following figure first and then introduced.

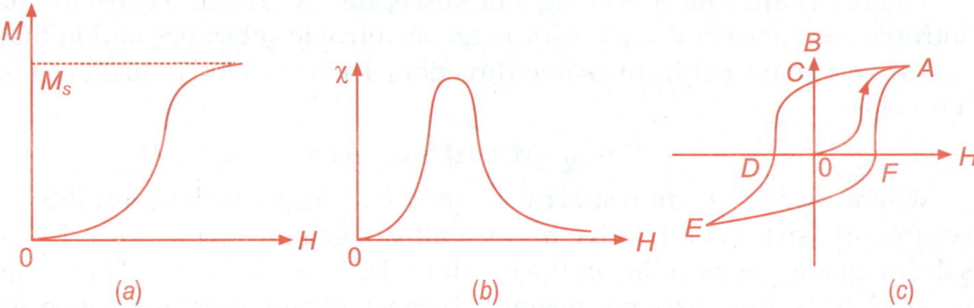


**Fig. 16.1** The variation of **M**, $\chi$ and magnetic induction **B** vs. magnetizing field **H** for ferromagnetic materials. (c) is the hysteresis loop

In the figure above, (a) represent first slow then rapid increase of magnetic field strength **M** with external magnetizing field **H** which tends to attain saturation. The variation of susceptibility $\chi$ with **H** is unlike pra and dia magnetic substance shows a maximum as it is not a constant for ferromagnetic and decreases rapidly as **M** attains saturation. The Fig. 16.1 (c) represents an important property, known as hysteresis. The magnetic induction **B** first increases as (a) with the increase of **H** attains a saturation and then do not follow the same route when H is decreased and follow A to C when **H** is decreased to zero. There is always a retained magnetization within the ferromagnetic equal to $OC$, which is known as 'magnetic retaintivity' and represents the remnant magnetic induction. On reversing the field **H**, **B** decreases and becomes zero at some negative **H**, $D$ which shows the coercive force to relieve the magnetization. Thus **B** completes a loop when **H** is varied from **+H** to **–H**. This loop formed is known as

'Hysteresis cycle' which represents the loss of energy due to magnetizing field cycle. Depending on the area of this loop, the ferromagnetic substances are classified as magnetic 'soft' and 'hard'. Soft magnetic materials can be easily magnetized and demagnetized without loss of much energy, where as 'hard' materials which have large areas enclosed by the loop are permanent magnets and difficult to magnetize and demagnetize.

When ferromagnetic substance is magnetized and demagnetized there is always fluctuation of the volume with frequency depending on the frequency of change in $H$. This phenomenon is known as 'Magneto striction'.

### 16.1.1 Theory of Diamagnetism

Diamagnetism has its origin in the change of orbital motion of atomic electrons due to applied magnetic field. We know that a loop of wire with a current through it produces a magnetic field. If $A$ is the area of the loop, the magnetic moment $m_1$ is given by $\mu_1 = Ai$, where $A$ the area is $\pi a^2$ and $i$ is the current. So, $m_1 = \pi a^2 i$, but as $i = ev$, where $n$ is the frequency $(= \omega/2\pi)$ of revolution and $e$ the charge of the electron. Hence the magnetic moment, $\mu_1$ for orbital motion of the electron is given as:

$$\mu_l = \pi a^2 \frac{e\omega}{2\pi} = \frac{e\,\omega a^2}{2} \text{ and}$$

the orbital angular momentum $p_l$ is given by: $p_l = m_e\,a^1 \cdot \omega$ then

$$\frac{\mu_l}{p_l} = \frac{e}{2m_e}, \text{ Provided both } p_l \text{ and } \mu_l \text{ are in the same direction.}$$

This ratio is known as 'Gyromagnetic ratio'.

We get from Bohr's theory the orbital angular momentum is quantized and is given as

$$p_l = l\hbar \qquad \text{where } l = 1, 2, 3, \dots.$$

From the above relation we get

$$\mu_l = \frac{e}{2m_e} p_l = l\frac{e\hbar}{2m_e} = l\mu_B$$

where $\mu_B = \dfrac{e\hbar}{2m_e}$ is known as Bohr magneton which is the basic unit of atomic magnetic moment. ($\mu_B = 9.2741 \times 10^{-24}$ J/Tesla, 1 Tesla $= 10^4$ Gauss).

The current loop due to electron circulating in an orbit thus behaves like a tiny magnet of atomic size. We know that when a bar is placed in a magnetic field and is free to oscillate, it oscillates back and forth before it can get settled in the direction of the external magnetic field. The atomic magnet however behaves differently; its magnetic moment vector $\mu_1$ precesses about the field direction making a definite angle like a rotating top precesses around the earth's magnetic field.

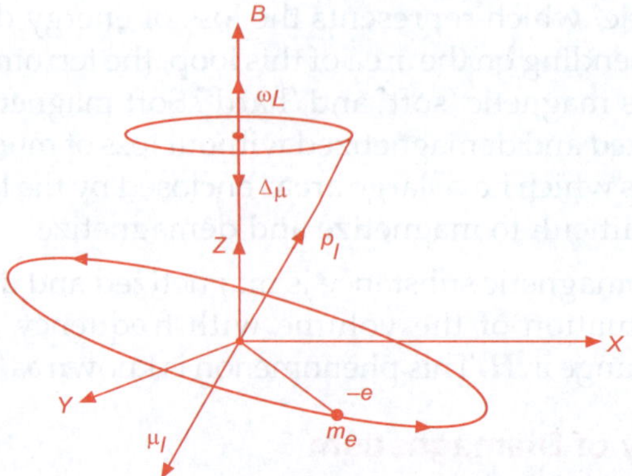

**Fig. 16.2** Larmor precession of orbital electron under magnetic field

In a magnetic field $B(= \mu_0 H)$, the torque acting on $\mu_l$ is $\mu_l \times B$ and this is equal to the rate of change of angular momentum $p_l$ and that is

$$\frac{d}{dt} p_l = \mu_0 (\mu_l \times H) = -\frac{\mu_0 e}{2m_e} (p_l \times H). \qquad \ldots(16.1)$$

The negative sign is due to the fact that electrons are negatively charged.

Now, the change of angular momentum can be written as

$$\Delta p_l = \omega_L \times p_l \, \Delta t$$

Where, $\omega_L$ is the 'Larmor precessional angular velocity'.

Therefore, $\qquad \dfrac{d}{dt} p_l = \omega_L \times p_l.$

Then, comparing with equn.(16.1) $\omega_L = \dfrac{\mu_0 e}{2m_e} H$ and Larmor precessional

frequency $v_L (= \omega/2\pi)$ is given by : $v_L = \dfrac{\mu_0 e H}{4\pi m_e}$.

Due to precession of electron orbit in the applied field there is induced magnetic moment $\Delta\mu$ whose direction is opposite to the direction of $B$ and is given by

$$\Delta\mu = -\frac{e}{2m_e} m_e \omega_L \left\langle \rho^2 \right\rangle = -\frac{e\omega_L}{2} \left\langle \rho^2 \right\rangle$$

where $\left\langle \rho^2 \right\rangle$ is the mean squared radius of the projection of the electron orbit in the plane perpendicular to $B$ and $m_e \omega_L \left\langle \rho^2 \right\rangle$ is the change in the angular momentum, due to precessional motion induced.

Now, if there are $n$ atoms per unit volume and there are $Z$ electrons per atom.

Then:

$$M = n\,Z\,\Delta\mu = -n\,Z\frac{e}{2}\frac{\mu_0\,e\,H}{2m_e}\langle\rho^2\rangle = -nZ\frac{\mu_0\,e^2 H}{4\,m_e}\langle\rho^2\rangle \qquad \ldots(16.2)$$

We consider spherical symmetrical charge distribution so that

$$\langle x^2\rangle = \langle y^2\rangle = \langle z^2\rangle = \frac{1}{3}\langle r^2\rangle.$$

Now, taking **B** in the direction of $z$-axis, the projection of electron orbit $r$ in the $x - y$ plane is given by

$$\langle\rho^2\rangle = \langle x^2\rangle + \langle y^2\rangle = \frac{2}{3}\langle r^2\rangle$$

Then $\qquad M = -nZ\dfrac{\mu_0\,e^2}{6\,m_e}\langle r^2\rangle\ \textbf{\textit{H}}.$

Hence the Diamagnetic susceptibility $\chi(= M/H) = -\mu_0\,nZ\dfrac{e^2}{6m_e}\langle r^2\rangle.$

Therefore, the susceptibility $\chi$ for diamagnetic substance is negative.

## 16.2 THEORY OF PARAMAGNETISM

In classical Langevin's theory of paramagnetism it was assumed that the inclination of atomic dipoles varies continuously in the direction of applied magnetic field. This was later found from Zeeman's effect is not a correct assumption as under the action external field an atom gives its characteristic super fine spectral structures. According to Sommerfeld's space quantization, the resultant angular momentum vector $J\,h$ can only align in $(2J + 1)$ special directions so that its components in the external field direction assume the discrete set of values $m_J = J, J - 1, J - 2, \ldots - J$.

Resultant magnetic moment of an atom is given by

$$\mu_J = \sqrt{J(J+1)}\,g\mu_B \qquad \ldots(16.3)$$

where $\mu_B$ is the Bohr magneton introduced above and $g$ is known as Lande's $g$-factor and is given by :

$$g = 1 + \frac{J(J+1) + S(S+1) - L(L+1)}{2\,J(J+1)}.$$

As discussed before **L** and **S** represent respectively orbital and spin angular momentum vectors of the atom comprising many electrons and **J** is their resultant and it will precess around the direction of applied magnetic field $\mu_L$ and $\mu_S$ represent the magnetic moments due to orbital and spin motions of the electron are aligned antiparallel to the vectors **L** and **S** and they also precess around the magnetic field. $\mu_J$ the total magnetic moment it is given as: $\mu_J = \mu_L + \mu_S$.

Now, if $\mu_H$ is the component of magnetic moment $\mu_J$ along the field direction then $\mu_H = -\mu_J\, g\, \mu_B$. The mean value of $\mu_H$ is given as

$$\langle \mu_H \rangle = \frac{\sum\limits_{-J}^{J} m_J\, g\, \mu_B \exp(m_J\, \mu_B\, B_0 / kT)}{\sum\limits_{-J}^{J} \exp(m_J\, g\, \mu_B\, B_0 / kT)}. \qquad \dots(16.4)$$

The magnetization of the substance is then given by

$$M_H = n <\mu_H> = \frac{n \sum\limits_{-J}^{J} m_J\, g\, \mu_B \exp(m_J\, \mu_B\, B_0 / kT)}{\sum\limits_{-J}^{J} \exp(m_J\, g\, \mu_B\, B_0 / kT)}. \qquad \dots(16.5)$$

At relatively high temperature i.e. room temperature we can assume $\dfrac{m_J\, \mu_B\, B_0}{kT} \lll 1$ and so in this case the exponential can be expanded as:

$$\exp\left(\frac{m_J\, \mu_B\, B_0}{kT}\right) \approx 1 + \frac{m_J\, \mu_B\, B_0}{kT}.$$

Then
$$M_H = \frac{n\, g\, \mu_B \sum\limits_{-J}^{J} m_J (1 + m_J g \mu_B B_0 / kT)}{\sum\limits_{-J}^{J} (1 + m_J\, g\, \mu_B\, B_0 / kT)}$$

$$= \frac{n\, g\, \mu_B \left\{ \sum\limits_{-J}^{J} m_J + g\, \mu_B\, B_0 \sum\limits_{-J}^{J} (m_J^2)/kT \right\}}{(2J + 1) + g\, \mu_B\, B_0 (\sum\limits_{-J}^{J} m_J)/kT}.$$

Now, as $\quad \sum\limits_{-J}^{J} m_J = 0 \quad$ and $\quad \sum\limits_{-J}^{J} m_J^2 = \dfrac{J(J+1)(2J+1)}{3}$

$$M_H = \frac{n\, g^2\, \mu_B^2\, B_0}{kT} \frac{J(J+1)(2J+1)}{3(2J+1)} \quad \text{and the } n \text{ finally, we get}$$

Susceptibility, $\quad \chi = \dfrac{M_H}{H} = \dfrac{\mu_0\, n\, g^2\, \mu_B^2\, J(J+1)}{3\, k\, T}.$ $\qquad \dots(16.6)$

If now the paramagnetic susceptibility $\chi$ is calculated using the above equn. (16.6) using Bohr magneton and $n \sim 10^{28}/m^3$, and we can get

$\chi \sim \dfrac{3}{100\,T}$ and for $T = 300\ K$ (room temperature) $\chi$ is of the order of $10^{-4}$.

This shows that for paramagnetic substances the susceptibility though positive but its magnitude is very small and it gets still smaller values at higher temperature.

The other approach to the susceptibility derivation of paramagnetic substances from the statistical concept of taking the electrons as free electron obeying Fermi statistical distribution is already discussed in Chapter 14 (Sec. 14.3).

## 16.3  ORIGIN OF FERROMAGNETISM

The theory of ferromagnetism was first introduced by P. Weiss in 1907 on the basis of molecular field and is based on the following two assumptions:

1. There exist some microscopic domains within the ferromagnetic material which are spontaneously magnetized. The magnetization of the whole body is the resultant of the magnetic moments of these individual domains.

2. The spontaneous magnetization of the microscopic domains arises due to existence of a molecular field within the material. Due to this field all the magnetic dipoles within a domain are aligned parallel to one another.

This Weiss original theory was based on Langevin's classical theory, however, we will discuss here from the stand point of quantum theory.

We have seen above in paramagnetism and recalling the equn.( 16.5) above as

$$M_H = \frac{n\sum\limits_{-J}^{J} m_J\, g\,\mu_B \exp(m_J\,\mu_B\, B_0\,/kT)}{\sum\limits_{-J}^{J}\exp(m_J\,g\,\mu_B\, B_0\,/kT)}.$$

This can be evaluated in a general case as:

$$M_H = n\, g\, J\, \mu_B\, B_J\,(\alpha) \qquad\qquad \ldots(16.7)$$

where $B_J(\alpha)$ is known as Brillouin function and $\alpha = Jg\, m_B\, B_0/kT$.

This is a general case and now introducing Weiss's assumption that the total magnetic acting on the dipoles within a ferromagnetic material is $H_m = H + gM$. Here $H$ is the external field, $M$ the magnetization of the material and $\gamma$ is a constant known as Weiss constant. Using this result and as $B_0 = \mu_0\, H_m$ we get

$$\alpha = J g \mu_B \mu_0 (H + \gamma M)/kT \text{ and so}$$

$$M = \frac{\alpha kT}{\gamma g \mu_0 \mu_B J} \qquad \qquad ...(16.8)$$

Now, if we use two equn. (16.7) applied for general case and (16.8) for ferromagnetic materials and plot the graph for the variation of $M$ with $\alpha$, we get the following Fig. 16.3.

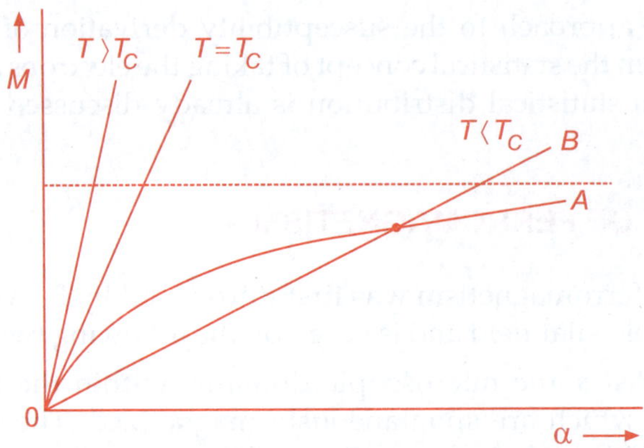


**Fig. 16.3**  Variation of magnetization $M$ with the function $\alpha$

In the figure the curve $A$ is drawn using the equn. (16.7) and curve $B$ using the equn. (16.8).The point of intersection shown by the dot gives the spontaneous magnetization $M$ at temperature $T$. The slope of the curve $A$ plotted using equn. (16.7) gives the limiting temperature $T_C$, above which spontaneous magnetization is not possible. For $T < T_C$, spontaneous magnetization is possible and this $T_C$ is known as ferromagnetic Curie temperature.

As per the domain theory of ferromagnetism, in the absence of any external field, the different domains are oriented at random so that the resultant magnetization is zero. In this case the free energy of the specimen is the minimum. By the application of a magnetic field the specimen is magnetized with the orientation of the domain so that the net magnetization vector is not cancelled and the resultant is oriented in the direction of external field. In 1928, *W.* Heisenberg proposed a quantum mechanical model of the molecular fields in ferromagnetic materials. According to his theory in ferromagnetic materials the spins of the electrons of the neighbouring atoms which are antiparallel align themselves in the parallel direction due to a force known as 'exchange force'. This exchange force results when two electrons between the neighbouring atoms are exchanged to form a symmetrical spatial state and thus results an attractive force between them when they are brought at some internuclear distance otherwise when they form antisymmetrical distribution they experience a repulsive force. As this symmetrical spatial state arises when the electrons align their spin direction parallel to each other, the total energy of the system decreases. This lower energy state is more stable and is the state when ferromagnetism arises. Whether the exchange interaction force will be positive or negative depends on the lattice parameter a and the diameter d of the partially filled electronic shell. If this exchange force given an integral $J_e$ is plotted against $a/d$, we get the following graph Fig. 16.4. In the graph the positions α-Iron, Nickel and Cobalt are in the positive side of the plot and γ-Fe, Mn are in the negative side.

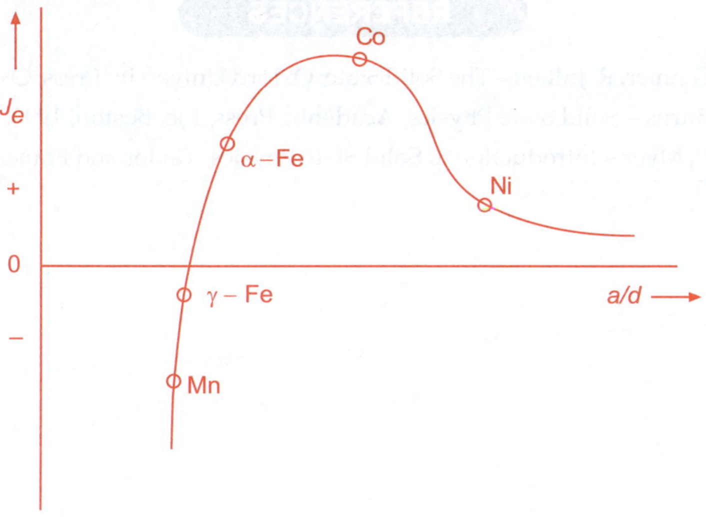


**Fig. 16.4** The plot of Exchange force (Integral) vs. *a/d*

## 16.4 ANTI-FERROMAGNETISM AND FERRIMAGNETISM

If exchange force given by the integral $J_e$ is negative the spin vectors of the atoms in the neighbouring positions in the lattice are aligned antiparallel and they do not show any spontaneous magnetization. Such demagnetization occurs at 0°K and this phenomenon is known as anti-ferromagnetism. If temperature of the substance is increased, this antiparallel alignment of the magnetic moment of the neighbouring atoms will be disturbed and magnetization increases. Above a specific temperature known as Neel temperature the substances show paramagnetism. An antiferromagnetic crystal may be regarded as composed of two sub lattices with opposite directions of magnetization. An example of MnO is an antiferromagnetic material.

If in some substances oppositely aligned magnetizations of two sub lattices are there but their magnetizations are not equal and so they can not compensate each other and the material as a whole show spontaneous magnetization. The property of these materials is known as Ferrimagnetisms. The materials are known as ferrites. As an example if in a magnetic crystal say FeO, $Fe_2O_3$ if the $Fe^{++}$ ions are substituted by some divalent metallic ions like Mg, Ni, Co, Mn etc., then we get a ferrite. The most important characteristic of these ferrites are that they have very low conductivity and high resistivity much higher than iron. But they have similar magnetic properties like iron. This has resulted in its indispensability in the high frequency technology where eddy current is an disadvantage.

## REFERENCES


1.  A. Guinier, R. Jullien – The Solid State, Oxford University Press, Oxford 1980.

2.  G. Burns – Solid State Physics, Academic Press, Inc, Boston, 1990.

3.  H. P. Myer – Introduction to Solid State Physics, Taylor and Francis, London 1991.

# CHAPTER 17

# Superconductivity

## 17.1 INTRODUCTION

The basic property of superconductivity is the existence of a persistent current or zero electrical resistance on cooling the material below a critical temperature. This very important property is however not a general property exhibited by all solids but is exhibited by elements, alloys and oxides with a range of compositions and structures. The sharp transition between the normal and the superconducting state below a temperature known as critical temperature, $(T_c)$ is shown in the following Fig. 17.1.

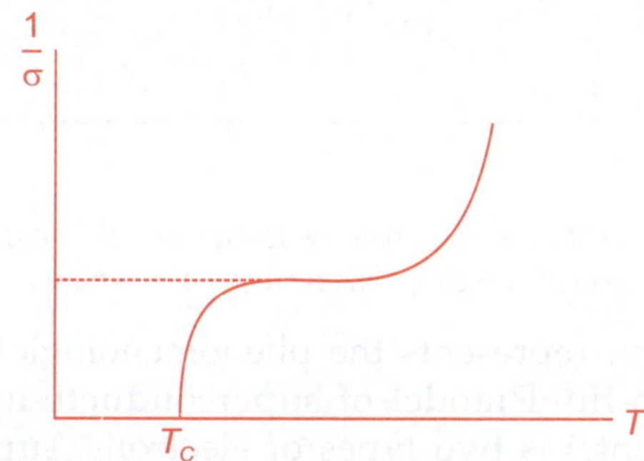

Fig. 17.1 Temperature dependence of resistivity $\rho(= 1/\sigma)$ for superconducting metal

The condition of perfect conductivity in terms of finite current density $j$ which flows indefinitely in the superconducting state such that the Ohm's Law gives

$$E = (1/\sigma)j = \rho j = 0. \qquad \ldots(17.1)$$

Substituting this result in Faraday's Law as introduced in Chapter. 6 (equn. 6.29) and writing it in differential form we get

$$\frac{\partial}{\partial t}B = -\nabla \times E = 0 \text{ or, } B = \text{const.} \qquad \ldots(17.2)$$

This means that inside a conductor $B$ cannot change with time. Now, if an electrical conductor which is cooled below $T_C$ in an external magnetic field $B_0$ and then becomes a perfect conductor, we can expect that the magnetic field within the samples would remain even after switching off the external magnetic field and that is due to induced current. However, it has been found experimentally that if a superconductor is cooled below $T_C$ in an applied magnetic field, the magnetic flux is expelled out of the material instead of the field be 'frozen' inside the material. Therefore, in superconducting state it is always required that: $B = 0$. This result which is not predicted by equn. (17.2), is known as Meissner Effect. This is a state of perfect diamagnetism and is shown in the following Fig. 17.2. For a perfect conductor even at $T_C$, the retained magnetic field within the material and then on further cooling it is expected to decrease linearly but for superconductors it simply drops down to zero. Therefore, two properties of perfect conductivity and perfect diamagnetism are mingled together in superconductors.

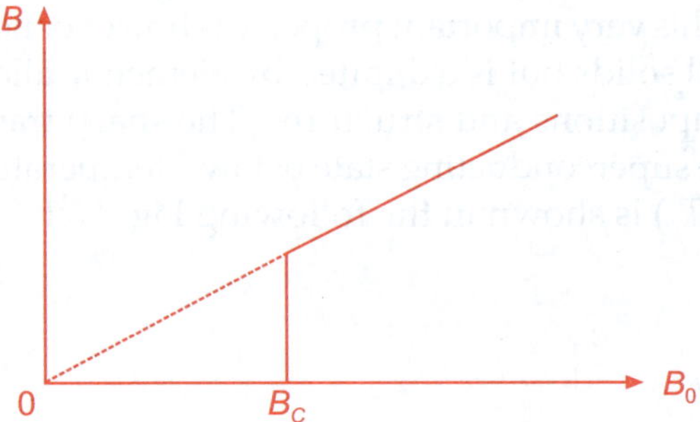


**Fig. 17.2** Magnetic behaviour of superconducting material. The bold line shows the superconducting material and the dotted line represent the behaviour of perfect conductors

This observation represents the phenomenological basis of London Theory or the two-fluid model of superconductivity. It assumes that superconductor contains two types of electrons, superconducting and normal, with densities $n_S$ and $n_N$ and velocities $v_S$ and $v_N$ respectively. The normal electrons obey the established current density equations as:

$$j_N = -n_N e v_N = s_N E.$$

However, for superconducting material the damping force arising due to resistance from the equation of motion can be dropped out and may be written as:

$$m_e \frac{d}{dt} v_S = -eE.$$

The current density of the superconducting electrons takes the form:

$$j_S = -n_S e v_S.$$

Now, combining these two equations we get

$$\frac{d}{dt} j_s = \frac{n_s e^2}{m_e} E. \qquad \qquad ...(17.3)$$

This equation is known as First London Equation.

Now, if we take curl of this equn. (17.3), we have:

$$\nabla x \frac{d}{dt} j_s = \frac{n_s e^2}{m_e} \nabla x\, E = -\frac{n_s e^2}{m_e} \frac{\partial}{\partial t} B = -\frac{\partial}{\partial t}\left(\frac{n_s e^2}{m_e}\right) B$$

or, $$\nabla x\, j_s = -\frac{n_s e^2}{m_e} B \qquad \qquad ...(17.4)$$

This is known as Second London Equation for a perfect diamagnetic substance.

## 17.2 COOPER PAIRS

Inside a superconductor the behaviour of electrons is vastly different. The impurities and lattice are still there, but the movement of superconducting electrons through the obstacles is quite different. They bump into nothing and create no friction and transmit electricity with no appreciable loss in current and no loss of energy. The early researches concluded that it is the low temperature which decreases the lattice vibration substantially is the cause of this superconductivity. But this decrease of temperature predicts a slow decrease of resistivity rather than abrupt loss of it, when the temperature is brought down to a critical value as shown in Fig. 17.1. Therefore, it was established that no classical and existing ideas could explain this phenomenon. It was in 1957, three American physicists namely J. Bardeen, L. Cooper and J. Schrieffer through their theory of superconductivity known as BCS Theory could explain the superconductivity phenomenon at temperature close to absolute zero.

According to the theory as one negatively charged electron passes by positively charged ions in the lattice of the superconductor, the lattice distorts. This distortion of the lattice creates phonons to be emitted which in turn form a trough of positive charge. Now, before the electron passes by and before the lattice springs back to the original position, another electron is drawn towards this trough and the process is repeated. These two electrons instead of repelling each other are linked up by the forces exerted by the phonons. This pair is known as Cooper Pair. The net effect is that when one electron of the pair emits a phonon due to shrinkage and subsequent release of the lattice, the other pair absorbs that phonon. It is this exchange that keeps the two negative electrons together and electron moving ahead makes the movement of the follower electron easier. However, though it is also important to remember that the pairs are constantly breaking and reforming, they can be considered as permanent pairs as electrons are indistinguishable.

The qualitative pictorial demonstration can be seen from the following (Fig. 17.3).

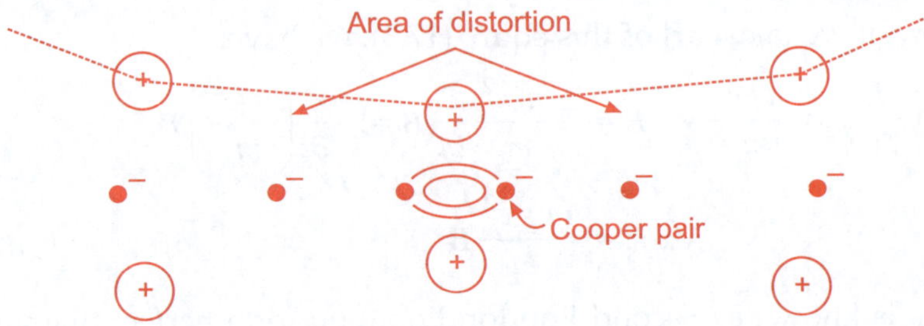


**Fig. 17.3**  Cooper pairs and their flow through lattice

This electron pairing is favourable because it has the effect of placing the material at low energy state. When the temperature is very low, the Cooper pairs remain intact due to reduced lattice vibration and with the increase of temperature the lattice vibration increases and the pairs are broken. As the pairs are broken, the superconductivity decreases and at a temperature above a critical temperature, $T_c$ the material is transformed into simple conductors. The superconductors made from different materials show different critical temperature. Among ceramic superconductors, $YBa_2Cu_3O_7$ has $T_c$ about 90 K. The following Fig. 17.4 shows such variation.

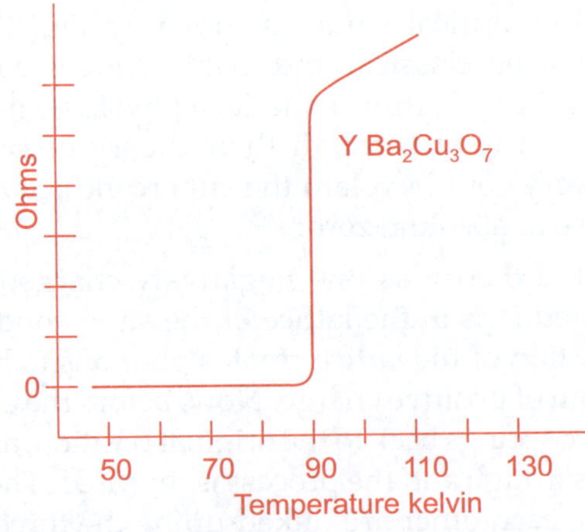


**Fig. 17.4**  Resistance *vs.* temperature of $Y Ba_2Cu_3O_7$

Since there is no loss of electrical energy due to Joule's heat when current flows in superconductors, relatively narrow wires of superconducting material can be used to carry huge currents. However, there is some maximum limit of the current giving rise to Critical current density, $J_C$

above which the superconductors loose their superconducting behaviour and return to normal state even if the Critical temperature, $T_C$ is not crossed.

For practical applications, $J_C$ values in excess of 1000 amperes per square millimeter are preferred. As an electric current in a wire produces a magnetic field around it and as the field is proportional to the current strength, superconductors which are capable of carrying huge current, are very suitable for making strong electromagnets When superconductors are cooled below their transition temperature and a magnetic field is increased around it, the magnetic field remains around the superconductors. The superconductor pushes the external field out of it if the temperature is below $T_C$ and then it behaves as a perfect diamagnetic material. This is because the superconductor (below $T_C$) creates a surface current in itself which in turn produces an opposite magnetic field and cancels all magnetic fields in its interior. This flux exclusion is an important property of superconductors and is known as 'Meissner Effect'. While existence of a critical temperature ($T_C$) and Meissner effect are macroscopic characteristics of the superconductors, the tunneling characteristic of superconductors has quantum origin.

## 17.3 JOSEPHSON EFFECT

It has been introduced in the earlier chapters that the tunneling has quantum origin and depends on the wave nature of the electrons. The tunneling of a pair of electrons between superconductors separated by an insulating barrier was first discovered by Brian Josephson in the year 1962 and the effect is known as Josephson Effect. Josephson discovered that if two superconducting metals were separated by a thin insulating barrier such as an oxide layer of 10 to 20 angstroms thick, it is possible for the electron pairs (Cooper pairs) to pass through the barrier without any resistance. This observation is in contradiction to what happens in ordinary material, where a potential difference must exist for a current to flow. As long as the current is below the critical current for the junction, there will be zero resistance and no voltages drop across the junction. This effect has many potential applications as switching device in computer over ordinary semiconductor switches.

Now, let us have now a consolidated view of this property of materials. It has been stated above that superconducting state is defined by the three very important factors like: (1) The critical temperature ($T_C$) (2) The critical magnetic field ($H_C$) and (3) The critical current density ($J_C$). Each of these parameters however are not independent but are mutually dependent parameters. Therefore, the superconducting state requires that the magnetic field, current and as well as the temperature must remain below the critical values and these values depend on the material. Figure 17.5 demonstrates the relationship between $T_C$, $H_C$ and $J_C$. When $H_C$ and $J_C$ are zero, the $T_C$ attains its maximum value and together they constitute a critical surface shown shaded in the Fig. 17.5. The states pointing towards zero is superconductors and outside it is either normal or mixed state.

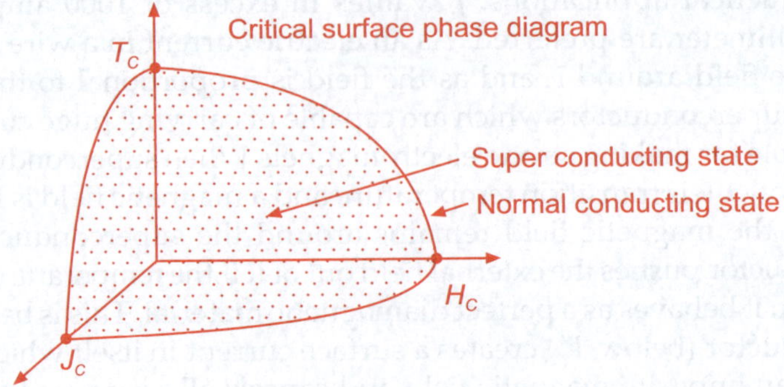


**Fig. 17.5** Critical surface phase diagram: Within the shaded surface the material possesses superconducting state and outside it behaves as normal conductors

## REFERENCES


1. S. B. Palmer and M.S. Rogalski - Gordon and Publishers, USA, 1996.

2. G. Burns – High Temperature Superconductivity, An Introduction, Academic Press, London, 1992.

3. K. Joos- Theoretical Physics, Dover Publication Inc, New York, 1986.

# 18

# Nuclear Physics

## 18.1 NUCLEAR STRUCTURE: AN OVERVIEW

It was long been established that an atom being electrically neutral must has within its volume equal quantity of positive and negative charges. J. J. Thomson first put forward a model of nucleus, according to which an atom was thought to be a positively charged matter within which equal quantity of negatively, charged electrons are embedded like plums in pudding. The model proposed was known as Thomson's plum-pudding model. The fact that an atom has a massive center called nucleus, was first established in 1911 by Rutherford by his famous alpha particle scattering experiment. He also proposed a planetary model of atom, having the positive massive center known as nucleus and surrounding it are negative electrons revolving in different orbits. It was Niels Bohr, who from the study of spectral lines emitted by gases, established in 1913, his model of atom, introducing the concept of quantization of angular momentum of electrons in stationary orbits. An atom according to Bohr has a nucleus which is positively charged and surrounding it in definite stable orbits are electrons revolving round it. The arrangement of electrons in stable orbits having different quantum numbers was introduced to explain the spectral lines having fine and superfine structures. In 1920, Rutherford advanced an hypothesis that an electron and a proton together form a composite nuclear particle but 1932 Chadwick proposed the existence of a neutral particle, named as neutron from its emission from nucleus. After this Chadwick's discovery W. Heisenberg proposed in 1932 that the nuclei are made up of protons having positive charge and neutrons which are electrically neutral. In this picture of nucleus, a nucleus having mass number A and atomic number Z consists of Z protons and A – Z neutrons, so that the total number of particles within a nucleus is maintained as A. Since the mass of protons and neutrons are almost same and equal to unity, the mass of an atom will be close to its mass number. The nucleus so composed will obviously possesses Z units of positive charge. The term nucleon is now used either for proton or neutron. The mass of the atoms or nucleons are expressed in the 'atomic mass unit' (a.m.u.) which is the mass of oxygen- 16 taken as 16 a. m .u. Thus the unit of a.m.u. in physical scale is one-sixteenth of mass of oxygen – 16 and that is: 1 a.m.u. = $1.6598 \times 10^{-24}$ gm. However, in 1961 onwards $1/12^{th}$ of the mass of $^{12}_{6}C$ is accepted as unit of nuclear mass and is known as "Unified atomic mass unit". According to this unit, 1 mole of $^{12}_{6}C$ has mass 12 gm = $12 \times 10^{-3}$ kg and contains $6.02205 \times 10^{23}$ (= $N_0$ Avogadro no.) atoms and so, the mass of a $^{12}_{6}C$ atom = $12 \times 10^{-3}/(6.02205 \times 10^{23})$ and the unit of mass in this unit scale is:

$$1\,u = \frac{1}{12}\frac{12 \times 10^{-3}}{N_0} = 1.660566 \times 10^{-27}\ \text{kg}$$

However, continuing with atomic mass unit (1/16 of $^{16}O$) and using this, the mass of :

**Proton :** 1.00759 a.m.u. = $1.6734 \times 10^{-24}$ gm

**Neutron :** 1.00898 a.m.u.

It has been stated above that as mass of protons and neutrons are almost same and can be approximately taken as one a.m.u · but there exists some difference between the mass of an atom $M$ with the mass number or number of particles in the nucleus $A$. This difference is stated by Aston in 1927 as mass defect.

Mass Defect : $\Delta = M - A$ and another parameter introduced by Aston as:

Packing Fraction: $p = \Delta/A$ which is the mass defect per nucleon.

Now, the variation of this packing fraction, $\Delta$ with mass number $A$ by Aston gave a qualitative indication of the 'Packing' of nucleons within the nucleus and that was perhaps the beginning of the study of nuclear structure and binding of nucleons.

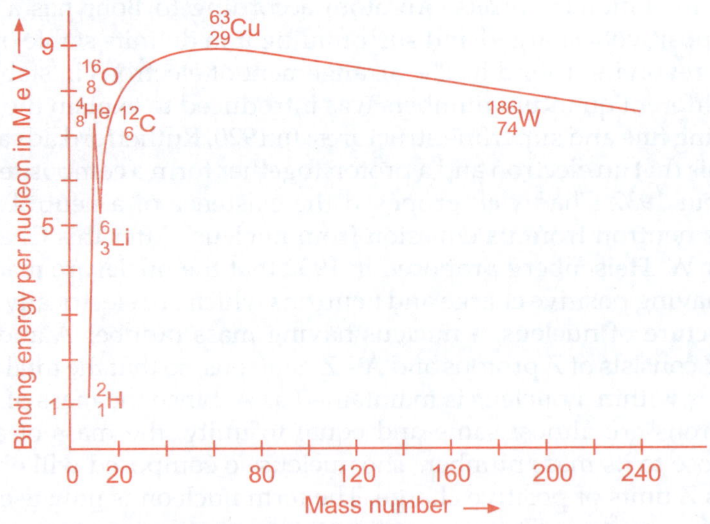


**Fig. 18.1** The binding energy per nucleons Vs. the mass number

When a nucleus of an element is formed by joining the nucleons, then it is found that the sum total of individual nucleons like proton and neutrons is found to be more than the mass of the nucleus of the element. This shows that there is always certain loss of mass while forming the nucleus and this loss of mass known as 'Mass Difference' $\Delta m$ results into creation of energy from $E = \Delta mc^2$. This energy equivalent of the loss of mass or mass difference is responsible for binding the nucleus and is known as binding energy. In Fig. 18.1 above the binding energy per nucleons Vs. mass number is shown. It can be seen from the figure that binding energy per nucleons for $He\,^4_2$ is much higher than hydrogen $H^2_1$. $O^{16}$, $Cu^{63}$ occur at the top of the curve and indicate that these are very stable nuclei. Now, using the mass difference $\Delta m$, which results into energy equivalent from Einstein mass energy relation as: $E = mc^2$ we get for $^7_3Li$ : $Z = 3$ (number of protons), $A = 7$ (number of nucleons including 3 protons and 4 neutrons). The sum of the constituents' mass (rest mass) is:

$$3 \times \text{Mass of Proton} = 3 \times 1.00759 \text{ a.m.u.} = 3.02277$$

$$4 \times \text{Mass of neutron} = \underline{4 \times 1.00898 \text{ a.m.u.} = 4.03592}$$

$$\text{Total mass in a.m.u.} = 7.05869$$

Now, the mass of one $^7_3Li$ atom is 7.01818. If we subtract from this the total mass of 3 electrons of the orbits, then we would get the mass of the nucleus from second approach.

$7.01818 - 3 \times m_e$ (where $m_e$ is the mass of an electron)

$$= 7.01653 \text{ a.m.u.}$$

Then the mass difference $\Delta m = 7.05869 - 7.01653 = 0.04216$ a.m.u.

But we know that 1 a.m.u. = $1.6598 \times 10^{-24}$ gm. $\cong 1.6598 \times 10^{-24} \times (2.9978 \times 10^{10})^2$ erg = $1.492 \times 10^{-3}$erg.

Therefore, 1 a.m.u. is equivalent to energy = $1.492 \times 10^{-3}$ erg

Now, we know that $1\,eV = 1.601 \times 10^{-12}$ erg and so as:

$$1 \text{ a.m.u.} \equiv \frac{1.492}{1.601} \times 10^9 \text{ eV} = 931 \text{ MeV.}$$

This is an important relation between 1 a.m.u. of mass and the equivalent amount of energy into which the mass is converted.

The mass difference $\Delta m$ for $^7_3Li$ atom nucleus is $0.04216 \times 931$ MeV = 39.3 MeV.

This equivalent energy is reasonably high and is used in binding the nucleons closely tight together and is called the source of binding energy. As this energy is reasonably high, the $^7_3Li$ nucleus as a verification is also found to be of very stable structure.

## 18.2 SEMI-CLASSIC MODELS OF NUCLEUS

A nucleus of an element having Atomic no. Z and Mass no. A is designated by the symbol $_Z^A X$. The number of neutrons within it is given by $N = A - Z$. The elements having same atomic number Z (same element) but different N and thus different mass number A are known as "Isotopes", those with same N and different Z are known as "Isotones" and those of the same mass number A, but different Z and N are said to be "Isobars".

Specific approximate descriptions, called "Models" have been developed in the case of nuclei, each of them only being appropriate for a limited range of nuclear properties. A general understanding of systematic trends found in the time-independent properties such as mass, size, charge can be obtained from semi classical models which give description of the nuclear phenomena without considering the inside details of the nucleus. It has been seen from Fig. 18.1 that binding energy per nucleons of most nuclei is about 8 M eV and approximately independent of mass number A. This implies that a nucleon in a large nucleus is not bound to more nucleons than in a small one. Hence nuclear forces have a range which is of the order of the diameter of one nucleon. The saturation of the curve indicates the effects which keep nucleons apart from each other. The nuclei are most tightly bound near A = 60, where the binding energy per nucleons is also maximum. In light nuclei, a single nucleon is attracted by a few other nucleons, thus the nucleon separation is large and stability is reduced. In heavy nuclei the decrease of stability is due to coulomb repulsion between protons, which becomes important for large Z and hence large A.

It is then concluded that energy can be released using nuclei situated at the both ends of the curve, either by combining light nuclei into heavier nuclei (nuclear fusion) or by breaking heavy nuclei into lighter nuclei (nuclear fission).

### *Liquid Drop Model for Nucleus*

Saturation properties of the nuclear forces are very similar to the properties of the intermolecular forces in a liquid. A formula for binding energy of the nucleus can be derived on the basis of liquid drop analogy for nuclear matter. It is considered in this model that matter within the nucleus is incompressible like liquid. If the nucleus is regarded as a spherical assembly of A nucleons, its volume must be proportional to A:

$$\frac{4\pi}{3} R^3 \sim A \text{ or,} \quad R = R_0 A^{1/3}$$

where the experimental value of $R_0 \cong 1.2 \times 10^{-15} \, m$. This liquid drop model was proposed by Bohr in 1936.

Considered as a sphere, the volume V of the nucleus will be given by:

$$V = \frac{4}{3}\pi R^3 = \frac{4}{3}\pi (1.2 \times 10^{-15})^3 \, Am^3 = 7.24 \times 10^{-45} \, Am^3.$$

The mass of the nucleus is approximately $A$ times the mass of the proton i.e. $1.6 \times 10^{-27}\,A$ kg. Then the density of the nucleus is given as:

$$\rho = \frac{1.67 \times 10^{-27}\,A}{7.24 \times 10^{-45}\,A}\, kg/m^3 = 2.3 \times 10^{17}\, kg/m^3.$$

The density of the nuclear matter is therefore independent of the type of nucleus. Furthermore, it has a colossal value $2.3 \times 10^{17}$ kg./$m^3$ i.e. about $10^8$ ton/$cm^3$.

The following analogies can be found between the nucleus and a small drop of liquid:

1. The drop is spherical because of the symmetrical surface tension forces which act towards the centre. The nucleus is also assumed to be spherical.

2. The density of the spherical drop is independent of volume. This is also the case for nucleus. The only difference is that while the density of nucleus is independent of the type of the nucleus, the density of the liquid drop depends on the type of the liquid.

3. The molecules in a liquid drop interact over short ranges compared with the diameter of the drop. Like nucleons of the nucleus, the molecules of the liquid drop interact only with their immediate neighbours.

4. The surface tension effect on the surface of the drop of liquid may be compared with the potential barrier acting on the surface of the nucleus.

5. Molecules of the drop move over short ranges with thermal velocities but if the temperature is increased, the molecules escape from the surface in the form of vapours. Similarly if the energy is supplied to the nucleons by bombarding particle, a compound nucleus is formed which emits nucleons almost instantaneously.

6. If a drop of liquid is made to oscillates, it breaks up into two drops of almost same size. Similarly, by capturing a neutron by nuclei of certain heavy elements, the nuclei of the heavy element breaks up into almost two parts of roughly equal size. This phenomenon is known as nuclear fission.

## Fermi Gas Model of the Nucleus

Fermi gas model is a statistical model which assumes the nucleus to behave like Fermi electron gas in metals. (Discussed before in Ch. 12). According to this model the nucleus is a degenerate gas of protons and neutrons. It may be said again that a gas is called degenerate if the number of energy states that can be occupied is comparable to the number of particles. Since nucleons are spin 1/2 particles, they are Fermions. Hence the behaviour of protons and neutrons will be described by Fermi-Dirac statistical distribution. At $0°K$ all energy levels will be filled up accommodating two nucleons of opposite spins. The Fermi level energy $E_F$ is given in Chapter 12 (Equn. 12. 45) and may be recalled here as:

$$E_F = \frac{h^2}{8m}\left(\frac{3N}{\pi V}\right)^{2/3}. \qquad \qquad ...(18.1)$$

Here, $m$ is the nucleonic mass and volume $V$ now represent the nuclear volume which contains N number of nucleons. Actually there are two types of nucleons one proton having number $Z$ and the other neutrons having number $A - Z$. Assuming further that the number of nucleonic states to be equal to the nucleon number in each case, we get density of states for two gases (of protons and neutrons) as: $n_p = \dfrac{Z}{V} = \dfrac{Z}{\frac{4}{3}\pi r_0^3 A}$ , where $r_0$ is the radius

of the nucleons. And similarly $n_n = \dfrac{3(A-Z)}{4\pi r_0^3 A}$ , Now taking or assuming $N = A - Z = A/2$ and taking nuclear radius parameter $r_0 = 1.2\,fm = 1.2 \times 10^{-15}$m

We get: $n_p = n_n = \dfrac{3/2}{4\pi(1.2)^3} = 0.069$ nucleons per$/m^3$.

Therefore, total nucleon density as $n_t = n_p + n_n = 0.138$ nucleons per $m^3$.

Now, if this value is substituted in equn. (18.1) and taking into mind that each state is occupied by two nucleons with spin up and spin down. The Fermi energy of each of the of the above two gases (protons and neutrons) as:

$$E_F = \frac{\hbar^2}{m}\left(\frac{3}{2}\pi^2 n_p\right)^{2/3} = 21\,\text{MeV}. \qquad ...(18.2)$$

In an actual nucleus the number of protons $(Z)$ and neutrons $(N = A - Z)$ are not equal, $N$ being slightly greater than $A$, the Fermi energies of these two types of nucleons will also be different. Since $N > Z$ the depth of potential well for neutrons is more deep than that of protons. The below, Fig. 18.2 shows the energy states within the potential well in the nucleus.

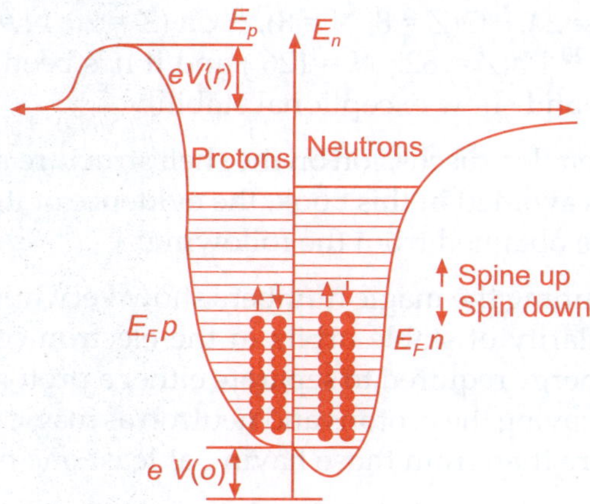


**Fig. 18.2** Potential well and discrete energy levels inside a nucleus

## 18.3  THE SHELL MODEL OF THE NUCLEUS

There is existence of enough evidences that there are periodicities in the nuclear binding energies which might be due the shell structure similar to the atomic shell structure. For neutrons closed shells have been associated with the magic numbers $N = 2, 8, 20, 28, 50, 82$, and $126$ and for protons with $Z = 2, 8, 20, 50$, and $82$, because the nuclei with one or both of the magic numbers shows a particular stability, when compared with other nuclei. The magic numbers can be derived using 'single particle shell model' where many nucleon problem is reduced to a single particle problem, under the assumption that despite the strong overall attraction between nucleons, which provides the nuclear binding energy each nucleon moves in a nuclear potential due to all the other nucleons. In brief, it may be said that the short range nuclear forces average out to create the nuclear potential and all the internucleon coupling can be neglected.

There are strong reasons to believe that as in the case of binding of orbital electrons in the atoms, the nucleons in the nuclei are arranged in certain discrete shells.

W. M. Elasser (1933) was first to introduce this concept of nuclear shells. Similar to the stable orbits in the electronic configuration round the nucleus, it was pointed out by some other physicists also that nuclei containing some definite numbers of neutrons and protons show very high stability. They are as introduced before, known as 'Magic numbers'. They are:

| Protons  | : | 2 | 8 | 20 | 28 | 50 | 82 |     |
|----------|---|---|---|----|----|----|----|-----|
| Neutrons | : | 2 | 8 | 20 | 28 | 50 | 82 | 126 |

Some nuclei contain magic numbers of protons and neutrons both. Examples:

$^4He$ ($Z = 2, N = 2$), $^{16}O$($Z = 8$, N = 8), $^{40}$Ca ($Z = 20$, N = 20), $^{48}$Ca($Z = 20$, N = 28), and also $^{208}$Pb($Z = 82$,  $N = 126$ ) and it has been found that these are doubly magic and show exceptional stability.

Though the detailed discussion on the shell structure as being quantum mechanical and is avoided in this book, the evidence of its existence inside the nucleus can be obtained from the following:

(a)   Nuclei containing the magic numbers show very high stability which shows  similarity of stable orbits in the electron configuration. The minimum energy required to separate either a proton or neutron from the nucleus having the protons and neutron as magic numbers is found to much more than from those having at least one number more.

(b)   The naturally occurring isotopes, whose nuclei contain magic numbers of neutrons or protons have generally greater relative abundance. This again because of their more stability.

(c)   The number of stable isotopes of an element containing magic number of protons is usually large compared to those for other elements. Example: $_{20}$Ca with $Z = 20$ has 6 stable isotopes compared to 3 and 5 for $_{18}$Ar and $_{22}$Ti.

(d)   The number of naturally occurring isotones with magic number of neutrons is usually large compared to those occurring nearby.

(e)   The stable end products of all the three radioactive series are the three isotopes of Pb i.e. $^{206}$Pb , $^{207}$Pb and $^{208}$Pb, which all have magic number of protons 82.

(f)   The neutron capture cross sections of the nuclei with magic numbers of neutrons are usually low. Since the neutron shells are filled up in these nuclei, the probability of capturing an additional neutron is small.

(g)   If the $\alpha$ and also $\beta$ disintegration energies are plotted for heavy nuclei as function of mass number $A$, for a given $Z$, then usually a regular variation is observed except sudden discontinuities are observed at the number neutron, $N$ is 126 which is the magic number of neutrons for $\alpha$ disintegration and also for $\beta$, similar discontinuities are also observed when neutrons or protons attain the magic numbers.

The experimental results summarized as above lend strong support to the proposition of shell structure for the nucleus and to develop a theory of the nuclear shell structure it is necessary to assume the existence of potential well within the nucleus and within this potential well there can exist a number of quantum states. However, the detailed quantum mechanical analysis of the quantum states of nucleons existing in the nucleus potential well is beyond the scope of this book.

## 18.4  RADIOACTIVE DECAY OF NUCLEUS

In the year 1896 Henry Becquerel discovered the emission of radiation from some double salts of uranium and confirmed that these radiations are spontaneous. This phenomenon known as Radioactivity was later extensively studied by Rutherford (1902) on thorium and its successive products. Rutherford studied the ionization power of the radiations given off by uranium using electroscope. He established that these rays are of two types : one called $\alpha$-rays and the other $\beta$-rays. A year later it was established that radium emits in addition to $\alpha$ and $\beta$-rays a third type of radiation known as $\gamma$-rays.

This third type ($\gamma$-rays) was found to be much more penetrating than $\alpha$ and $\beta$-rays. On application of a magnetic field it was found that this $\gamma$-rays were not affected by magnetic field, whereas, the $\alpha$-rays were deflected to a small extent in a direction confirming that they are positively charged and heavier than $\beta$-rays, which were deflected more and in opposite direction confirming that they are lighter and negative particles. The use of magnetic field to deflect the emitted radiation is shown in the following Fig. 18.3.

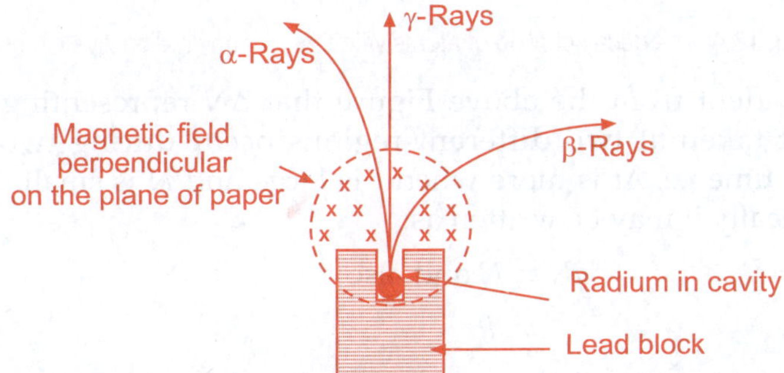


**Fig. 18.3** $\alpha$, $\beta$ and $\gamma$-rays emitting from radioactive radium ($Rd$ ), separated by magnetic field

Due to this emission of radiations from radioactive metals, the mother nucleus goes on disintegrating by emitting of these radiations like $\alpha$, $\beta$ and $\gamma$-rays and constantly goes to daughter nuclei and thus creating a series of elements in the series known as Radioactive series. There are several important such series called 'Uranium Series', 'Thorium Series' and 'Actinium Series' named after the mother element.

It was also established that these $\alpha$-rays emitted is ionized Helium atoms and $\beta$-rays are electrons and so $\alpha$-rays are composed of positively charged particles and much heavier than $\beta$-ray which is nothing but negatively charged electrons. The $\gamma$-rays which is not deflected by magnetic field is electromagnetic radiations. Therefore, the following radioactive reactions occur during $\alpha$ and $\beta$-emissions.

$$\ _Z^A X \ \rightarrow \ \ _{Z-2}^{A-4}Y + \alpha(_2^4He)$$

$$\ _Z^A P \ \rightarrow \ \ _{Z+1}^A Q + \beta(e^-) \qquad\qquad ...(18.3)$$

Due to this disintegration the number of nucleons in the radioactive nucleus goes on decreasing and it was found that with time the intensity of radiation emitted decreases. The plot of number of atoms with time from one initial takes the following nature:

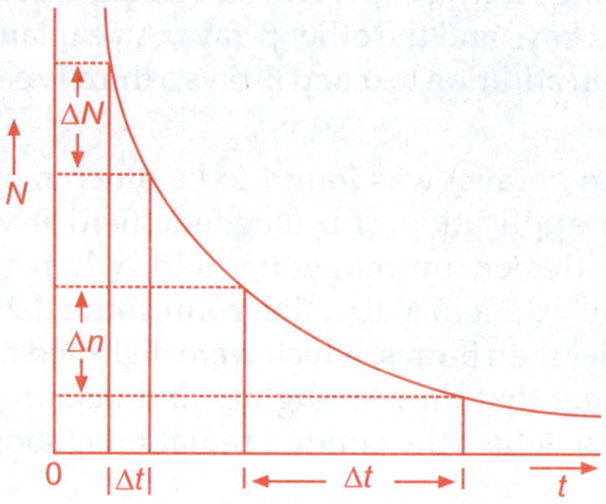

**Fig. 18.4** The decay of $N$ no of atoms with time $t$ of any radioactive element

It is evident from the above Figure that $\Delta N$ representing the same decrement taken at two different regions occur during two different interval of time i.e. $\Delta t$ is more when $t$ is large and $N$ is small. Therefore, mathematically it may be written as:

$$- dN \propto N \text{ and also}$$

$$\propto dt$$

or, $$\frac{dN}{N} = -\lambda dt.$$

Integrating we get:

$$\log_e N = -\lambda t + k, \text{ where } k \text{ is a constant.}$$

If $N_0$ is the number of atoms at the beginning, when $t = 0$ and $k = \log_e N_0$.

Therefore, $$\log_e \frac{N}{N_0} = -\lambda t$$

or, $$N = N_0 e^{-\gamma t}. \qquad\qquad ...(18.4)$$

This is famous Rutherford-Soddy equation of radioactive disintegration.

Half life period of any radioactive decay is the time elapsed from $t = 0$, so as to decrease the initial number of atoms from $N_0$ to $N_0/2$.

Then, $$\log_e 0.5 = -\lambda T, \text{ where } T \text{ is the half life.}$$

Then $$\lambda T = 2.303 \log_{10} 2$$

Therefore, $$T = \frac{0.693}{\lambda}.$$

This half life is an important parameter to know the activity of the radioactive elements and in a radioactive series at every step of the transformation, this half life being different and it changes. For example, in Uranium Series the half life changes from $4.5 \times 10^9$ year (for Uranium I) to as low as $1.6 \times 10^{-4}$ sec (Ra C$'$). If the half life is large, it shows that the element takes a longer time through slow decay to reach the next element in the series.

## 18.4.1 The Growth and Decay of Uranium X in Uranium

Uranium, having mass number 238 and atomic number 92 and noted as $^{238}_{92}U$ emits one $\alpha$ particle to disintegrate into $UX_1$. The disintegration reaction is given by:

$$^{238}_{92}U \rightarrow {}^{234}_{90}UX_1 + {}^{4}_{2}He\,(\alpha)\,T\,(\text{Half life}) = 4.5 \times 10^9 \text{ years.}$$

This $UX_1$ forms as a by product within the uranium. $UX_1$ the by product has a half life ($T$) only 24.1 days and emits one $\beta$ particle. The radioactivity is then due to the breaking of $UX_1$ to emit $\beta$ particles and when this $UX_1$ is separated from uranium it will decay by emitting $\beta$ particles and its activity will become half of its initial activity within 24.1 days.

Consider now the growth of $UX_1$ in freshly separated uranium, $^{238}_{92}U$. Let the number of uranium atoms at this time taken as $t = 0$ is $N_0$, which may be assumed to be sensibly constant for very long half life. The rate of formation of $UX_1$ atoms will therefore be constant and equal to $\lambda_1 N_0$, where $\lambda_1$ is the decay constant of uranium, $^{238}_{92}U$. If now the number of $UX_1$ at a time $t$ is $N$, the $UX_1$ will decay at time $t$ by the rate $\lambda_2 N$, where $\lambda_2$ is the decay constant of $UX_1$. The net increase of $UX_1$ atoms in uranium is given by:

$$\frac{dN}{dt} = \lambda_1 N_0 - \lambda_2 N$$

or, $$\frac{dN}{(\lambda_1 N_0 - \lambda_2 N)} = dt .$$

Integrating we get: $\ln(\lambda_1 N_0 - \lambda_2 N) = -\lambda_2 t + C$, where $c$ is a constant.

When $t = 0$, $N = 0$ as there was no $UX_1$ atoms and then $C = \ln \lambda_1 N_0$. Then the above equation can be written as:

$$\ln\left[\frac{\lambda_1 N_0 - \lambda_2 N}{\lambda_1 N_0}\right] = -\lambda_2 t$$

or,
$$1 - \frac{\lambda_2\, N}{\lambda_1\, N_0} = e^{-\lambda_2 t}$$

i.e.
$$N = \frac{\lambda_1\, N_0}{\lambda_2}\left(1 - e^{-\lambda_2 t}\right) \qquad \qquad \dots(18.5)$$

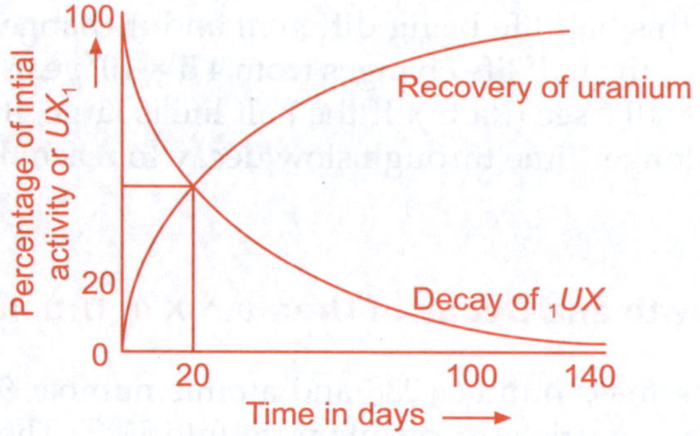

**Fig. 18.5** The decay of uranium $X_1$ and the recovery of uranium

Now, in the above disintegration, the activity demonstrated in intensities, which is measured in ionization chamber, the effect of the $\alpha$ particle emission by uranium is very small (Long half life). Thus the $\beta$ emission by $UX_1$ is only significant in creating ionization and so the rate of decay of the freshly separated $UX_1$ will be decided by

$$N' = N'_0\, e^{-\lambda_2 t} \qquad \qquad \dots(18.6)$$

Where $N'_0$ is the initial number of atoms of $UX_1$ at time zero and $N'$ is the number of these atoms at time $t$.

The recovery curve for the growth of $UX_1$ in uranium will depend on equn. (18.5). However, $N_0$ in equn. (18.5) is not same as $N'_0$ of equn. (18.6) because $N_0$ is the number of uranium atoms and $N'_0$ is the number of $UX_1$ atoms in the original uranium before the $UX_1$ is separated from it. However, a relation between $N_0$ and $N'_0$ can be derived because of the fact that uranium left for many months or years, will have reached a state of radioactive equilibrium. In that equilibrium state the rate of formation of $UX_1$ in it is equal to the rate of disintegration. This is:

$$\frac{dN}{dt} = \frac{dN'}{dt} \quad \text{and therefore,} \quad \lambda_1\, N_0 = \lambda_2\, N'_0$$

The equn. (18.5) can therefore, be written as:

$$N = N'_0\left(1 - e^{-\lambda_2 t}\right) \qquad \qquad \dots(18.7)$$

The above Fig. 18.5, the intensity equivalent of the equns. (18.6) and (18.7) are plotted.

### 18.4.2 A Comparative Study of $\alpha$, $\beta$-Particles and $\gamma$-Radiation and Artificial Radio Activity

In the natural radio activity described before, three types of radiations are found to emit from naturally radioactive atoms and they are $\alpha$ and $\beta$-particles and $\gamma$-radiations. A comparative study of these emitted particles and radiation is given below.

**$\alpha$-Particles**: They are by far most strongly ionizing particles and are readily absorbed by thin foils of metal and by the passage of few centimeters of gas at atmospheric pressure. They are as mentioned before are ionized helium atoms $^4_2\text{He}$. The velocities of these $\alpha$-particles depend on the nature of the emitting radioactive atoms and are of the order of $c/16$, where $c$ is the velocity of light. This velocity ranges between, highest from short lived radioactive element to lowest from atoms having long half life. For example, Th C' having half life T = $3 \times 10^{-7}$ sec emits $\alpha$-particles having $\alpha$ velocity $2.054 \times 10^9$ cm/sec and U1, half life, T = $4.5 \times 10^9$ yrs. emit $\alpha$ with velocity $1.42 \times 10^9$ cm/sec.

The $\alpha$-particle has specific charge e/m of 4,823 e.m.u./gm and a charge of $3.202 \times 10^{-20}$ e.m.u. which is twice the charge of an electron. This through the experiments of Rutherford it is established that $\alpha$-particle is a helium atom with two positive charges.

**$\beta$-Particles:** The experiments on the charge and specific charge established that $\beta$-particles are fast moving electrons. Being as light as that of electron, $\beta$-particles have ranges in materials of the order of twenty times those of a particles. A $\beta$-particle with an energy of 0.5 MeV, has a range in air at N.T.P. of 1 $m$ approximately. A continuous energy spectrum is observed to be occupied by $\beta$-particles emitted from the nuclei of radioactive elements.

**$\gamma$-Radiation:** This is a form of very high energy electromagnetic radiation of low ionizing power but having high penetration through materials. $\gamma$-Rays from radium are capable of traversing a thickness of steel up to 12 in. and more but have only 1% of ionizing power of that of $\beta$-particles. $\gamma$-ray spectrum is a line spectrum in the form of a number of discrete frequencies characteristics of the energy levels of the nucleus, they emit from.

### *Induced (Artificial) Radioactivity*

In 1933 Curie – Joliot discovered that when aluminium foil was bombarded with $\alpha$-particles from naturally radioactive Polonium, emission of neutrons took place. They also found that positrons were also emitted at the same time and this emission continued even after the radioactive source of $\alpha$-particle, Polonium was removed though the intensity of positron was found to decrease exponentially with time. To interpret their results, Joliots assumed that Aluminum on bombardment with a particles from Polonium source initiates a reaction as

$$^{27}_{13}\text{Al} + {}^{4}_{2}\text{He} \rightarrow ({}^{31}_{15}\text{P}) \rightarrow {}^{30}_{15}\text{P} + {}^{1}_{0}n. \qquad \qquad ...(18.8)$$

This ${}^{1}_{0}n$ is the neutron produced in the reaction and ${}^{30}_{15}p$ is radioactive and breaks up producing the decay

$$^{30}_{15}\text{P} \rightarrow {}^{30}_{14}\text{Si} + {}^{0}_{+1}e \text{ (positron).} \qquad \qquad ...(18.9)$$

The above reaction (18.9) has half life 2.2 min. Joliots confirmed the production of positron by studying the deflection in magnetic field. They further after separating the new radioactive product by radiochemical method established that positron emission took place from separated phosphorus. This induced reaction is known artificial or induced radioactivity. Joliots also observed similar phenomena with boron and magnesium ${}^{10}_{5}\text{B} + {}^{4}_{2}\text{He} \rightarrow {}^{13}_{7}\text{N} + {}^{1}_{0}n$ and ${}^{13}_{7}\text{N}$ being radioactive emits

$$^{13}_{7}\text{N} \rightarrow {}^{13}_{6}\text{C} + {}^{0}_{+1}e.$$

The discovery of this artificial radioactivity is of great importance and for this Joliots were awarded Nobel Prize in 1935. Most products of artificial transmutations of elements are radioactive. They decay mainly by ${}^{0}_{-1}e\,(\beta^{-})$ or by ${}^{0}_{+1}e(\beta^{+})$ or by orbital electron capture. In cases of heavy elements they are found to decay by a emission or by spontaneous fission.

The induced radioactivity may be Proton (${}^{1}_{1}\text{H}$) induced like $(p - \alpha)$, $(p - n)$ or $(p - \gamma)$ or even deuteron (${}^{2}_{1}\text{H}$) induced reactions. The detailed descriptions of these reactions are however out of the scope of this book. However, the neutron induced reactions which have some special importance is discussed later.

### 18.4.3 Discovery of Neutron

The discovery of neutrons is an important phenomenon and its detection was delayed because of the fact that they are electrically neutral and remains undeflected by electric field. It was however, certain that neutrons were being produced in experiments on artificial disintegration of nuclei from 1919 onwards but remained undetected until 1930 because of difficulty involved in interpreting fully the experimental results. In 1930, Bothe and Becker on bombarding beryllium with a particles found that a very penetrating radiation was produced and this radiation was found to be uncharged, it was reasonable at that time to assume it to be highly energetic $\gamma$-radiation.

$$^{9}_{4}\text{Be} + {}^{4}_{2}\text{He} \rightarrow ({}^{13}_{6}\text{C}) \rightarrow {}^{13}_{6}\text{C} + h\nu(\gamma) \qquad \qquad ...(18.10)$$

On measuring the absorption of this radiation in lead, it was found that the energy of the proposed $\gamma$-radiation to be 7 MeV which was greater than the energy of any $\gamma$-radiation known at that time. In 1932 Curie-Joliot measured the intensity of this radiation by ionization chamber and found that ionization had marked increase, if this radiation is passed first through paraffin or other hydrogen reach materials. They found that protons were emitted from paraffin on being bombarded with this radiation, measured the energy of the protons to be 4.5 MeV and considering this energy transfer to protons due to Compton recoil process found the energy of the proposed $\gamma$-radiation to be as high as 55 MeV. This energy of 55 MeV was however, much greater than the energy available from a reaction of the type given in equn. (18.10) calculated on a basis of mass energy difference of final and initial products. This result found inconsistent with the conservation of momentum and energy was explained by J. Chadwick of Britain in the year 1932, by proposing a different type of reaction as below:

$$ {}^{9}_{4}\text{Be} + {}^{4}_{2}\text{He} \rightarrow ({}^{13}_{6}\text{C}) \rightarrow {}^{12}_{6}\text{C} + {}^{1}_{0}n. \qquad \ldots(18.11)$$

This is the brief history of the discovery of neutral particle, neutrons theoretically predicted long back but experimentally detected after many years.

The general $\alpha$-particle induced reactions ($\alpha - n$) which emit neutrons can be written as:

$$ {}^{A}_{Z}\text{X} + {}^{4}_{2}\text{He} \rightarrow {}^{A+3}_{Z+2}\text{Y} + {}^{1}_{0}n + Q, \text{ where Q is the energy emitted.}$$

A few typical ($\alpha - n$) reactions are:

Lithium: $\quad {}^{7}_{3}\text{Li} + {}^{4}_{2}\text{He} \rightarrow {}^{10}_{5}\text{Be} + {}^{1}_{0}n$

Nitrogen: $\quad {}^{14}_{7}\text{N} + {}^{4}_{2}\text{He} \rightarrow {}^{17}_{9}\text{F} + {}^{1}_{0}n$

Aluminium: ${}^{27}_{13}\text{Al} + {}^{4}_{2}\text{He} \rightarrow {}^{30}_{15}\text{P} + {}^{1}_{0}n.$ $\qquad \ldots(18.12)$

### 18.4.4 Mass of Neutron

In the experiment of ejection of protons (hydrogen nuclei) from paraffin wax, Chadwick considered that this ejection was due to collision with neutron and applied the laws of mechanics for the elastic collision $\cdot$ Let $M$ and $V$ represent the mass and the maximum initial velocity of the neutrons. It suffers elastic collision with a nucleus of mass m at rest and produces a recoil velocity $v_R$ and its velocity decreases to $V_1$. Applying the conservation of momentum and energy we get:

$$ MV = mv_R + MV_1 \text{ and}$$
$$ 1/2 \, MV^2 = 1/2 \, mv_R{}^2 + 1/2 \, MV_1^2$$

From the momentum equation we get $V_1 = \dfrac{MV - mv_R}{M}$ and substituting this in the energy equation we get:

$$MV^2 = mv_R^2 + \frac{(MV - mv_R)^2}{M}$$

Which on algebraic simplification $Mv_R - 2MV + mv_R = 0$ and

$$v_R = \frac{2MV}{M+m}$$

This is a general expression, which can be written when neutron impinges on protons in paraffin wax we get simply by replacing $v_R$ by $v_H$

$$v_H = \frac{2MV}{M + m_H} \quad \text{and considering collision with}$$

nitrogen atoms in nitrogenous materials, we get

$$v_N = \frac{2MV}{M + m_N}$$

Dividing we get:

$$\frac{v_H}{v_N} = \frac{M + m_N}{M + m_H}$$

Measuring the recoil velocities of protons $v_H$ and nitrogen nuclei $v_N$, and as $m_H$ and $m_R$ are known from this we get the mass of neutrons. Chadwick after finding the maximum recoil velocities of hydrogen and nitrogen nuclei from cloud chamber calculated the value of mass of neutron as 1.17 a.m.u. This was the first value of neutron mass reported and is too high. Chadwick himself modified the process and using the methods of greater accuracy measured the value as 1.00898 a.m.u.

## 18.4.5 Disintegration of Nuclei by Neutron Bombardment

The interaction with nuclei with neutron result into a compound nucleus having same atomic number but only the mass number is increased by unity. The general form of this reaction can be written as:

$$_Z^A X + {}_0^1 n \rightarrow \left( {}_Z^{A+1} X \right)$$

The increased energy given to compound nucleus makes it unstable and it generally returns to a stable state by emitting any of the resulting products like $\gamma$-rays, $\alpha$ particles, protons $p$, neutrons etc. The reactions created are known respectively as: $(n - \gamma)$, $(n - \alpha)$, $(n - p)$ and $(n - n)$ etc. Some of the examples of these reactions are given below:

**(n – g) reaction**:

$$^1_1\text{H} + ^1_0 n \rightarrow (^2_1\text{H}) + h\nu$$

Deuterium $^2_1 H$ is formed from hydrogen.

**(n – α) reaction**

$$^6_3\text{Li} + ^1_0 n \rightarrow (^7_3\text{Li}) \rightarrow ^3_1\text{H} + ^4_2\text{He}$$

$^3_1 H$ is tritium an unstable isotope of hydrogen having half life $T$ = 12.4 years.

**(n – p) reaction**

$$^{14}_7\text{N} + ^1_0 n \rightarrow (^{15}_7\text{N}) \rightarrow ^{14}_6\text{C} + ^1_1\text{H}$$

$^{14}_6\text{C}$ (radio carbon) $\rightarrow ^{14}_7\text{N} + ^{\ 0}_{-1} e$ having half life $T$ = 5, 570 years.

**(n – n) reaction**

$$^{107}_{47}\text{Ag} + ^1_0 n \rightarrow ^{107}_{47}\text{Ag} + ^1_0 n$$

This $^{107}_{\ 47} Ag$ is a meta stable and it decays by emitting γ rays

$$^{107}_{47}\text{Ag} \rightarrow ^{107}_{47}\text{Ag} + h\nu.$$

## 18.4.6 Neutron Cross Section

The concept of cross section becomes a general and useful measure of many classes of "collisions". If two particles interact in more than one way, the probability of various interactions can be measured in terms of cross section of each.

For neutrons also when they are moving through the matter we find various kinds of inter actions and each has its own probability which is directly proportional to its cross section.

In a material if  $n$ = molecules per unit volume

$l$ = dimension of the square area

$dx$ = thickness of the of the volume element

$\sigma$ = Cross section of each molecule = $\pi(r_b + r_t)^2$

where $r_b$ represent the radius of say bullet and $r_t$ is the radius of the target molecule.

Then area of the molecular targets = $\sigma \cdot n \cdot l^2\, dx$. This per unit area gives $n\, s\, d\, x$ which is collision probability.

The probability of a particular neutron induced reaction including Fission for a neutron moving a distance $dx$ is $P_f = s_f N dx$, where $s$ stands for microscopic cross section and $N$ is the number of nuclei per unit volume.

When this microscopic cross section is multiplied by $N$ gives macroscopic cross section.

The total probability $P_t$ for a number of possibilities is written as:

$$P_t = (\sigma_f + \sigma_S + \sigma_a + \sigma_{r+} \ldots \text{ etc.}) \, Ndx$$

where, $\sigma_f$, $\sigma_S$, $\sigma_a$, $\sigma_r$ respectively represent the micro cross section for fission, scattering, absorption and radioactive absorption etc.

This concept as said above, is particularly important in dealing with various chances and probabilities of various reactions during a process where only one type cannot be singled out. This has special significance in Nuclear Fission that is being discussed in the following section.

## 18.5 NUCLEAR FISSION: AN INTRODUCTION

In 1934, Fermi during the course of his study suggested that the bombardment of $^{238}_{92}\text{U}$ might lead to the formation of transuranic elements having atomic number greater than 92 by following the reaction:

$$^{238}_{92}\text{U} + ^{1}_{0}n \rightarrow ^{239}_{92}\text{U} \rightarrow ^{238}_{93}\text{X} + ^{0}_{-1}e$$

The element $X$ would then be possible as suggested by Fermi a transuranic element and his suggestion seemed to be borne out by experiments in his laboratory in which $^{238}_{92}\text{U}$.

When bombarded by neutrons emitted negative β-particles. But in the year 1938, through the elaborate investigations of the products obtained through this neutron bombardment of natural $^{238}_{92}\text{U}$, Hahn and Strassmann found that the precipitate responsible for β emission was an isotope of Barium. Now, if one of the fragment is barium of mass number 56, the other must be 92 – 56 i.e. 36 which is characteristic of a gas krypton. Therefore it was proposed by them (Hahn and Strassmann) that the uranium nucleus was split up into two nuclei on irradiation with neutrons. However, this new process of nuclear reaction was named by Meitner and Frisch as " Fission of Nucleus".

The completed such Fission reaction of Uranium 235 by thermal neutron is:

$$^{235}_{92}\text{U} + ^{1}_{0}n \rightarrow ^{236}_{92}\text{U} \rightarrow ^{141}_{56}\text{Ba} + ^{92}_{36}\text{Kr} + 3\,^{1}_{0}n \qquad \ldots(18.14)$$

## 18.5.1 Energy Release Per Fission of Uranium Atom

Recalling the equn.(18.14) which is:

$$^{235}_{92}\text{U} + ^{1}_{0}n \rightarrow ^{141}_{56}\text{Ba} + ^{92}_{36}\text{Kr} + 3\,^{1}_{0}n$$

Total reacting mass:

$$^{235}_{92}\text{U} = 235.1175 \text{ a.m.u.}$$

$$^{1}_{0}n = 1.00898 \text{ a.m.u. (Neglecting the kinetic energy}$$
$$\text{of thermal neutron)}$$

$$\overline{236.1265 \text{ a.m.u.}}$$

Total resulting mass:

$$^{141}_{56}\text{Ba} = 140.9577 \text{ a.m.u.}$$
$$^{92}_{36}\text{Kr} = 91.9264 \text{ a.m.u.}$$
$$3 \times {}^{1}_{0}n = 3.02694 \text{ a.m.u. } (3 \times 1.00898)$$

$$235.911 \text{ a.m.u.}$$

Mass Difference ($\Delta m$) : 236.1265 - 235.911 = 0.2154 a.m.u.

$$1 \text{ a.m.u.} \cong 931 \text{ M eV.}$$

This mass difference is equivalent to 200 MeV. Therefore, this energy release per fission of uranium atom will appear mostly in the form of kinetic energy of the fragments and product three neutrons. Now in more practical form as 1 MeV is equivalent to $1.6 \times 10^{-6}$ erg, 200 MeV is equivalent to $3.2 \times 10^{-4}$ erg. This though appear very low but if we look into in more details we get:

There are $6.025 \times 10^{23}$ atoms per gm atom of $^{235}_{92}\text{U}$ and so the number

of atoms per gm of $^{235}_{92}\text{U} = \dfrac{6.025 \times 10^{23}}{235} = 2.56 \times 10^{21}$ atoms.

Now, if all these $^{235}_{92}\text{U}$ atoms undergo fission, the energy produced will be given by:

$$E = 2.56 \times 10^{21} \times 3.2 \times 10^{-4} = 8.2 \times 10^{17} \text{ erg}$$

But as $10^7$ erg = 1 joule = 1 watt-sec and 1 KWh = $10^7 \times 10^3 \times 3600$ erg

Therefore, energy released per gm of $^{235}_{2}\text{U}$,

$$E = \dfrac{8.2 \times 10^{17}}{10^7 \times 10^3 \times 3600} = 2.28 \times 10^4 \text{ KWh.}$$

Now, this energy release will appear substantial. It has been shown that on complete disintegration of 1 gm of $^{235}_{92}\text{U}$ th e energy release is equivalent to the energy supply of a power plant of 1 MW capacity for nearly one day. More interestingly, a pound of uranium-235 will supply power of 1 MW capacity for nearly a year.

## 18.5.2 Products of Fission

The fission reaction as stated in equn. (18.14) shows that 3 neutrons are produced per fission of $^{235}_{92}\text{U}$. As there exists other product possibilities, the general fission from $^{235}_{92}\text{U}$ with one slow, thermal neutron can be written as:

$$^{235}_{92}\text{U} + {}^{1}_{0}n \left({}^{235}_{92}\text{U}\right) \rightarrow {}^{A}_{Z}\text{X} + {}^{A1}_{92-Z}\text{Y} + \text{Prompt neutrons}$$

If there are two prompt neutrons then the mass number of the nuclei $Y$, $A1$ will be 234-A. If $_Z^A X$ is lighter nuclei then A can be anywhere in the range of 85 to 104 and the number $A1$ of $Y$ will be 149 to 130. For example, there is a possibility that instead of the reaction (18.14) the following reaction yielding two prompt neutrons may also result.

$$_{92}^{235}U + _0^1 n \rightarrow \left(_{92}^{236}U\right) \rightarrow _{54}^{140}Xe + _{38}^{94}Sr + 2\,_0^1 n + \gamma \qquad ...(18.15)$$

It then can be stated that fission products are then of a statistical nature. The following Fig. 18.6 shows the graph of the percentage of total number fission which gives rise fragments of certain mass number plotted against the mass number.

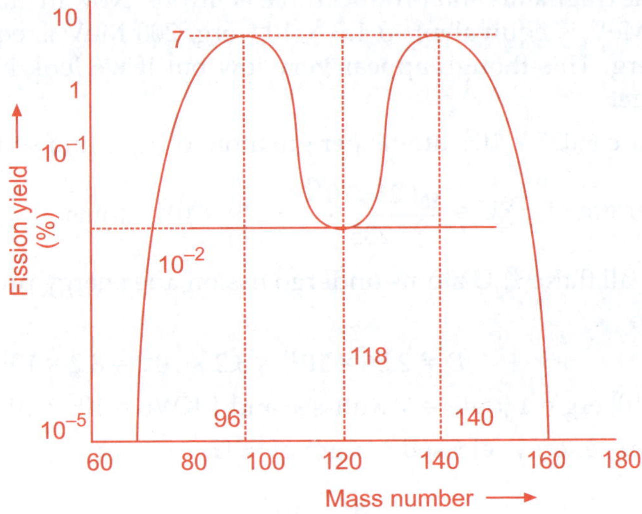


**Fig. 18.6**  Percentage of fission product with mass number for thermal fission of $_{92}^{235}U$

It can be seen that the probability of having fission product having mass number of both equal to 118 is only 0.01%. The most probable (7%) fission products are those having mass number 96 and 140 approximately. Therefore the reactions of the types given in equns. (18.14) and (18.15) have almost equal possibilities. Therefore, in our subsequent discussions it is said that 2.5 numbers of neutrons are produced per fission. However, all these neutrons are very fast and prompt neutrons as they carry a part of the energy liberated due to mass loss. One example of the enormous quantity of the energy to be released can be seen from this fission of 1 kg. of $_{92}^{235}U$. We know that:

$$235 \text{ gms of } _{92}^{235}U \text{ contains} = 6.025 \times 10^{23} \text{ atoms}$$

$$1 \text{ kg. of } _{92}^{235}U \text{ contains} = \frac{6.025 \times 10^{26}}{235} \approx 25 \times 10^{23} \text{ atoms}$$

Now assuming there is no loss of neutrons i.e. if each of those neutrons product take part in fission with another $^{235}_{92}$U atom and if there are n such fission reactions are necessary to react with all these $25 \times 10^{23}$ atoms, then:

$$(2.5)^n = 25 \times 10^{23}$$

and $n = 60$ approximately and the time required to consume all the atoms of 1 kg of $^{235}_{92}$U is only $60 \times 10^{-8}$ sec $\approx 10^{-6}$ sec. We have seen that 1 kg of $^{235}_{92}$U releases an energy of $2.28 \times 10^7$ KWh and that too in a span of time of one micro second. This release within such small time will lead to a rise of temperature of the order of $10^7$ °C.

It should be realized that this fission reaction has an enormous potentiality as a source of energy which should be used for peaceful purposes.

### 18.5.3 Neutron Cross Section and Multiplicity Factor

We have got introduced to neutron cross section in section 18.4.6 and now consider it in more details with giving emphasis on some important elements used in nuclear fission processes. The unit of the cross section of nuclear reaction is 'barn' which is $10^{-12}$ cm. This is because most of the nuclides have diameters about $10^{-12}$ cm. Neutrons are generally classified as fast as they are produced in fission reaction and are about 2 Mev and the energy of the neutrons which have obtained equilibrium with the molecular motion of the material is about 0.025 eV and are called thermal neutrons. The cross sections of the nuclear reactions are different for these two broad categories of neutrons. A brief prior knowledge can be obtained from the following table (Table: 18.1). It is noticeable that the cross sections which signify the probability of the reaction nature widely different for different materials, reaction natures and also on the energies of the neutron. For example, the absorption cross section for slow neutrons for Cadmium is very large and so this element has utilities as slow neutron absorber in controlled fission reaction in reactors, and the fission cross section $s_f$ for U-235 for fast neutron is only 1.3 whereas it is 580 for slow neutrons. This implies that U-235 undergoes fission only with slow thermal neutrons. For U-238 abundantly occurring in natural uranium the fission cross sections is negligible for slow neutrons and very low with fast neutrons. To sustain the fission reaction in uranium it is necessary to slow the neutron produced in fission as they are prompt and fast. We now introduce the concept of multiplicity factor $k$ for neutrons in fission reactions. If we could use an appropriate device to slow down the neutrons produced then the reaction can be sustained and then the sustained reaction process is called a "Chain Reaction".

**Table 18.1** Neutron cross sections in barns

| Elements/ compounds | $\sigma_t$ (slow) | $\sigma_t$ (fast) | $\sigma_s$ (slow) | $\sigma_a$ (slow) | $\sigma_f$ (slow) | $\sigma_f$ (fast) |
|---|---|---|---|---|---|---|
| $H_2O$ | 110 | | | | | |
| $D_2O$ | 14.5 | | | | 0 | 0 |
| B | 760 | | 4 | 755 | 0 | 0 |
| C | 4.8 | | 4.8 | 0.0032 | 0 | 0 |
| Zr | 8 | | 8 | 0.18 | 0 | 0 |
| Cd | 2560 | | 7 | 2550 | 0 | 0 |
| U (Natural) | | 7 | 4.7 (fast) | 2 (fast) | | 0.5 |
| U235 | 697 | 6.5 | 10 | 107 | 580 | 1.3 |
| U238 | 2.8 | | 0 | 2.0 | 0.0005 | 0.5 |
| Pu | 1075 | | 9.6 | 315 | 750 | |

**The multiplicity factor** is defined as the ratio of the number of neutrons in one generation to the neutrons in the preceding generation and if the loss of neutrons is prevented or can be neglected, instead of $k$, it is noted as $k_\infty$. The fast neutrons produced may be slowed down to thermal limit due to their passage through the bulk uranium 235 and the minimum size necessary to maintain the chain reaction is known as "Critical Size". It can then be concluded that above the critical size $k$ is greater than unity and the neutron population increases exponentially with time, at critical size $k$ is equal to unity and so the neutron population is constant and if it is less than unity the neutron population exponentially falls down with time and the fission can not be sustained.

Now, the calculations are sufficient to show this conclusion that the neutrons may produce fission for which cross section is 0.5 barns for U-238. These neutrons are productively absorbed and tend to maintain chain reaction. The probability that one neutron will produce another fission is the microscopic fission cross section divided by total cross section.

$$P_f = \frac{\sigma_f}{\sigma_a + \sigma_f} \approx \frac{0.5}{2 + 0.5} = 0.2.$$

Thus one out of five neutrons may produce fission and since 2.5 number of neutrons are produced on the average then $0.2 \times 2.5 = 0.5$ the number neutrons produced by each neutrons. If we start with 10 neutrons in one generation, then $10 \times 0.2 = 2$ neutrons will be available for fission. In the next generation then $2.5 \times 2 = 5$ neutrons will be available and out of that 5 only $5 \times 0.2 = 1$ will only available. Thus the reaction cannot be maintained and dies down very fast. This shows that in natural uranium U-238 is about 99.3% whereas the U-235 is only 0.7%. Therefore, the natural neutron can not chain react and enrichment with U-235 is essential. This can be shown as follows:

$$k_\infty = 2.5 \frac{N_5 \sigma_{f5} + N_8 \sigma_{f8}}{N_5 (\sigma_{a5} + \sigma_{f5}) + N_8 (\sigma_{a8} + \sigma_{f8})}$$

Putting the values from Table 18.1 where suffixes 5 and 8 represent U-235 and U-238, we get:

$$= 2.5 \frac{1.3 N_5 + 0.5 N_8}{(0 + 1.3) N_5 + (2 + 0.5) N_8} .$$

If we put $N_5 = 0$, we get our just earlier calculation of $k_\infty$ and if we put $N_8 = 0$ ( pure U-235), we get our second result. Now, if we put $k_\infty = 1$ and calculate $N_8/N_5$ we get 1.5, which says that a large body of uranium would be critical it must contain about 40 per cent of U-235 and for a finite size where there are losses, the enrichment of U-235 must be more than 40%. As the separation of U-235 from natural uranium is a very costly process by gaseous diffusion of $UF_6$, the cost of pure U-235 is enormous compared to U-238.

Now, the question which may puzzle the young readers that even if the content of U-235 in the uranium ore of the mine is only 0.7% and U-235 which is an isotope of uranium, in the mine the total quantity of U-235 is definitely enormous so U-235 present may cause a fission and emit fast neutrons and these neutrons while moving through the huge quantity of ore may be thermalized and may cause further fission. But this does not occur and a Uranium mine never explode. The reason is the fast neutrons produced in any fission and the cause is that the natural thermalization is not possible. As the neutrons produced decrease their energy due to collision and as soon as it reaches 7 eV, the neutrons suffer resonance capture by plenty of U-238 present and it is not possible for the neutrons go below 7 eV. to attain the probability that they can create fission with U-235 present. This resonance capture in U-238 is shown in the following Fig. 18.7(a) and 18.7(b) shows the high fission cross section in U-235 at thermal neutrons at 0.025 eV.

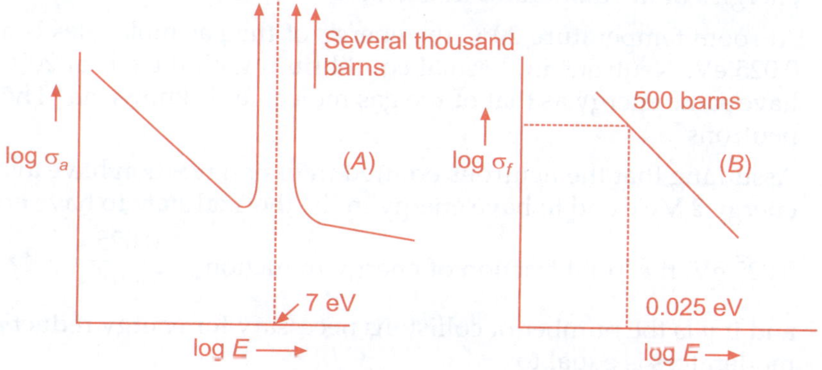


**Fig. 18.7** (a) The absorption cross section in U-238 showing resonance capture at 7 eV and (b) The fission cross section in U-235 for thermal neutron

### 18.5.4 Some More Aspects of Critical Size

It has been stated before that a critical size of the fissionable material is necessary for sustaining the chain reaction. There are some important aspects which should be attended to before designing the nuclear reactors for the production of electric power as one of the most important peaceful use of fission process. These aspects are stated below:

1. As fission occurs throughout the volume of the fissionable material there is fair chance that the neutrons which are produced in the fission reaction may escape from the fuel surface as they are fast and prompt. Therefore, to prevent such loss of neutrons the determination of size of the fission material (fuel in reactor) is of immense importance.

2. As the size of the fission material is changed, its volume also changes. The former is proportional to the square of the dimension and the latter is proportional to the cube of the dimension.

3. The ratio of the neutron production to the loss due to leakage then varies as the first power of the dimension.

4. To prevent $k_\infty$ to be greater than one, a smaller size is required compared to that for the material to be just critical with $k_\infty$ equal to unity. Below this size $k_\infty < 1$.

5. For $k_\infty > 1$, neutron population increases exponentially.

6. For $k_\infty = 1$, neutron population is constant.

7. For $k_\infty <$ , neutron population decreases exponentially.

8. Natural thermalization is not possible as the neutrons at 7 eV energy will suffer resonance capture by $^{235}_{92}U$ and so mixture of $^{235}_{92}U$ and $^{235}_{92}U$ cannot sustain the fission reaction by $^{235}_{92}U$.

9. For slowing down the neutrons with average energy 2 MeV, gathered during fission, elastic scattering in 'Moderators' is much more suitable.

10. No moderators can however, reduce the energy of neutrons below energies of the moderator molecules.

11. At room temperature 20°C the energy of the gas molecules is about 0.025 eV. Neutrons in thermal equilibrium with the gas at 20°C will have same energy as that of the gas molecules is known as "Thermal neutrons".

12. Assuming that the neutrons emitted in fission reaction have average energy 2 MeV and to have energy in the thermal state to have energy 0.025 eV, the Total fraction of energy reduction $= \dfrac{0.025}{2 \times 10^6} = 1.3 \times 10^{-8}$

    and if $n$ is the number of collisions necessary for energy reduction in moderators is equal to:

    (Average fractional energy loss per collision)$^n = 1.3 \times 10^{-8}$.

## 18.5.5 Four Factor Equation

It has been seen that the working value of $k_\infty$ is very important to decide whether the fission reaction can be self sustained or even whether the reaction will die down or will be uncontrollable. The following schematic diagram, Fig. 18.8, throws some light on it.

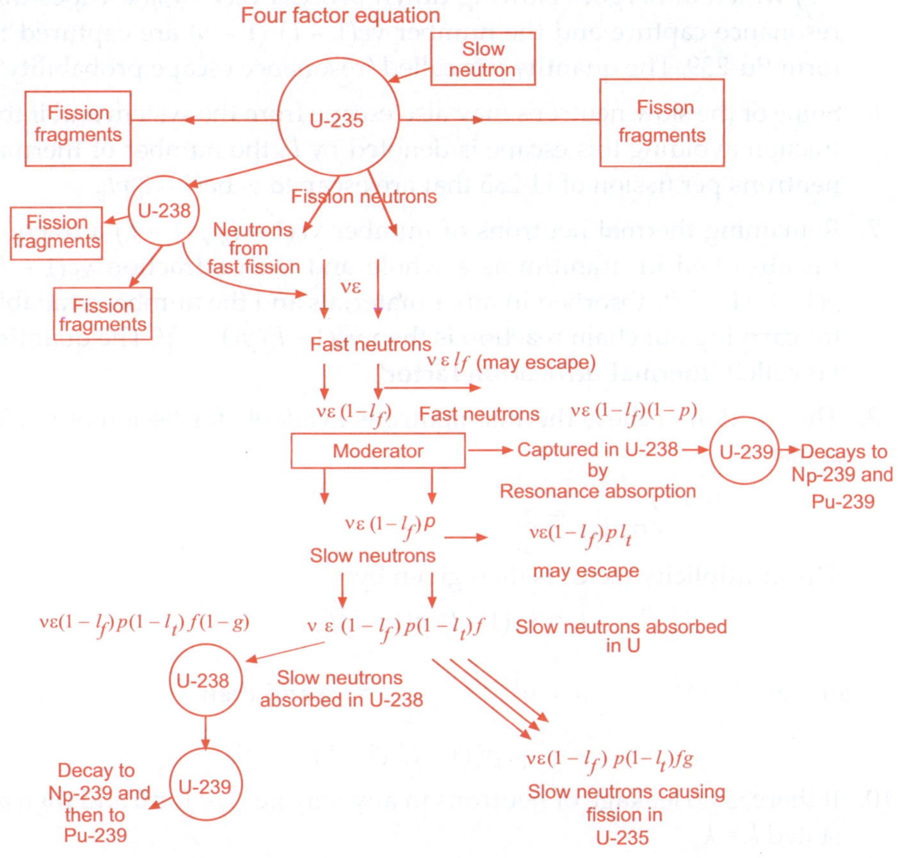


**Fig. 18.8** The chain reaction in fission with almost all possibilities

The above Fig. 18.8, schematically represents the various factors involved in determining effectively the multiplicity factor. However, this can be explained as follows:

1. The process starts with fission of U-235 rich uranium with a thermal neutron and as a result it produces fission fragments along with ν number of fast neutrons

2. Some of these fast neutrons collide with U-238 present and can cause fissions resulting in the increase of fast neutrons by say ε times more.

3. The total number of fast neutrons then increases from ν to νε. This ε is called 'fast fission factor'.

4.  $v\varepsilon$ Numbers of fast neutrons now move through the 'pile' (fuel, moderator, assembly) most of them are slowed down but a fraction $l_f$ escape and so number of neutrons available remains as $ne(1 - l_f)$.

5.  During the slowing down process some neutrons will be captured by U-238 to form U-239 which decays to Pu-239. The number $ne(1 - l_f)$ which undergoes slowing down process $v\varepsilon(1 - l_f)p$ escapes this resonance capture and the number $v\varepsilon(1 - l_f)(1 - p)$ are captured to form Pu-239. The quantity $p$ is called 'resonance escape probability'.

6.  Some of the slow neutrons may also escape from the system and if the fraction avoiding this escape is denoted by $l_t$, the number of thermal neutrons per fission of U-235 that are escaped is $ne(1 - l_f)pl_t$.

7.  Remaining thermal neutrons of number $v\varepsilon(1 - l_f) p(1 - l_t)$ a fraction $f$ is absorbed in uranium as a whole and then a fraction $v\varepsilon(1 - l_f) p(1 - l_t)(1 - f)$ is absorbed in other materials and the number available for carrying out chain reaction is then $v\varepsilon(1 - l_f) p(1 - l_t)f$. The quantity $f$ is called **'thermal utilization factor'**.

8.  The fraction of these thermal neutrons available for fission of U-235 is just

$$\sigma_f(U) \Big/ \sigma_a(U) = g.$$

9.  The multiplicity factor is then given by:

$$k = v\varepsilon(1 - l_f) p(1 - l_t)fg$$

and let

$$\eta = v \,\sigma_f(U) \Big/ \sigma_a(U) = vg \text{ and then}$$

$$k = \eta\varepsilon\, pf (1 - l_f) (1 - l_t).$$

10. If there is no leakage of neutrons in any way i.e., $l_f = l_t = 0$ and then as stated $k = k_\infty$

$$k_\infty = \eta\,\varepsilon\,pf$$ this is known as 'Four Factor Equation' which is an essential theoretical basis in designing the nuclear reactor and includes the factors already introduced and are as:

$v$ = fast neutrons produced per fission of thermal neutron.

$\varepsilon$ = fast fission factor (more fast neutron from U-238).

$p$ = resonance escape probability.

$f$ = thermal utilization factor.

## 18.6  NUCLEAR FUSION

Fusion process is exactly opposite to that of fission. In fission the heavy nuclei break up into fragments and thus the release energy (Section 18.5.1). In fusion to very light nuclei combine together to form a heavier nucleus and release energy.

$$4\,{}^{1}_{1}\text{H} \rightarrow {}^{4}_{2}\text{He} + 2\,{}^{0}_{+1}e\,(\beta^{+}) + 24.7\ \text{MeV} \qquad ...(18.16)$$

Now, these two positrons ($\beta^{+}$) immediately after their creation will combine with two negative electrons and release energy due to their annihilation. The energy that will be released in addition is of the order 2 M eV. The total energy released per fusion of hydrogen nucleus will be then 26.7 M eV.

In 1939 Bethe proposed a set of reactions that occurs in sun and named it as 'Carbon Cycle'. The steps of that set of reactions proposed by Bethe are:

$${}^{12}_{6}\text{C} + {}^{1}_{1}\text{H} \rightarrow {}^{13}_{7}\text{N} + Q$$

$${}^{13}_{7}\text{N} \rightarrow {}^{13}_{6}\text{C} + {}^{0}_{+1}e$$

$${}^{13}_{6}\text{C} + {}^{1}_{1}\text{H} \rightarrow {}^{14}_{7}\text{N} + Q$$

$${}^{14}_{7}\text{N} + {}^{1}_{1}\text{H} \rightarrow {}^{15}_{8}\text{O} + Q$$

$${}^{15}_{8}\text{O} \rightarrow {}^{15}_{7}\text{N} + {}^{0}_{+1}e$$

$${}^{15}_{7}\text{N} + {}^{1}_{1}\text{H} \rightarrow {}^{12}_{6}\text{C} + {}^{4}_{2}\text{He}$$

---

$$4\,{}^{1}_{1}\text{H} \rightarrow {}^{4}_{2}\text{He} + 2\,{}^{0}_{+1}e\,(\beta^{+}) + 24.7\text{MeV}$$

To produce these reactions on earth, the carbon and nitrogen nuclides must be bombarded with accelerated protons. But the temperature at the center of the sun is so great that some thermal protons at the high energy end of the Maxwellian distribution of velocity are found able to react. Considering the fact of presence of enormous quantity of high energy protons resulting to produce this reaction, enormous quantity of solar energy is produced. From presence of the absorption spectra of hydrogen in the continuous radiation emitted and reaching earth, it is established that the fusion reaction of hydrogen actually takes place within sun. It is also now known that when all hydrogen will be converted into helium, the radiation will cease to be produced. It will take, however, 30 billion years to consume all the hydrogen present within the sun. The sun is in equilibrium because even though fission increases the temperature of the core of sun and thus increases the energy of the protons and increases the fusion rate along with temperature, on the other hand this probable increase of temperature would expand the sun volume and thus decrease the neutron concentration. These two opposite effects are balanced and sun steadily emits energy. There is thus ample reason o believe that the sun is not unique; but the other stars also follow the nuclear fusion reactions and emit light. The sun is the source of energy in the form of heat and light, so life on earth is possible.

It may be stated again that the nuclear power is a chain reaction which maintains itself. While neutrons are the 'link' in maintaining fission chain reaction, the 'link' in fusion is the protons. The major difference between these to chain reactions is that in fission thermal neutrons at room temperature are responsible, in fusion protons in fusion are at millions of degree. This makes the possibility of making fusion chain on earth bleak. Only hope to maintain fusion on earth lies on the following reactions:

$$_3^6\text{Li} + {}_0^1 n \rightarrow {}_2^4\text{He} + {}_1^3\text{H}$$

$$_1^3\text{H} + {}_1^2\text{H} \rightarrow {}_2^4\text{He} + {}_0^1 n$$

The first reaction produces a tritium ($_1^3H$) and if this tritium is struck by deuterium, it produces a neutron and this neutron may initiate the first reaction and maintain the chain. Once the fusion is started, the fusion itself maintains the temperature to keep the process going. The energy liberated is a function of amount of fusion material present and there is no theoretical limit. In fission bomb, the parts before detonation must be smaller than the critical size otherwise the bomb will explode automatically, in fusion there is nothing like critical size and any amount would be safe to be preserve until it is ignited. The constructive use of the energy release from fusion is now a major field of research. Fusion in sun is 'contained' by tremendous gravity of sun but on earth for peaceful purpose not only the production of high temperature but also to contain the reaction at this temperature many orders of magnitude which is much above the temperature of vaporization of earth materials. However, calculations have shown that it may be possible to maintain fusion continuously at a temperature as low as 45 million degrees because the charged particles may be given this kinetic temperature by accelerating them under an electric field of 4000 volt. At this temperature and electric field, the gases will be ionized and as this state of the gas is called 'Plasma', the problem of plasma is to contain them within an enclosure whose surface must not be struck by the ions of plasma to prevent evaporation of the container. There is possibility to contain this plasma without allowing the ions to react with the material wall of the container by applying a strong magnetic field, which may be able to fcontain the plasma in a volume less than the volume of the container or even without any container. The detailed discussion of the processes possible is beyond the scope of this book and as most the aspects are yet to be fully standardized.

## 18.7  NUCLEAR REACTORS AND SOURCE OF ENERGY

The nuclear reactors are basically of two types (i) Research Reactors whose power output is insignificant and are mainly used in making radio isotopes and (ii) Power Reactors whose most important function is to produce electric power.

As there are three components of a reactor (i) Fission Fuel (ii) Moderator (iii) Absorber and (vi) Safety rods. The reactors are again classified as (i) Heterogeneous Reactor where all the components are individually placed and (ii) Homogeneous Reactor where the components (fuel and moderator) exist in a homogeneous mixture within the pile.

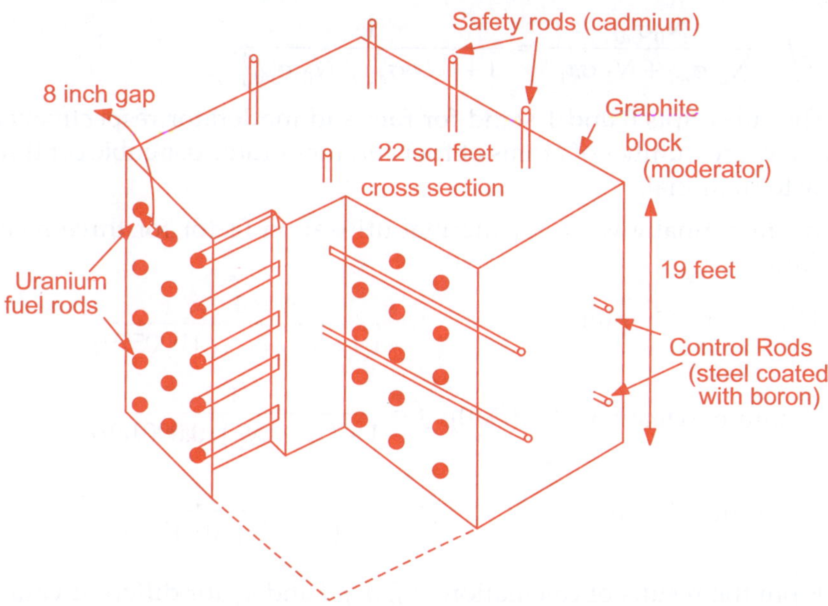


Fig. 18.9  Uranium reactor pile of heterogeneous reactor type

The heterogeneous thermal reactor at Chicago University, reached its criticality on December 2, 1942 and the schematic diagram of the nuclear pile is shown in the above Fig. 18.9. The power level achieved in this reactor was initially 1,000 kW which was doubled later. This was the first demonstration of successful attempt to control nuclear energy. The fuel (6.2 tons of Uranium metal and Uranium oxide) in aluminum sealed casings were placed at regular intervals within the graphite block (used as moderator) of 9 inch. The fast neutrons produced in the fission in uranium have to travel this distance of 9 inch to reach next fuel rod and during this travel through graphite are thermalized as the slowing down length in graphite is 7.4 inch. Strips of Cadmium which has high absorption cross section for thermal neutrons were inserted at regular intervals. They are noted in the figure as 'Control rods'. For further safety so that just critical condition is always maintained and the reaction rate is kept constant, cadmium 'Safety rods' are inserted under gravity from the top as and when the situation demands. The total pile (graphite in layers) along with fuel, control and safety rods was shielded by cubical shaped 7 ft. thick concrete wall, so that there is no radiation hazard to outside operators.

The homogeneous reactors are a mixture of fuel and heavy water ($D_2O$) which acts as moderator. The control and safety rods are introduced separately. The first of such system is known as 'Swimming Pool' type.

Now, if we consider an assembly consisting of only fuel and moderator, we may write the thermal utilization factor $f$, introduced in section 18.5.5 as:

$$f = \frac{N_0 \sigma_{a0}}{N_0\, \sigma_{a0} + N_1\, \sigma a_1} = \frac{1}{1 + \left[ N_1\, \sigma_{a1}\, /\, N_0\, \sigma_{a0} \right]}.$$

The subscripts 0 and 1 stand for fuel and moderator respectively and $N_1$ and $N_0$ are number of atoms of fuel and moderator per cubic centimeter of reactor material.

We may finally write for thermal utilization factor for three reactors as follows:

Uranium – Graphite :
$$f = \frac{1}{1 + (N_1\, /\, N_0)\,(0.000579)}$$

Uranium - Heavy water ($D_2O$):
$$f = \frac{1}{1 + (N_1\, /\, N_0)\,(0.000169)}$$

Uranium – Water :
$$f = \frac{1}{1 + (N_1\, /\, N_0)\,(0.0432)}$$

From the results of calculation of $f$, $p$, $p\,f$ and $k_\infty$ for different values of $N_1/N_0$ which is the ratio of number of moderator atoms per number of Uranium atoms, it is seen that

1.  For small values of $N_1/N_0$, $f$ approaches unity while $p$ is small.
2.  An arrangement which increases $f$ should decrease $p$ and the product $p\,f$ should pass through maximum as $N_1/N_0$ is varied.
3.  The maxima for $k_\infty$ is found to be quiet flat occurring in the ranges of $N_1/N_0$ equal to 4 to 10 for $H_2O$, 150 to 500 for $D_2O$ and 300 to 600 for graphite as moderators.
4.  The important observation is that $k_\infty > 1$ can be attained in a homogeneous mixture containing natural Uranium and heavy water, but not with homogeneous mixtures of natural uranium–graphite or natural uranium–ordinary water.
5.  When uranium enriched in $^{235}U$ is available, a chain reaction $k_\infty$ can be achieved with either graphite or water as moderator.

The following Fig. 18.10 represents a schematic diagram of a Power Reactor where, high temperature, high pressure steam is produced to run the turbine to generate electric power, which is much cheaper than using conventional fossil fuels like coal or petroleum and there is by large no danger if the reactor's safety devices function effectively. However, there exist many more safety measures and monitoring devices to prevent any disaster. However, it may be mentioned that the disposal of used fuels which remains very radioactive is an imposing problem.

## Nuclear Power Reactor

In some other kinds of reactor the Breeder Reactor needs mentioning. In this reactor the primary and produced fuels are same. For example $^{239}_{94}\text{Pu}$ undergoes fission as fuel and the neutrons are absorbed by $^{238}_{92}\text{U}$ inserted in the reactor and get converted into $^{239}_{94}\text{Pu}$ which is continued to be used as fuel. As $^{238}_{92}\text{U}$ enrichment from natural uranium is a very costly process and so production of electricity by using $^{238}_{92}\text{U}$ as fuel will not be cost effective. $^{239}_{94}\text{Pu}$ used as fuel produces further $^{239}_{94}\text{Pu}$ and can be used as fuel so the Breeder Reactor is more cost effective.

It may be stated here that the problem of safe disposal of radioactive wastes from fission reactors could be totally avoided if the fusion reactor in an established scale could be effectively implemented as none of the fusion products like helium is not radioactive. However, it still remains as a dream.

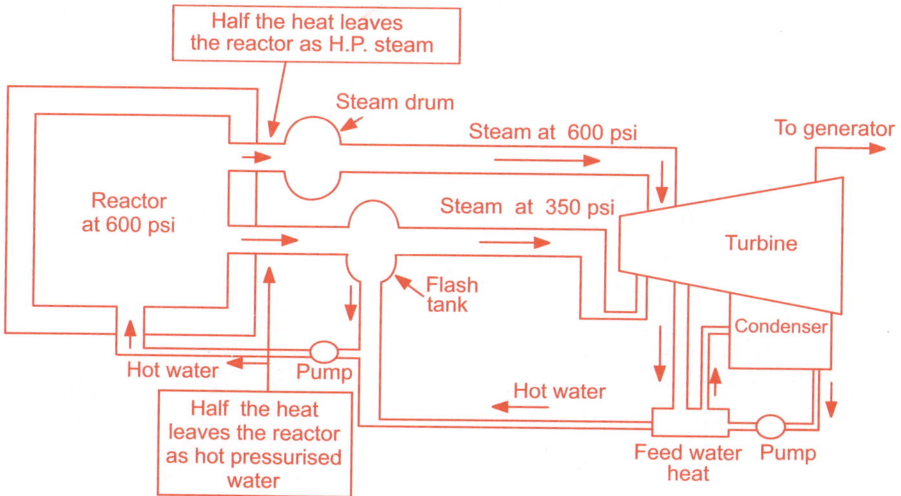


**Fig. 18.10** Schematic diagram of a boiling water power reactor

# REFERENCES


1. K. S. Krane – Introductory Nuclear Physics, John Wiley and sons, New York 1988.

2. J. Yarwood – Atomic Physics (Vol. II), Oxford University Press, London, 1965

3. M. R. Wehr and J.A. Richards- Physics of the Atom, Addison-Wesley Pub. Co. Reading, Mass, 1964.

4. I. Kaplan - Nuclear Physics, Addison Wesley Pub. Co. (Indian Student Edition), Narosa Publishing House, New Delhi, 1995.

5. S. N. Ghoshal – Atomic and nuclear Physics (Vol. I and Vol. II), S. Chand & Co., New Delhi, 1991.

# The Universe: A Brief Discussion

From the time unknown, human beings have looked upon the sky and get amazed by finding moon and sparkling stars during night and sun during day. They were so amazed by their presence in the sky that they later wrote many mythological scripts to associate them with God and worshipped them. What the human used to see at that time and what we see today during night on clear sky are only a small part of that, we call universe.

The universe as it was thought just a century ago, extends just up to the disc of stars what we call 'Milky Way' and beyond it exist only black empty space. Our sun and the solar systems are only a part of this Milky Way. But now we realize that this Milky Way galaxy is only one of the billions of galaxies just strewn across the distances of billions of light years. One light year is the distance covered by light (c = 3 × 10$^8$ m/sec) in one year and what we 'see' today was existing billions of years ago and what exists today, the information can only be known billions of years after! Such is the amazing vastness of the universe. The study of the objects in the galaxies is known as Astronomy and the study and formulation of the natural laws is Astrophysics. The science of the universe as a whole is known as 'Cosmology'. Let us start our very brief discussion on universe with an humble introductory study of our own home galaxy, the Milky Way and then peep into the other normal and unusual galactic objects.

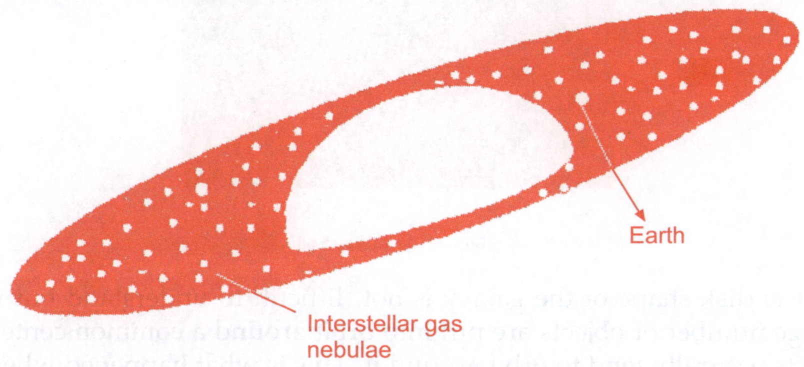


**Fig. AI 1** An artist's impression of Milky way galaxy of which our sun is just member. It is a disc shaped collection of hundreds of billions of stars

In studying the Milky Way, we may explore the universe as we instead of looking at any particular star, we look at a system of stars, and we consider the overall arrangement and history of a huge stellar community of which our sun is a member. For over a century, the prevailing idea was that our sun and its planets lie on the centre of the galaxy but now we know that Milky Way does not only contain stars but also copious interstellar dust. These dusts hide most of the galaxy stars from view and thus forbid us to make idea of the true dimension of the galaxy. There are three major components of our galaxy first a thin disk mainly consists of interstellar dusts with metal rich stars second a central bulge and third the halo consists of metal poor stars. The bulge at the centre is a mixture of these two types of stars.

Because the stellar dusts obscured the detailed observation of the structure of the galaxy by light telescopes, the knowledge about its structure had to wait until the discovery of radio astronomy. Both the optical and radio observations reveal that the star forming regions of the galaxy has 'Spiral arms'. The process of tracking the hydrogen element, the main constituent of the galaxy by the observation of the radio waves emitted by it is used for this purpose. Interstellar dusts and hydrogen gas mostly make up the interstellar medium of the disk. These constituent materials are not evenly spread but during the period of over billion years is churned up by passing stars and supernovae, have resulted it into non uniformly spread lumpy and frothy. The space missions indicated that Sun lies inside one of such regions, where interstellar medium is hot and thin. The figure below shows the structure of the spiral arms of the galaxy.

**Fig. AI 2** The spiral structure of Milky way galaxy

The disk shape of the galaxy is not difficult to understand but when a large number of objects are put into orbit around a common center, the objects naturally tend to orbit around it. This is what happened when our solar system was formed from solar nebula. The giant cloud of objects organizes themselves to form planets and all of them orbit in nearly same plane. To understand the formation of spiral arm of the galaxy, let us consider the following figure.

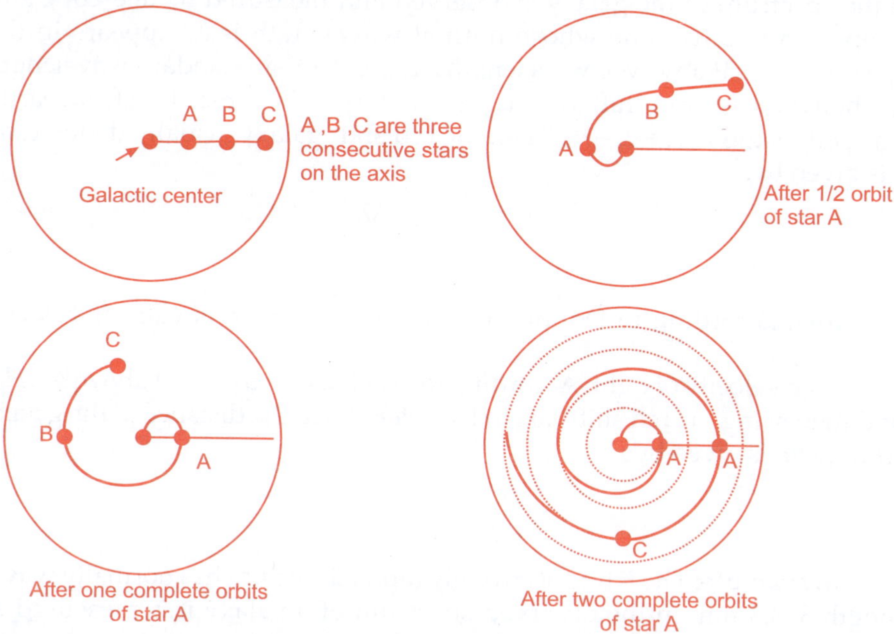


**Fig. AI 3** Shows that the spiral arm of our galaxy is caused by systematic movement of stars

The figure above shows that as star A completes one orbit and more the spiral continues to tighten up. Now, the question arises why then not the spiral wing of the galaxy does collapse? If the spiral arms are caused by mass movement, then it has been calculated that to collapse or wind up, it would need only few hundred million years which is a very brief time compared to the age of the galaxy. In 1940 astronomers predicted that instead of matter movement the spiral structure is caused by creation of mass density wave, whose crest passes out making spiral configuration. This latter prediction is more likely to have caused the spiral structure.

As mentioned before, this Milky Way is not the only one galaxy present. The boundary of the universe is not limited but is continued so as to enclose several other galaxies of different natures. The famous astronomer Hubble classified these galaxies into four major categories like: **Spirals, Barred Spirals, Ellipticals and Irregulars**. The disks of spiral and barred spiral galaxies are sites of active star formation. The elliptical galaxies do not contain interstellar gases and cannot favor any star formation. Irregular galaxies are not well defined and have asymmetrical shape.

The distance between the galaxies can be approximately measured from the linear relationship defined by **Hubble Law**. This is $v = H_0 d$, where $v$ is the recessional velocity of the galaxy, $d$ is the distance of the galaxy from earth and $H_0$ is a constant known as Hubble constant.

## Application of Hubble Law

If the spectrum of the galaxy is observed and measured by telescope and suppose a spectral line whose normal wavelength is $\lambda_0$ appears in the spectrum as $\lambda$. If observed wavelength $\lambda$ is greater than standard wavelength $\lambda_0$, then there is a red shift indicating that the galaxy is receding from earth. The shift is then $\Delta\lambda = \lambda - \lambda_0$. Then the red shift galaxy, usually denoted by $z$ is given by:

$$z = \frac{\lambda - \lambda_0}{\lambda_0} = \frac{\Delta\lambda}{\lambda_0}.$$

Now, according to Doppler effect $\dfrac{\Delta\lambda}{\lambda_0} = \dfrac{v}{c}$ which is valid if velocity $v \ll c$, the velocity of light. According to Hubble law, as stated above $v = H_0 d$ and then using the red shift due to Doppler effect, the distance of the galaxy from earth is given by:

$$d = \frac{zc}{H_0}.$$

**An example**: The $K$ line of strongly ionized calcium has normally wavelength 393.3 nm. In the observed spectrum of a galaxy it is measured as 401.8 nm. The red shift of the galaxy is then $z = \dfrac{(401.8 - 393.3)\ \text{nm}}{393.3\ \text{nm}} = 0.0216$.

The said galaxy is then moving away from earth with a speed of

$$v = zc = 0.0216 \times 3 \times 10^5 \text{ km/s}.$$

The value of the Hubble constant varies between 50 to 80 km/s/Mpc [1 pc(parsec) = $3.09 \times 10^{13}$ km = 3.26l $y$( Light year ).

1l $y$ = 3600 × 30 × 12 × 3 × $10^5$ km/s = 9.46 × $10^{12}$ km

1 Mpc = $10^6$ pc

Any of light year or parsec can be used to measure distances in space]

Using the above expression $d = \dfrac{zc}{H_0}$ and using the values of zc and the constant $H_0$ as 75 km/s/Mpc, we get the distance of the galaxy from earth as 87 Mpc.

This is equal to 280 million light years.

The determination of temperature on the surface of a star is determined by using Stefan-Boltzmann law (Described in Ch. 4) which is

$E = \sigma T^4$, where $E$ is the radiation energy emitted per sec per square meter of a black body. The luminosity of a star at a temperature $T$ is given by:

$$L = 4\pi R^2 \sigma T^4$$

The ratio of the star's radius to the sun's radius can be obtained from a related expression as:

$$\frac{R_{Star}}{R_{Sun}} = \left(\frac{T_{Star}}{T_{Sun}}\right)^2 \sqrt{\frac{L_{Star}}{L_{Sun}}}\,.$$

## Classification of Stars and Stellar Energy

The details of the star's spectrum reveal whether it is a giant, a white dwarf or a main sequence star. Hydrogen lines are good indicator of the luminosity as they are affected by both density and pressure. Higher the density and pressure, the collision between hydrogen atoms and other atoms are more frequent and this results into more ionization and broaden the hydrogen spectral lines, otherwise not. The density and pressure in the atmosphere of a **luminous giant star** is low as the star's mass is spread over a larger volume and so collisions are less frequent resulting into narrow hydrogen Balmer lines. A **main sequence star** is however more dense and compact than giant or super giant stars, the density of hydrogen is more and the collisions are more frequent. This results in the broadening of the hydrogen spectral lines. The stars are classified on the basis of their luminosity and as this is a continuous process spread over hundreds of billion years, there is a luminosity scale for this classification. **White dwarfs** are not given a luminosity class and they represent the final stage of stellar evolution in which no thermonuclear reactions take place. In the understanding of the stellar evolution which includes the questions of how the stars are born, live their lives and finally die, we search the answers from the realm of Physics. Our sun the basic cause of our existence on earth emits energy due to the thermo nuclear reaction at its center. The reaction converts hydrogen into helium by fusion and as a process it consumes $6 \times 10^{11}$ kg of hydrogen per sec. It has been mentioned in Chapter 18, that when this hydrogen will be exhausted, the sun will die. This fusion is commonly called 'Hydrogen burning'. There is a sequence of this hydrogen burning process. It first starts at the core of the star and slowly goes to the outer part as the density of hydrogen at the center decreases. As this thermonuclear fusion of hydrogen into helium ceases due to exhaust or decrease of density of hydrogen, a second type of thermonuclear reaction starts at the core, where two helium atoms combine to form Be and then carbon and oxygen. (Please visit Bethe's carbon cycle, Chapter 18. Sun's source of energy is due to hydrogen burning core with a diameter about $10^5$ km as it is today. In about 5 billion years, the sun will draw energy from hydrogen burning shell which will surround a compact helium rich core. The helium core will have its diameter about 30,000 km and this condensed helium core will initiate 'Helium burning' by converting gravitational energy into thermal energy. The helium burning occurs in two steps. First two helium atoms combine to form an isotope of $_4Be^8$

$$_2He^4 + {}_2He^4 = {}_4Be^8$$

At the core, this beryllium atom before it decays is attacked by third helium atom and forms carbon and release energy

$$_4Be^8 + {}_2He^4 = {}_6C^{12} + \gamma$$
$$_6C^{12} + {}_2He^4 = {}_8O^{16} + \gamma$$

This Carbon is the main constituent of human body and the oxygen is what we breathe.

## The Life Time and the Death of a Star

During hydrogen burning (Fusion) a part of the star's mass is converted into energy and if f is the fraction, then using Einstein's mass energy relation, we get

$E = fMc^2$. Now if $L$ is the star's luminosity i.e. energy released per unit time and if t is the total life time then

$$L = \frac{E}{t} \qquad So, E = Lt$$

and
$$Lt = fMc^2$$

or,
$$t = \frac{Mc^2}{L}$$

The luminosity is roughly proportional to 3.5 power of mass and so we can write

$$t \propto \frac{1}{M^{2.5}} = \frac{1}{M^2 \sqrt{M}}$$

Thus a star's life time is inversely proportional to 3.5 times of its mass.

We 'see' sometimes in the clear night sky a distant star suddenly increases it luminosity and this may be the beginning of the end of the star or sudden lack of balancing reactions stated above. A high mass star dies in a violent cataclysm in which its core collapses and most of its matter is ejected into space with high speeds. This results into sudden increase of luminosity by a factor of $10^8$ and during this explosion supernova is produced. The matter ejected glows as nebula.

## Neutron Star

The neutrons are heavier than proton but unlike protons which contain positive charge, neutrons are electrically neutral. Protons and electrons may combine with each other under high pressure and if that happens in a dying star, it then transforms into a neutron star. Neutron stars are inconceivably dense and its compactness may be imagined as a mass having equal to the mass of sun (solar mass = $1.989 \times 10^{30}$ kg) and packed in a sphere of diameter only few km (solar radius = $6.9599 \times 10^5$ km). They are actually the collapsed cores of massive stars that have perished in cataclysmic supernova explosions. The material within these neutron stars is very different from ordinary materials. These materials are super fluid i.e. they flow without any friction and they also conduct electricity without offering any resistance and so these are super conducting materials. The pulsating radio signals received on earth from these neutron stars is due to their powerful magnetic field which sweeps radiation across the sky. Neutron stars rotate around an axis and as they are also strong magnets having their magnetic axis is tilted with respect to their axis of rotation, the magnetic field sweeps around in space and as it reaches the earth we receive the pulsar (on and off pulse).

## Black Hole and its Formation

If the burnt out matter presses inexorably and if it is too heavy and beyond the Chandrasekhar limit resulting into a density more than a nucleus. If any object approaches close to it, it is attracted by the star and to escape the object needs a velocity more than the velocity of light, which is not possible. Therefore nothing can emerge out of this and it is so called as Black hole. There is compelling evidence that the black holes do exist and it has been discovered that black holes more than a million solar masses may lie at the center of many galaxies including our own, Milky Way. The behavior of the black holes is explained by Einstein's General theory of Relativity. Due to enormous mass, according to this theory the space around this black hole is curved.

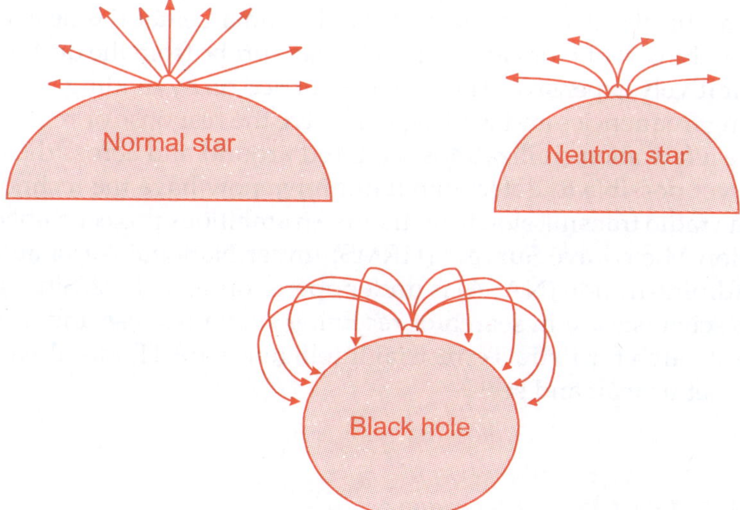


**Fig. AI 4** A comparison showing how the light beam emerges out from the normal star, the Neutron star and black hole. Note that in black hole all the radiation including photons to bend back to the black hole due to enormous gravity pull.

The curvature of space around normal star is so small that photons emitted almost move out straight from the surface. At the surface region of a collapsing neutron star the space is more curved and hence the photons emitted from surface of the star follow curved path.

When and as the collapsing process continues in a neutron star, the curvature of the space near its surface increases and the photons begin to follow a trajectory so that they return back to the surface. When this neutron star condenses below a critical size, the curvature of space around becomes so large that nothing can come out. A black hole is thus in fact, an 'information sink'. For anything to come out of a black hole, it must have its escape velocity more than the velocity of light, a condition which is impossible.

## In Search of Extra Terrestrial Life

The heavens we see a minute part of it in the sky above earth raise in our mind many profound questions. The series of questions that we contemplate includes the berth, life and the death and nature of stars, black holes and above all the possibility of life in other worlds. The questions in our mind that often make us curious include: Are we alone? Is there a chance that someday we will meet an alien civilization?

Unfortunately there exist no answers to these questions though the chemical building blocks of life on our earth are found throughout the space. Even, based on our present technology, it seems impracticable to send unmanned space probes to other stars. Firstly it would take a century or less to reach as the distance of the nearest star; Proxima Centauri is nearly 4.22 *ly*. Secondly, the cost of fabrication of such space probe from the design would be prohibitively expensive. The search is carried on by sending radio signal of varying frequencies and waiting to receive the response or reply. Even if there are a few alien civilizations scattered around and across the Galaxy, it is not yet possible to detect one though we now have the technology to detect the radio transmission from them. An ambitious project named 'High Resolution Microwave Survey' (HRMS) under National Aeronautics and Space Administration (NASA) is under operation since 1992. Since last few decades scientists are in search of receiving such response and a separate institute 'Search for Extra Terrestrial Intelligence' (SETI) has already been formed. Let us wait and see!

# Some Solved Problems

## Solved Problems On Classical Physics (Chapter 1)

1. If two forces $F_1$ and $F_2$ act on an object of mass 5.00 Kg. If $F_1$ and $F_2$ are of magnitudes 20.0 $N$ and 15.0 $N$ respectively and act in the directions a) 90° and b) 60° with each other, then find accelerations.

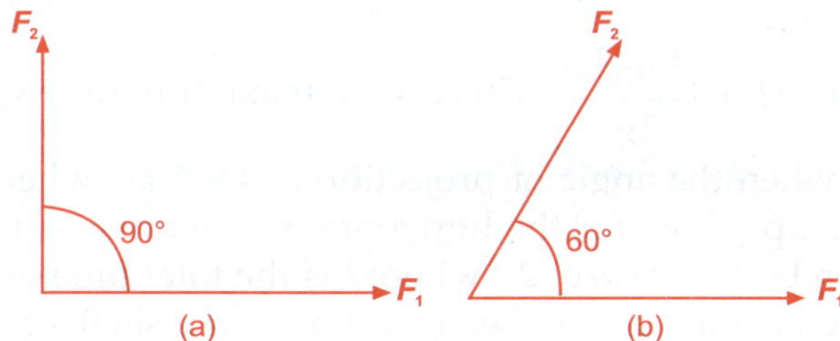


**Solution:**

a) As two forces are perpendicular to each other, the resultant force will be

$F = (F_1^2 + F_2^2)^{1/2}$ that is $F = (20^2 + 15^2)^{1/2} = (400 + 225)^{1/2} = 25$ N

Acceleration (a) = $F/M$ = 25/ 5 = 5 m/sec$^2$

b) Using the relation $R^2 = F_1^2 + F_2^2 + 2 F_1 F_2 \cos 60°$

$$= 20^2 + 15^2 + 2 \cdot 20.15 \cdot \frac{1}{2} = 400 + 225 + 20.15$$

$$= 400 + 225 + 300 = 925$$

$$R = 30.41$$

$$a = 30.14/ 5 = 6.028 \text{ m/sec}^2$$

Alternatively:

For (a)

$F_1 = iF_1$ and $F_2 = jF_2$ Resultant force $F = iF_1 + jF_2$ and $\tan \alpha = F_2/F_1$ where $\alpha$ is the angle between resultant and $F_1 \cdot |F| = (F_1^2 + F_2^2)^{1/2}$

$$F = (20^2 + 15^2)^{1/2} = 25 \ N.$$

For (b)

Total $\qquad F_1(x) = iF_1 + iF_2 \cos 60°$

$\qquad F_1(x) = 20 + 15 \cos 60 = 27.5$

Total $\qquad F_2(y) = jF_2 \sin 60°$

$\qquad F_2(y) = 15. .866 = 12.99$

$$R = (27.5^2 + 12.99^2)^{1/2} = 30.41 \ N$$

$$a = 30.41/5 = 6.08 \ \text{m/sec}^2.$$

2. Show for what value of the projection angle, a projectile motion will have maximum range and also maximum height.

   The velocity $v$ with which a body is projected making an angle $\theta$ with the horizontal will have two mutually perpendicular components. The component responsible for vertical height to be attained is equal to $v$ $\sin\theta$ and direction vertically up and against gravity. The maximum height to be attained is then using: $v^2 = u^2 - 2\,gH$ where $H$ is the maximum vertical height. Which is then $0 = v^2 \sin^2\theta - 2\,gH$.

   Therefore, $H = \dfrac{v^2 \sin^2\theta}{2g}$. This can be maximum for a given value of velocity when the angle of projection $\theta$ is 90° i.e. when it is thrown vertically up. Now, for the horizontal range traversed by the body, $R$ is given by $R = v \cos q, 2\,t$ where $t$ is the total time of travel which is obtained from $v = u - gt$ i.e. $0 = v \sin\theta - gt \ v \sin\theta - gt$ and the total time of travel (up and down) is given by $2t = \dfrac{2v\sin\theta}{g}$. Now, putting this time in the expression for range $R$ we get:

   $R = 2v^2 \sin\theta \cos q \ \dfrac{1}{g} = \dfrac{v^2 \sin 2\theta}{g}$. This range $R$ will therefore be maximum when angle of projection $\theta$ is 45°.

   Therefore, for a projectile motion the vertical height will be maximum when the angle of projection $\theta$ is 90° and horizontal range will be maximum when this is equal to 45°.

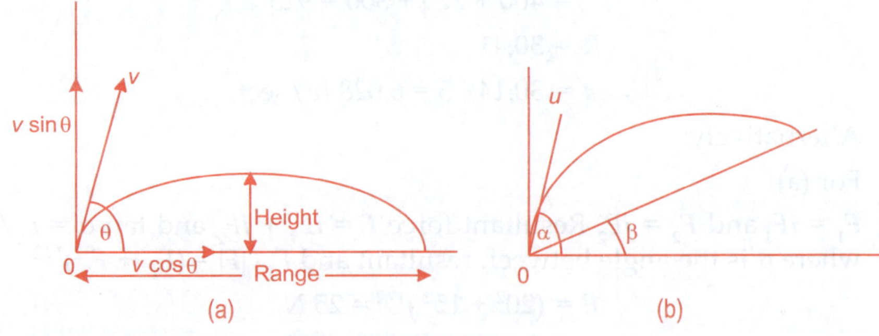

3. A particle is projected at an angle $\alpha$ with the horizontal from the foot of a plane, whose inclination to the horizontal is $\beta$. Show that it will strike plane at right angles if $\cot \beta = 2 \tan (\alpha - \beta)$. Fig. (b) above.

If $u$ be the velocity of projection, then $u \cos(\alpha - \beta)$ and $u \sin(\alpha - \beta)$ be the initial velocities respectively parallel and vertical to the inclined plane. The acceleration of the particle along these parallel and perpendicular directions will be respectively as $-g \sin \beta$ and $-g \cos \beta$. Then we can write taking the total time $t$ as the time to move up and down making net displacement as zero:

$$0 = u \sin(\alpha - \beta) . t - \frac{1}{2} g \cos \beta\, t^2 \text{ and } t = \frac{2u \sin(\alpha - \beta)}{g \cos \beta}.$$

Now, as required, the direction of motion at the instant when the particle strikes the plane is perpendicular and so the component parallel to the inclined plane vanishes. Thus, $u \cos (\alpha - \beta) - g \sin \beta . t = 0$.

$$\therefore \quad \frac{u \cos(\alpha - \beta)}{g \sin \beta} = t = \frac{2u \sin(\alpha - \beta)}{g \cos \beta}$$

$$\therefore \cot \beta = 2 \tan(\alpha - \beta).$$

4. As in the following diagram find the relative acceleration between $A$ and $B$ when a force of 5 Kg wt. is applied on the body $A$

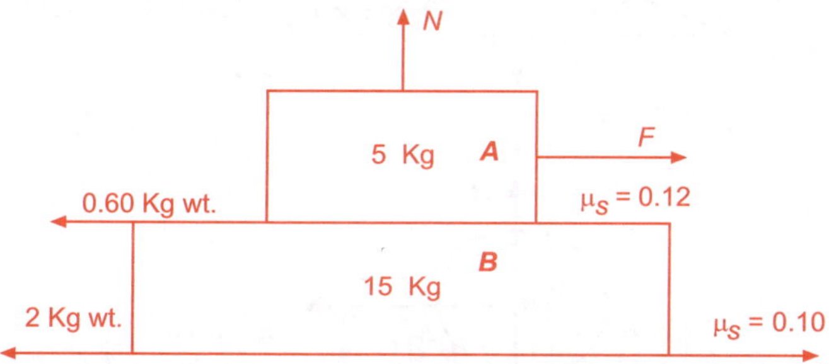


When a force $F$ of 5 Kg wt. is applied in horizontal direction on A

The frictional resistance applied by B on A through the surface between $A$ and $B$ is:

$$f_s = \mu_s N = 0.60 \times 5 \text{ Kg wt.} = 3 \text{ Kg wt.}$$

Net force on $A$ = (5 – 3) Kg wt. = 2 Kg wt = 19. 6 N

Acceleration of $A$ with respect to $B$ = 19.6/ 5 m/sec$^2$ = 3.92 m/sec$^2$

Reaction force acting on $B$ due to friction on $A$ = 0.60 Kg wt.

But frictional resistance on $B$ because of friction between $B$ and ground is

$$f_s = \mu_s N = 0.01 \times (5 + 15) \text{ km wt.} = 2 \text{ Kg wt.}$$

As this is exactly equal to the effective force on $B$, there will not be movement of $B$ and so the relative acceleration between $A$ and $B$ will remain as 3.92 m/sec$^2$ in the direction of force $F$ = 5 Kg wt.

5.  If in the above problem the force of 5 Kg wt. is applied on the lower body B, then find the relative acceleration.

Total frictional force on $B$:

$f_S = (\mu_s \times 5 = \mu_s \times 20) = 0.12 \times 5 + 0.10 \times 20 = 2.6$ Kg wt.

Net effective force on $B$: $(5 - 2.6)$ Kg wt = 2.4 Kg wt.

Acceleration of $B$ in the left to right direction: $(2.4 \times 9.8)N/20$ Kg = 1.176 m/sec$^2$.

Pseudo force acting on $A$ is $(1.176 \times 5)$ N = 5.88 N in the opposite direction and the frictional resistance on it $(A)$ is also 60 Kg wt. = .60 $\times$ 9.8 = 5.88 N

Therefore, effective force acting on it in the opposite direction is zero.

Therefore,

Net relative acceleration of $A$ towards left: 1.176 m/sec$^2$.

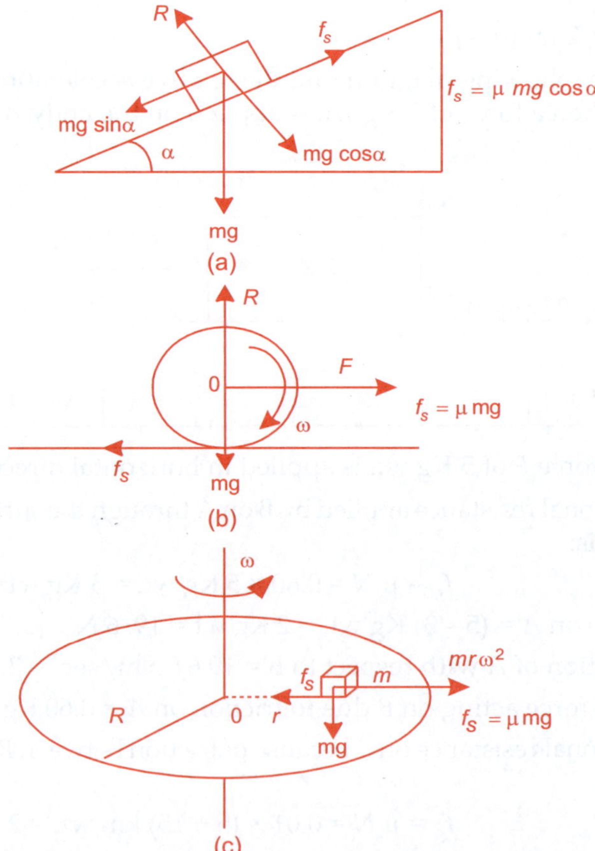


**Fig. AII 1 (a)** Mass on an inclined plane **(b)** Rotating sphere or spherical ball **(c)** Rotating disc

6. A wedge shaped inclined plane of inclination $\alpha$ is shown in the figure below (Refer to figure AII. 1 (A), a mass $m$ is rested on the plane of inclination. The angle of inclination is $\alpha$. Study the characteristics of its motion on the inclined plane.

   The weight vertically down $mg$ has two components $mg\cos\alpha$ acting normal on the inclined plane and $mg\sin\alpha$ acting along the plane and is responsible for downward sliding motion of the body. The frictional force $f_S$ opposes the motion and is equal to $\mu\, mg\cos\alpha$. Now, two possibilities are there in the first:

   $mg\sin\alpha > \mu\, mg\cos\alpha$ and then the net force responsible for sliding down the inclined plane will be $mg\sin\alpha - \mu\, mg\cos\alpha$ and the acceleration a will then be: $g(\sin\alpha - \cos\alpha) = a$. The second case when $mg\sin\alpha = \mu\, mg\cos\alpha$, the body will remain at rest. The case when $\mu\, mg\cos\alpha > mg\sin\alpha$, interesting situation results. As the under this condition the body develops a tendency to move up under the resultant force $\mu\, mg\cos\alpha - mg\sin\alpha$. But as the friction force is a self adjusting force it changes (reverses) its direction and acts downwards along with $mg\sin\alpha$ and then the body reverses the tendency of movement and tries to move down and then the frictional force acts upwards. This swinging tendency continues and the body will remain in its state of rest.

7. When a football is kicked by a footballer, it first slide over the ground and after sliding through certain length it starts rolling. Find the distance the ball must slide before rolling.

   When a ball is kicked with force $F$ say it is resisted by the sliding friction which is $\mu mg$ where $\mu$ is the frictional coefficient and m is the mass of the ball. Then the initial velocity V will decrease due to friction and let $S$ be the distance that the ball has to slide before it assumes a velocity say $V_1$. Then:

   $$V_1^2 = V^2 - 2\,\mu mg\, S.$$

   And this velocity $V_1$ is such that $V_1 = R\omega$, where $R$ is the radius of the ball and $\omega$ is the pure rotational velocity. Then the above equation can be written as:

   $$R^2\omega^2 = V^2 - 2\,\mu mgS$$

   Then after sliding over a distance $S$ on the ground the ball would start pure rotational velocity.

8. A circular plate is capable of rotating about the axis. A mass $m$ is placed over the plane at a distance $r$ from the centre. Find the minimum rotational speed of the circular plate that would initiate the sliding motion of the mass over the plane.

   The mass will experience centrifugal force due to the rotational motion of the circular disc and that would be under equilibrium condition $m\, r\,\omega^2 = f_s = \mu mg$ the minimum rotational speed would be:

$$\omega = \sqrt{\mu g / r}.$$

9. A particle slides down a smooth curve from a vertical height $h$ which is sufficient so that it can make a complete round. Prove if the inside radius of the circular path is $r$ then $2h$ is greater than $5r$.

   P.E. at the vertical height h is $mgh$ and the kinetic energy at the bottom is say $\dfrac{1}{2}mv^2$ and $mgh = \dfrac{1}{2}mv^2$ and then $v = \sqrt{2gh}$.

   Again the forces acting on the point vertically up the circular path will be :

   Weight of the particle and the centrifugal force

   $$mg = \frac{mu^2}{r} \; i.e. \, u^2 = gr.$$

Now, P. E. at $h$ = K. E. at vertically down the circular path = P. E. + K. E. at vertical top point on the circular path i.e.

$$mgh = \frac{1}{2} mu^2 + mg \times 2r \;\; \text{and the condition of the body to make just one}$$

complete round on the vertical circular path

$$mgh > \frac{1}{2} mu^2 + 2mgr$$

$$2\, gh > u^2 + 4\, gr$$

$$2\, gh > gr + 4\, gr$$

$$\therefore \qquad\qquad 2\, h > 5r.$$

10. A cylindrical steel drum of diameter 20 cm rotates about its axis which is vertical. A small steel body can be kept stuck i.e. in unmoved state on the inner wall of the cylinder if the cylinder rotates with 200 revolutions per minute and falls if the revolution decreases. Find the coefficient of friction of the inner wall.

    When the rotational speed is $\omega$ and it is sufficient to keep the body of mass $m$ stuck on the inner wall of the cylinder, then the normal reaction on the wall of the cylinder

    $$N = m\omega^2 r \text{ and } \mu N = mg, \therefore \; \mu = \frac{mg}{N} = \frac{mg}{mw^2 r}$$

    Putting $g = 980 \text{ cm/s}^2$, $\omega = 2\pi \times \dfrac{200}{60}\pi \; rad/s$ and $r = 10$ cm we get

    $$\mu = 0.223.$$

11. A small body starts sliding off the top of a smooth sphere of radius $r$. Find the angle corresponding to the point at which the body losses contact with the sphere. Find also the break off velocity of the body.

   Let the line joining the body of mass $m$ and the centre makes an angle $\theta$ with the vertical when it gets detached from the surface of the sphere and the break off velocity is $v$ then the component of the weight of the particle towards the centre will be $mg\cos\theta$ and the centrifugal force on the particle at that point is $\dfrac{mv^2}{r}$, the force tangential responsible for downward sliding will be $mg\sin\theta$. The potential energy of the particle mass m at the vertically up point on the spherical surface will be $mg$ and at the point of having detachment $mg - mg\cos\theta = mg(1 - \cos\theta)$. From the principle of conservation of energy we get:

$$\frac{1}{2}mv^2 = mg\,r(1 - \cos\theta) \text{ i.e. } v^2 = 2\,gt(1 - \cos\theta)$$ and equating the force towards the centre $mg\cos\theta$ with the centrifugal force we get:

$$mg\cos\theta = \frac{m^2 gr(1 - \cos\theta)}{r} \text{ and } \cos\theta = \frac{2}{3}$$

Again the break off velocity will be:

$$v^2 = 2\,gr(1 - \cos\theta) = 2gr\left(1 - \frac{2}{3}\right) = \frac{2}{3}\,gr$$

$$\therefore \qquad v = \sqrt{\frac{2}{3}\,gr}$$

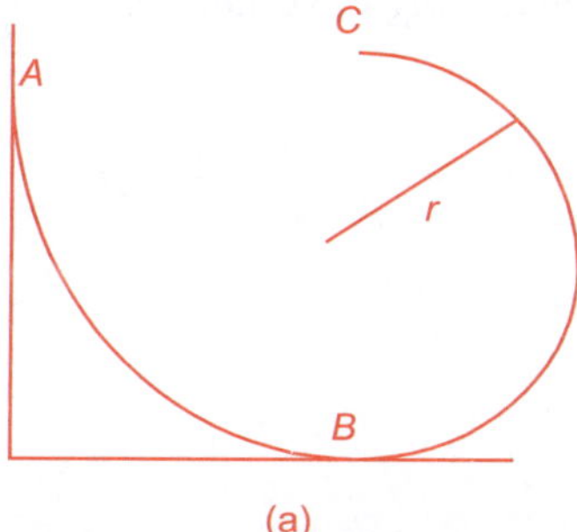


(a)

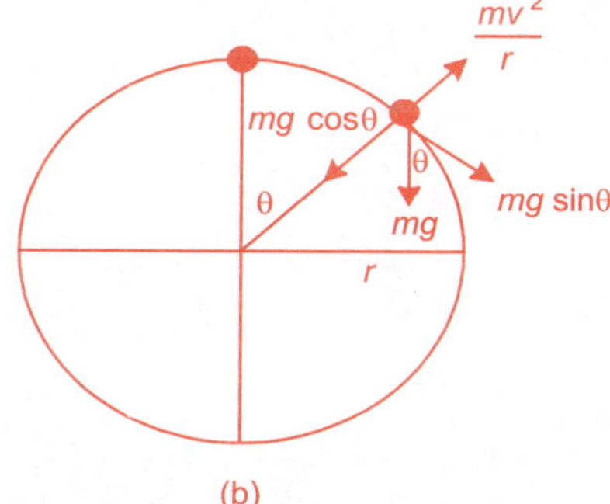


(b)

**12.** A bob of mass $m$ hangs from a string. It is released from a position where it is displaced at angle 90° from vertical and released. Now, where the sting would snap if it can withstand the tension equal to only the double of the weight of the bob.

Let the string snaps when the bob descended from the vertical height to a position making an angle $\theta$ with the vertical. If $u$ is the velocity of the bob at that point then the tension $T$ is given by:

$T = mg\cos\theta + \dfrac{mu^2}{l}$ where $l$ is the length of the string.

P.E. at the vertical top position = P. E. + K.E of the bob at the position concerned.

Then $mgl = mgx + \dfrac{1}{2}mu^2$, where $x$ is the vertical distance of the bob from bottom position and this $x$ is given by : $l - l\cos q$ and then: $u^2 = 2gl - 2g(1 - \cos\theta)l$ and thus,

$T = mg\cos\theta + \dfrac{m}{l} \times 2gl\,(1 - 1\cos\theta) = 3\,mg\cos\theta.$

The maximum tension that the string can with stand is $2\,mg$ and so

The string not be snapped

$T_{max} = 2\,mg \geq T$ i.e. $2\,mg \geq 3\,mg\cos\theta$ i.e. $\cos\theta \leq \dfrac{2}{3}.$

# Index